PERFORMANCE EVALUATION AND DESIGN OF FLIGHT VEHICLE CONTROL SYSTEMS

PERFORMANCE EVALUATION AND DESIGN OF FLIGHT VEHICLE CONTROL SYSTEMS

ERIC T. FALANGAS

IEEE PRESS

WILEY

Published by John Wiley & Sons, Inc., Hoboken, New Jersey. All rights reserved.
Published simultaneously in Canada.

For general information on our other products and services or for technical support, please contact our Customer Care Department within the United States at (800) 762-2974, outside the United States at (317) 572-3993 or fax (317) 572-4002.

Wiley also publishes its books in a variety of electronic formats. Some content that appears in print may not be available in electronic formats. For more information about Wiley products, visit our web site at www.wiley.com.

Library of Congress Cataloging-in-Publication Data is available.

ISBN: 978-1-119-00976-4

10 9 8 7 6 5 4 3 2 1

CONTENTS

PREFACE

The purpose of this book is to assist analysts and students toward developing dynamic models, analyzing and controlling flight vehicles of various blended features comprising aircraft, launch vehicles, re-entry vehicles, missiles, and spacecraft. The vehicle models are for simulations, control design, and stability analysis purposes. The vehicles are controlled by different types of effectors, such as aerosurfaces, thrust vector control, throttling engines, reaction control jets, control moment gyros, and reaction wheels. It is intended for flight control designers, dynamic analysts, and in general, for engineers and students who design and analyze atmospheric vehicles with asymmetric blended features using different types of actuators. It presents a unified approach in modeling, effector trimming, and in combining multiple types of flight vehicle control effectors.

The book begins with the rigid-body nonlinear equations of motion used in generic flight vehicle 6-DOF simulations and it gradually includes additional dynamic effects and details, such as propellant sloshing, structural flexibility, aeroelasticity, actuators, and tail-wags-dog dynamics. It presents some commonly used actuator models and describes in detail the dynamic coupling between actuators and flexibility of the supporting structure that often causes flexure oscillations. It also presents a method for augmenting the dynamic model to include structured parameter uncertainties used for analyzing the control system robustness using μ-analysis. It describes the design of a mixing-logic matrix that combines various types of effectors together, maximizes control efficiency, and reduces dynamic coupling between the control axes. It highlights two of the most commonly used flight control design methods, the LQR and H-infinity, and includes numerous design examples in detail. It describes methods of effector trimming and presents interactive trimming methods for balancing the vehicle moments and forces along a trajectory by adjusting the positions of the effectors.

It develops criteria for analyzing the static performance of a generic flight vehicle and presents simple graphical methods for evaluating static stability, performance, and control authority prior to detailed modeling, analysis, and simulations. It concludes with design examples of two flight vehicles and a space station that demonstrate the methodology described in detail.

ERIC T. FALANGAS

ACKNOWLEDGMENTS

I would like to acknowledge Dr. Aditya Paranjape for his contribution on spin susceptibility, my colleagues John Harduvel for the derivation of conditions on slosh stability, Viet Nguyen for educating me on reentry dynamics and static performance analysis, Dale Jensen for his inputs on flexibility and launch vehicle dynamics, and Don Edberg for reviewing this book. I would also like to thank NASA for allowing me to download illustrative pictures from their website, and also my family.

INTRODUCTION

The stability and performance of a flight vehicle can be analyzed from two distinct perspectives: static and dynamic. The static analysis is described in the second half of this book and it deals with the capability of the vehicle controls to balance the steady-state aerodynamic moments and forces and also having sufficient control authority to counteract the expected wind shear along a trajectory. Dynamic analysis evaluates the behavior of the closed-loop vehicle and control system together by using linear control analysis and requires dynamic models of the vehicle consisting of differential equations. In aerospace vehicle development, one of the central features is the simulation model of the vehicle dynamics. This model is a mathematical representation of its expected behavior and dynamic response to the control commands and to wind disturbances. This math model is used by the stability and control engineer to develop control laws that allow a human pilot or an autopilot to maneuver the vehicle and to perform its mission. It is necessary to understand the physics of how forces cause objects to move in order to be able to predict the resulting motion of a flight vehicle to excitations. This field of study is called "dynamics" and it examines the behavior of structures under the effects of external forces and torques. Equations of motion describe how the vehicle will move in response to applied forces. For example, simple equations describe how a rocket will accelerate when a constant thrust is provided by the rocket's engine. More difficult equations describe how the sloshing of propellant in a rocket's tank will cause the rocket's structure to vibrate or throw the rocket off course. Another type of modeling would be to predict, in a mathematical equation,

Performance Evaluation and Design of Flight Vehicle Control Systems, First Edition. Eric T. Falangas.
© 2016 by Eric T. Falangas. Published 2016 by John Wiley & Sons, Inc.

how an aircraft will respond to hitting an updraft in the atmosphere, or how the aircraft will respond to the deflection of the control surfaces at different airspeeds. These equations are differential equations, in which the rate of change of some variables, called states, is described as being dependent upon inputs and other states. The set of mathematical equations that describe these motions are collectively called a math model or simulation model of the vehicle, and they can range in complexity from a single equation to a complex set of equations. Complex vehicle models are, for convenience, broken down into subsystems that deal with different sets of dynamics, such as vehicle, sensor, and actuator subsystems.

The dynamic behavior of flight vehicles, such as launch vehicles, re-entry vehicles, and high performance aircraft, is generally studied in two distinct phases. The first phase deals with orbital mechanics and flight trajectory optimization assuming that the vehicle can be perfectly steered. This analysis is usually referred to as *"long period dynamics."* The second study deals with variations of the vehicle from its flight trajectory and they have a relatively shorter period. The dynamic equations describe how a vehicle model will behave in response to a combination of applied forces and torques and they are expressed in two forms: (a) the nonlinear large-angle equations suitable for six-degrees-of-freedom (6-DOF) simulations and (b) the linearized equations that capture the shorter period dynamics and they describe small variations of the vehicle from its trim condition. Both types of equations are used by different groups of people for flight vehicle design to address different issues. The 6-DOF equations are used to simulate the nonlinear response of the vehicle to commands and to disturbances along the entire flight. The 6-DOF simulations, however, usually include only the rigid-body motion, they are complex because they contain a lot of operational details and logic, and they cannot be used for control design and stability analysis.

The linear models are used for modeling the short period dynamic behavior of flight vehicles and for analyzing stability and control system performance. The performance analysis includes robustness against uncertainties and the vehicle response to commands and to wind-gust disturbances. The coefficients of the linear equation are functions of vehicle parameters at fixed flight conditions, which are typically *"time-slices"* for rockets or Mach versus alpha for aircraft. The parameters include mass properties, aerodynamic coefficients, trajectory data, slosh parameters, and structural modes. The vehicle data are obtained from 6-DOF simulations, wind tunnel aerodynamic data or computational fluid dynamic (CFD) models, mass properties, propellant slosh models, and finite element structural models, such as NASTRAN®. The complexity of the dynamic model varies according to the maturity of the project. An analyst typically begins with a simple rigid-body model originating from a fixed flight condition that is gradually updated, as more data become available, to a complex model that includes high-order dynamics, such as detail actuators, tail-wag-dog effects, hinge-moments feedback, wind gusts, propellant sloshing, flexibility, and aeroelasticity, including hundreds of structural modes.

The control system is typically designed based on linear state-space models that correspond to fixed flight conditions and the gains are interpolated in-between the design points. The linearized models are usually valid for small variations relative to the trajectory flight condition that it originated from. The control system, therefore, must stabilize the vehicle not only at the design points but also be capable of providing a certain amount of robustness to variations. It must also have the control authority and performance to correct small departures from the target trajectory caused by wind gusts or other dispersions. The assumption is that if the flight control system can provide an acceptable performance, stability margins, and robustness to uncertainties at multiple flight conditions along a trajectory (or Mach versus alpha) this will obviously be a good indication that the vehicle can be successfully guided without deviating from the trajectory due to instability or due to its inability to respond to guidance. This assumption is generally acceptable when the variation of vehicle parameters occur at rates considerably slower than the time constants associated with the vehicle dynamics. In launch vehicles, control design and analysis is typically performed at critical flight conditions which are lift-off, maximum dynamic pressure, maximum slosh, before and after staging, and main engines cut-off. In aircraft the control design and analysis is performed for a wide range of Mach versus angle of attack conditions and the gains are interpolated as a function of Mach and alpha. The control design, however, is finalized with a detailed 6-DOF nonlinear simulation that includes multiple parameter dispersion runs, various types of winds, and other variations.

Launch vehicles are controlled by small deflections of the thrust-vectored-control (TVC) engines relative to their trim positions. Trim positions are the gimbal angles required to balance the aerodynamic moments. Engine throttling is also used in some cases to control the vehicle speed and attitude by varying the engine's thrust relative to the nominal thrust T_e. That is, $T_e \pm \delta T_h$, where δT_h is the thrust variation control. Aircraft, gliders, and re-entry vehicles are controlled by deflecting the control surfaces relative to their trim positions. The dynamic models used in the analysis and control design are in state-space form which is the standard mathematical representation used for control system analysis such as the singular values (SV), the linear quadratic regulator (LQR), and the H-infinity control design methods, which are described in Chapter 9. The inputs to the dynamic models are control surface and engine deflections, accelerations about a hinge or gimbal, thrust variations or throttle control inputs, and wind-gust velocity in (ft/sec). Wind-gust inputs are disturbances which are defined as variations in wind velocity and they affect the aerodynamic angles and therefore the aeroforces on the vehicle. The vehicle model outputs are attitude, rates, accelerations, the aerodynamic angles (α, β) relative to a nonmoving air mass, and vane sensors that measure (α_w, β_w) relative to a moving air mass. The outputs include flexibility at the sensor or gimbal locations.

This book is focusing on modeling, trimming, and analyzing vehicles with blended features of aircraft, launch vehicles, re-entry vehicles, and spacecraft controls. It includes modeling of high-order dynamics which are typically left out in rigid-body analysis. It describes a methodology of combining various types of effectors together for the purpose of optimizing the control effectiveness and decoupling the control axes. It also describes two of the most commonly used flight control design methods, the LQR and H-infinity, and it includes several detailed design examples. It finally provides graphical methods for evaluating vehicle performance and controllability. The following topics are discussed in the upcoming sections.

- Chapter 2 describes the nonlinear rigid-body 6-DOF equations of motion for a generic type of atmospheric vehicle controlled by aerosurfaces, TVC, throttling engines, and RCS jets.
- Chapter 3 presents the linearized 6-DOF equations of motion of a generic flight vehicle relative to the trim conditions. The equations include flexibility, aeroelasticity, propellant sloshing, and tail-wags-dog (TWD) dynamics. They can be used to model the dynamics of an aircraft or a space shuttle type of vehicle from lift-off to re-entry and to perform linear control analysis and design.
- Chapter 4 describes standard electromechanical and hydraulic actuator models that provide the torque to rotate the TVC engines or aerosurfaces about their hinges against aerodynamic loads. It also describes the coupling between vehicle dynamics and the actuator models and the related dynamic effects of "tail-wags-dog" and "load-torque feedback." The detailed modeling of an electromechanical actuator is included as an example.
- In Chapter 5, a method is presented for combining multiple types of effectors as a system. The method uses a pseudoinverse algorithm to derive a mixing-logic matrix that efficiently combines aerosurfaces, gimbaling engines, throttling engines, and RCS jets together to increase controllability, provide redundancy, and decouple the control axes. The mixing-logic matrix is included in the flight control law. It connects the flight control system output and the effector inputs and converts the roll, pitch, and yaw flight control demands to effector commands that drive the actuator subsystems according to their individual capabilities. Examples are also included.

The second part of the book deals with evaluating the performance of a flight vehicle concept based on static analysis alone. This is done in the early phase using preliminary vehicle data and before getting involved with dynamic modeling, control design and analysis. It is possible to predict the vehicle stability and performance based on parameters that can be calculated from the vehicle data as a function of the trajectory. Flight vehicles must satisfy static stability, controllability, and maneuverability performance requirements in order to be certified for operations. A vehicle's ability to meet those requirements is often limited by the amount of control authority available. Thus, it is essential for designers to evaluate the control authority of candidate concepts early in the conceptual design phase. Normally, the designer considers numerous possible configurations before the stability and control group begin their analysis. An early evaluation of the concept before a detailed control design begins makes the design process much more efficient.

- Chapter 6 describes the effector trimming process used for balancing the moments and forces along a trajectory by adjusting the positions of vehicle effectors. Trimming of the control effectors is an important process that is included in the static analysis. A vehicle is trimmed when its effector positions balance the external torques due to aerodynamics and thrust, and sometimes also the linear steady-state accelerations according to a prescheduled trajectory. An interactive trimming method is presented that allows to trade control authority among effectors when multiple control effectors are available.

- Chapter 7 presents a method for quickly evaluating the performance of a conceptual flight vehicle against the design requirements imposed by the vehicle mission which are captured in a predefined trajectory. The vehicle performance is measured in terms of parameters which are calculated directly from vehicle data. It bypasses the time-consuming control design and analysis until an acceptable vehicle concept is recognized. The performance parameters evaluate, among other parameters, static stability, control authority, and maneuverability.

- In Chapter 8, the performance parameters are presented in contour plots as a function of Mach versus alpha. Contour plots are graphical tools that provide a wider viewpoint of vehicle performance in the entire operational range of Mach versus alpha. They help locate undesirable or favorable flight conditions and provide insight on how to reshape the trajectory in order to improve the vehicle performance. Vector diagrams are used for analyzing the vehicle control authority against wind disturbances by comparing the control moments and forces against the moments and forces produced by wind disturbances which are defined in terms of angles of attack and sideslip dispersions from trim.

- Chapter 9 presents the LQR and H-infinity design methodologies with examples. It also describes a method for creating uncertainty vehicle models and analyzing robustness.

- The overall methodology described in this book is illustrated by several design examples of re-entry and launch vehicles, including Space Shuttle and Space Station designs, which are presented in Chapters 9 and 10. The methodologies described are also implemented in a software program that can be accessed from Flixan.com together with the data and work-files.

1

DESCRIPTION OF THE DYNAMIC MODELS

The dynamic models described in this book consist of three rotational (roll, pitch and yaw), and three translational equations along x-, y- and z-axes. The vehicle forces and moments generated are calculated with reference to the body axes system. The reference axes are shown in Figure 1.1. The x-axis is aligned with the fuselage reference line and its direction is pointing forward. The z-axis is positive downward, and the y-axis is perpendicular to the x- and z-axes. The Euler angles (φ, θ, ψ) define the vehicle attitude with respect to the inertial reference axes. In a launch vehicle the attitude reference is usually measured with respect to the launch pad with the Euler angles initially at $(0°, 90°, 0°)$, respectively. The equations of motion include dynamic coupling between the pitch and lateral axes that can occur due to lack of symmetry. For example, thrust mismatch in the thrust-vectored control (TVC), cross-products of inertia, Y_{CG} offset, structural lack of symmetry, a nonsymmetrical aerodynamic shape that causes coupling between pitch and yaw such as pitching moment due to sideslip $C_{m\beta}$, or yawing moment due to alpha $C_{n\alpha}$. The dynamic models require that the x, y, z coordinates of important vehicle locations such as the engine gimbals, the control surface hinges, gyros, accelerometers, slosh masses, the center of mass (CG), the moments reference center (MRC) to be defined with respect to the vehicle reference axes. Vehicle data such as trajectory, mass properties, structural models are often defined by their authors in different coordinate frames and units and they must be transformed to a common set of units and directions.

Performance Evaluation and Design of Flight Vehicle Control Systems, First Edition. Eric T. Falangas.
© 2016 by Eric T. Falangas. Published 2016 by John Wiley & Sons, Inc.

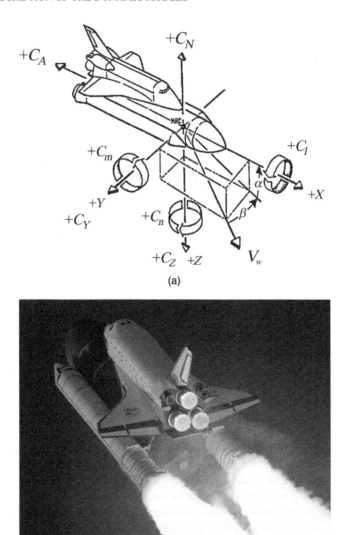

(a)

(b)

FIGURE 1.1   (a) Vehicle Axes and Directions of the Aerodynamic Forces and Moments; (b) Space Shuttle Vehicle Showing the TVC and the Control Surfaces

1.1   AERODYNAMIC MODELS

The aerodynamic coefficients used in the equations of motion are based on complex aerodynamic models. An aerodynamic model describes how the vehicle will respond to forces caused by motion of the vehicle through the atmosphere, and predicts the effects of each control surface (such as the flaps, rudders, ailerons) upon the motion of the vehicle. Aerodynamic models are often complex and are usually based on wind-tunnel data, in which the forces and moments exerted on a scaled model are measured at various speeds, flow angles, and with combinations of control surface deflections, until enough data is available to predict the forces and moments that will act on the full-scale vehicle. Increasingly more data is being added by using a technology called computational fluid dynamics (CFD) in which the same forces and moments are predicted in a computer program, using the geometrical shape of the vehicle in a virtual wind tunnel. The resulting aerodynamic subsystem model will predict what are the forces and moments on the vehicle as a result of any combination of control surface deflections, thrust settings, and flight conditions. The equations of motion in Chapters 2 and 3 use aerodynamic coefficients and derivatives at steady-state conditions, which are extracted from these complex aerodynamic models.

1.2   STRUCTURAL FLEXIBILITY

Structural flexibility is a very important issue in flight control system design and stability. Flight vehicles are designed with minimum weight objectives and hence in some cases their structures exhibit a considerable amount of flexibility, requiring the development of flexible structure equations to account for motion of various parts of the vehicle in relation to other parts. Long and slender vehicles made of lightweight materials require some attention to this aspect of modeling. Some parts of the vehicle can develop considerable amounts of displacement and acceleration as a result of excitations and structural flexibility in addition to the displacement and acceleration that arise owing to the rigid-body motion. The structural dynamics should be considered as an integral part of the control loop. If the deformation characteristics are ignored, the flight vehicle may not be properly controlled and in many instances it may exhibit self-excited divergent oscillations that can be destructive. Thus, the control system designer must be aware that divergent structural feedback can occur and must ensure that the flex phenomena are properly modeled and analyzed. Flexibility also limits the control system bandwidth, affects vehicle performance, and often requires filter design in the control system. In launch vehicles the main source of flex mode excitation is the TVC. In aircraft, the acceleration of the vehicle in combination with the aerodynamic forces can excite the structure into flexure oscillations, especially in the wings and the tails. This causes significant aeroelastic phenomena to occur which may have a serious impact on vehicle stability and performance.

The elastic behavior of a vehicle structure can be defined by the superposition of multiple flex modes that are excited by the actuator forces and moments, slosh forces, aerodynamic coupling, and other disturbances. The term "elastic modes" refers to the normalized mode shapes of the flight vehicle in "free–free" vibration. The mode shapes and frequencies of the "free–free" vehicle (often referred to as the modal data) are obtained from a finite elements modeling program such as NASTRAN®. Each bending mode j is defined by a second-order transfer function that calculates its generalized modal displacement η_j. Each mode is defined by a low damping coefficient ζ_j and a resonance frequency ω_j (rad/sec). The modal equations are excited by forces and moments applied on the vehicle structure in different locations, and also by aerodynamic forces (aeroelastic terms). The coefficients of each bending mode are derived from the finite elements model. The modal data for each mode consist of the mode frequency ω_j, the generalized mass $m_{g(j)}$, and generalized mode shapes $\varphi(j,k)$, and slopes $\sigma(j,k)$, at different locations k on the vehicle (nodes). In most applications 20–80 modes are sufficient to create an accurate representation of the structural flexibility. Sometimes as many as 400 modes may be included for verification analysis. When a sensor is mounted on the structure, in addition to the rigid-body motion, it also measures a linear combination of generalized modal displacements η_j from all structural modes which are observable at this sensor. The elastic modes produce high frequency oscillations superimposed on the rigid-body measurements.

The aeroelastic coupling between aerosurfaces and flexibility is implemented in the equations using two different approaches. The first approach is easier to model but less accurate. It assumes that the structure is excited only by the aerodynamic forces and torques at the hinges generated by the deflections and accelerations of the aerosurfaces, which are assumed to be rigid, and they couple with the vehicle structure as separate bodies. It requires modal shapes at the aerosurface hinges. The mass properties of the surfaces, gimbaling engines, and slosh masses should not be included in the finite elements model because they couple with the vehicle by the forces generated by the pivoting bodies as described in the equations. The second method is more efficient because it uses aeroelastic data created from an aeroelastic model such as a CFD used for flutter analysis. It captures the dynamic coupling between structural flexibility and the aerodynamic forces and moments created by the vehicle motion due to: (α, β, p, q, r) variations, and also due to the deflection of the aerosurfaces δ_{as}. This method requires, in addition to the modal data, two additional sets of coefficients: the "generalized aero force derivatives" (GAFD) data and the "inertial coupling coefficients" or "h-parameters." Hinge moment equations calculate the moments at the hinges of the aerosurfaces which are generated by the vehicle motion, flexibility at the hinge, and surface deflections. The "inertial coupling coefficients" define the dynamic coupling between structural flexibility and the angular accelerations of the engines or control surfaces. The finite elements model in this case must include the control surfaces or the engines rigidly attached at the hinges. The surfaces are released in the equations of motion by the inertial coupling coefficients, as we shall describe in Section 3.9. Both approaches require a finite elements model that is "free–free," that is, not clamped on either side.

1.3   PROPELLANT SLOSHING

The dynamic behavior of propellant sloshing inside the tanks is a very important issue in launch vehicle stability and design. Sloshing is defined as the periodic motion of the free surface of a liquid in a partially filled tank or container. Typical fuels and oxidizers used in launch vehicles are liquid oxygen (LOX) liquid hydrogen (LH$_2$), peroxide (H$_2$O$_2$), hydrazine, etc. Slosh frequencies depend on the tank size and acceleration, and they typically range between 2.5 and 6 (rad/sec). Propellant slosh frequencies are usually lower than the structural natural frequencies but they are sometimes near the control system bandwidth and may cause problems when they become unstable. Sloshing is excited by variations in normal and lateral accelerations due to maneuvering, thrust variations, wind gusts, etc. If the liquid is allowed to slosh freely, the uncontrolled oscillations can produce disturbance forces that cause additional accelerations on the vehicle. These oscillations are at low frequency and often near the control system bandwidth. They are sensed and responded to by the flight control system, forming a closed loop that may degrade performance or lead to an instability. It is also possible for the slosh resonances to interact with flexibility and to cause even further deterioration in stability and to excite oscillations. Most often slosh modes are passively phase stable, meaning that the closed-loop control system is attenuating the mode. When phase unstable, the control system destabilizes the mode and may require lead-lag phase compensation. In general, propellant sloshing is an undesirable effect on the vehicle, not only in terms of potential instability and attitude oscillations but it can also cause other problems and hardware malfunctions. The most commonly used mechanical solution for dampening slosh instabilities is to include mechanical baffles in the interior of the tank. There are situations, however, where slosh instabilities may be acceptable. For example, if the duration of the instability is relatively short or if the slosh mass is small in comparison with the total vehicle mass causing, therefore, a small disturbance or limit cycle. In reality, the amplitude of the slosh mass oscillation inside a tank will not grow forever, but it is limited by the tank walls. When slosh instability occurs engineering intuition is required to determine the severity of the problem. One has to consider what would the impact on the vehicle be if the slosh mass would be allowed to limit cycle at 3/4 tank radius. It may be acceptable if the mass is small.

The liquid motion inside the tank can be analyzed using pendulum or spring–mass analogy models. The sloshing propellant is often approximated with an oscillating spring–mass system with very low damping coefficient. One end of the spring is attached to a point in the tank centerline below the surface and the other end of the spring is attached to a mass that represents the sloshing part of the propellant. The slosh mass has two translational degrees of freedom and generates normal and lateral forces on the vehicle (along the z- and y-axes). It is constrained to oscillate along a plane that is perpendicular to the acceleration vector. The spring–mass analogy linear model is shown in Figure 3.11. It is acceptable for linear stability analysis but it is too conservative for larger slosh mass deflections because it does not limit the amplitude of the slosh mass. The pendulum model, however, shown in Figure 2.4, is more realistic for larger amplitudes because it bounds the slosh deflections and it is better suited for time domain simulations. The parameters required to implement the spring–mass slosh model are: the slosh mass (m_s), the slosh frequency (ω_s) in (rad/sec) which is a function of the vehicle acceleration, the damping coefficient (ζ_s), and the location of the slosh mass (x_s, y_s, z_s) at zero deflection, in vehicle reference axes. The slosh parameters are usually obtained from experiments. The slosh damping coefficients (ζ_s) are usually very small and they vary from 0.0002 for a tank without baffles to 0.002 for a tank with baffles. Note that the slosh masses should not be included in the calculations of the vehicle mass properties (mass, inertias, and CG).

The effects of the slosh masses on the vehicle are captured by the applied forces which are presented by the equations of motion described in Sections 2.6 and 3.8.

1.4   DYNAMIC COUPLING BETWEEN VEHICLE, ACTUATORS, AND CONTROL EFFECTORS

The inertia forces produced by the accelerating motion of gimbaling engines or control surfaces may also be a source of dynamic instability. The "tail-wags-dog" (TWD) is a torque created by the swiveling accelerations of the TVC engines or the control surfaces and these accelerations are further intensified by the local flexibility at the effector support structure. On a flight vehicle controlled by gimbaling engines, an excitation frequency exists at which the magnitude of the engine inertia reaction force is equal and opposite to the magnitude of the lateral component of thrust, or for a pivoting control surface, the force at the hinge due to the surface inertia is equal to the lateral component of the aero force. Below this so-called "TWD" frequency, the resultant lateral force at the gimbal is predominantly due to thrust being in phase with the gimbal angle. That is, an increase in gimbal angle results in an increase in lateral force. Above this frequency, the engine inertia forces produce the dominant lateral force which is in phase with the gimbaling acceleration, or 180° out of phase with the gimbal angle. The TWD introduces a complex pair of zeros in the transfer function "$\theta(s)/\delta(s)$." A 180° phase reversal occurs at frequencies greater than the TWD frequency and a system designed without the TWD consideration may perform unsatisfactorily above the TWD frequency. In particular, some of the higher frequency flex modes may be driven into divergent oscillations by this phase reversal if adequate structural damping or filter attenuation is not present. The TWD frequency should, therefore, be higher than the control system bandwidth. Fortunately, the TWD phenomenon provides significant amount of attenuation at around the TWD frequency which helps the flex mode attenuation.

Oscillations due to TWD phenomena have occurred during staging when the engines thrust decays and for a few seconds an active control system exists with very low or zero thrust, causing the TWD frequency to drop considerably. The drop in the control force phase reversal then drives some flex modes to instability causing divergent oscillations until the hydraulic pressure in the actuators is depleted and the oscillations cease due to lack of actuation. This type of problem is fixed by phasing down the control system gains during staging. TWD type of instabilities have also occurred in aircraft exciting control surface oscillations at low speeds or during ground testing when the control system is turned on. The aerosurfaces oscillate at their local natural frequency. These types of problems are usually alleviated by including filters in the control loop and stiffening the backup structure of the control surfaces in combination with detailed control analysis and simulations.

The dynamic coupling between the engines or surfaces, the actuators, and the effector backup structure play a critical role in the control system stability and in actuator performance. It often causes "TWD" type of oscillatory instabilities if not properly designed. Oscillations may occur due to unaccounted interactions between actuator dynamics and the supporting structure of the control surface or gimbaling engine. Sometimes all three deformations contribute to instability: actuator flexibility, local, and vehicle body deformations. They often couple with the actuator servo nonlinearities and they cause local control instability. This type of instability sometimes is not predicted in the preflight analysis due to lack of modeling details. A filter is often included either in the control system or in the actuator control loop to eliminate the instability problem.

Another effect that involves the actuator and the effector backup structure and requires proper modeling is the "load torque," which is an external loading torque at the hinge or gimbal and it is caused by local acceleration. When the vehicle accelerates, as a result of maneuvering or due to the stimulus of disturbance forces, the accelerations create an external loading torque on the nozzle or control surface that may react against the control torque provided by the actuator. The load-torque dynamics is described in Section 3.10 and in the actuators Section 4.2. Initially, the TWD dynamics and load-torque feedback are not included in the preliminary modeling and analysis phase, often due to lack of effector data. They are incorporated later as the design matures. Vehicle and actuator parameters such as hydraulic fluid compressibility, nozzle or control surface flexibility, backup structure stiffness, Coulomb/Dahl friction at the gimbal, load inertia play critical roles in the control system performance, and they are usually captured in a separate actuator model that is preferably nonlinear.

To properly model the TWD and load-torque dynamics, a dedicated actuator model is required for every control surface hinge or gimbaling engine rotation that is included in the vehicle model. The actuator model is more than just a simple transfer function, but it has two inputs: a deflection δ-command and an external load-torque input. It also has three outputs: engine or surface deflection, rate, and acceleration. The actuator inputs and outputs couple with the vehicle model to create the TWD and load-torque dynamics, as described in Section 4.2 and shown in Figure 4.7 for three aerosurfaces. Actuator models similar to those described in Section 4.1 are used to rotate each aerosurface or TVC engine. The models are applicable to both effector types but the actuator parameters may be different. Notice that when the engines and aerosurfaces are implemented as separate bodies interacting with the vehicle, their masses and moments of inertia should not be included in the vehicle mass, moments of inertia, and CG calculations.

1.5   CONTROL ISSUES

Feedback from the Inertial Measurements Unit (IMU), the rate gyros, accelerometers, or vane sensors is used to control the vehicle attitude and flight direction. Feedback from the accelerometers or vanes is often used for "load relief," especially in launch vehicles operating at high dynamic pressures, in order to reduce the normal and lateral aerodynamic loading on the structure. Normal acceleration feedback is also used to control the rate of descent of a re-entry vehicle. The equations of motion also include variations in altitude and in velocity (δV, δh) variables that characterize the

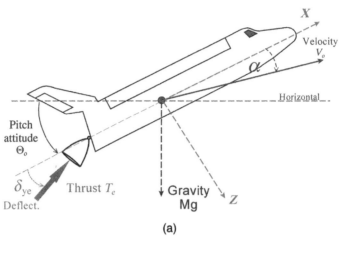

(a)

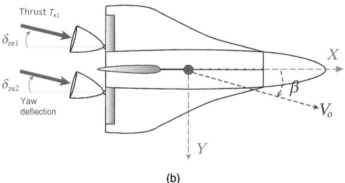

(b)

FIGURE 1.2   Vehicle Axes, Aerodynamic Angles, and Directions of Gimbal Rotations

phugoid dynamics. These variables are often used to design feedback control laws for longitudinal guidance in aircraft or in re-entry vehicles, especially at high angles of attack, and using speed brake, variable thrust, or alpha control as means to regulate the speed and altitude. Cross-range velocity is also included in the equations and it is used by lateral guidance to control heading. In some cases some of the state variables are not required in the dynamic model, especially in control design models, and they should be taken out in order to minimize the state vector. For example, the variation in altitude and velocity $(\delta h, \delta V)$ states are not useful in a launch vehicle having a fixed thrust and flying at zero (α). The x-acceleration output may not also be needed because it is not controlled from a TVC input. It is a good practice to reduce the vehicle model by removing the noncontributing state variables, especially when creating models for control synthesis, because most design algorithms require minimal state-space realizations.

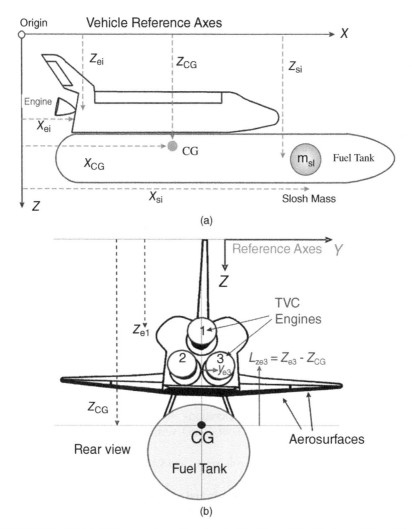

FIGURE 1.3   Vehicle Reference Axes Showing the Positions of the Engines, CG, and Slosh Mass

1.6   COORDINATE AXES

The axis used in this book to describe the equations of motion and to define the forces and moments is the body axes system. For a typical flight vehicle such as the Space Shuttle shown in Figure 1.1, the x-axis is along the vehicle structural symmetry pointing forward, the z-axis is pointing toward the floor, and the y-axis is perpendicular to the other two, according to the right-hand-rule, which is, toward the right wing. The origin of the body axes system for Guidance, Navigation and Control (GN&C) people is usually the CG. Other points are also used by different people such as the MRC which is a fixed point used by the aerodynamics groups because the CG varies. A fixed point outside the vehicle body is often used for reference by mass properties, or structural dynamics groups, see Figure 1.3. The stability axis system that has the x-axis pointing along the velocity vector is also used in Section 3.11.

The aerodynamic moments and forces on a flight vehicle are produced by its relative motion with respect to the moving air mass and its orientation with respect to the airflow V_w. Two orientation angles with respect to the relative wind are needed to specify the aerodynamic forces and moments, the angle of attack α_w and the angle of sideslip β_w. They are known as the "aerodynamic angles," and they are shown in Figure 1.1. Figure 1.2 shows the direction of the velocity vector relative to the body axes and defines the angles of attack and sideslip. The aerodynamic forces along the body $-x$-, y-, and z-axes are defined by the force coefficients: C_A, C_Y, and C_Z, and the aerodynamic moments about the x-, y-, and z-axes are defined by the moment coefficients C_l, C_m, and C_n. The aerodynamic equations are described in Section 2.5. Reference axes such as the one shown in Figure 1.3, are used for defining the locations of various parts of the vehicle such as the TVC gimbals, the CG, slosh masses. The reference frame is also used for defining rotational directions of aerosurfaces and nozzle gimbals see Figure 1.2. A surface deflection is positive when the rotation is clockwise about the hinge vector.

NOMENCLATURE

Symbol	Description
A_X, A_Y, A_Z	Vehicle sensed accelerations along x-, y-, and z-axes (ft/sec^2)
A_{XI}, A_{YI}, A_{ZI}	Vehicle inertial accelerations along x-, y-, and z-axes (ft/sec^2)
U_o, V_o, W_o	Vehicle forward, lateral, and vertical velocities along x-, y-, and z-axes (ft/sec)
P_o, Q_o, R_o	Nominal vehicle roll, pitch, and yaw rates about x-, y-, and z-axes (rad/sec)
Φ_o, Θ_o, Ψ_o	Euler angles (roll, pitch, and yaw) in radians
u_g, v_g, w_g	Wind-gust velocity components along vehicle x-, y-, and z-axes (ft/sec)
I_{xx}, I_{yy}, I_{zz}	Vehicle moments of inertia about the vehicle CG (ft-lb-sec^2)
I_{xy}, I_{xz}, I_{yz}	Vehicle products inertia about the vehicle CG (ft-lb-sec^2)
$\Delta_E(k), \Delta_Z(k)$	Pitch and yaw trim angles of the kth engine nozzle or jet in radians (elevation and azimuth angles measured from the vehicle $-x$-axis, see Figure 3.3)
$\delta_{ya}(k), \delta_{za}(k)$	Pitch and yaw gimbal rotations of the kth engine nozzle from trim due to actuator deflections alone in (rad), excludes bending at the gimbal
$\delta_{ye}(k), \delta_{ze}(k)$	Total deflections about the y- and z-axes, respectively, of the kth engine nozzle in radians (due to actuator deflection including bending)
$\delta_{cs}(k)$	Rigid deflection at surface (k) due to actuator rotation in radians

$\delta_{fcs}(k)$	Total deflection at surface (k) about the hinge (includes rotation due to structure flexibility) in radians
T_{Lyk}, T_{Lzk}	Pitch and yaw load torques at engine (k) gimbal (ft-lb)
$L_{hs(k)}$	Moment arm between the hinge line and the control surface CG $(X_{sk} - X_{sk\,cg})$ in feet
l_{ek}	Distance between the engine gimbal and the engine CG, $(X_{ek} - X_{ecgk})$ in feet
$l_{xek}, l_{yek}, l_{zek}$	Distance from vehicle CG to engine (k) gimbal along x, y, and z in feet
M_v	Mass of the vehicle (engines and slosh masses excluded) in lb-$\sec^2$/ft
m_{si}	Slosh mass (i) in slugs
$l_{sxi}, l_{syi}, l_{szi}$	Distance of the ith slosh mass from the vehicle CG along x, y, and z in feet
m_{ek}	Mass of the kth engine (slug)
T_{ek}	Thrust of the kth engine (lb)
δT_{ek}	Change in thrust of the kth engine from its nominal value (lb)
X_{cg}, Y_{cg}, Z_{cg}	Location of the vehicle center of gravity (ft)
X_{ek}, Y_{ek}, Z_{ek}	Location of the kth engine nozzle gimbal (ft)
X_{gi}, Y_{gi}, Z_{gi}	Location of the ith gyro (ft)
X_{ai}, Y_{ai}, Z_{ai}	Location of the ith accelerometer (ft)
X_{si}, Y_{si}, Z_{si}	Location of the ith slosh mass (ft)
X_{vk}, Y_{vk}, Z_{vk}	Location of the kth vane sensor (ft)
ϕ, θ, ψ	Small changes in vehicle attitude from nominal (roll, pitch, yaw) in radians
p, q, r	Changes in vehicle body rates (rad/sec)
α, β	Changes in the angles of attack and sideslip (rad)
$C_{z\alpha}, C_{y\beta}$	Normal and lateral force aerodynamic derivatives due to α and β (1/deg)
$C_{m\alpha}, C_{l\beta}, C_{n\beta}$	Aerodynamic moment derivatives (1/deg)
C_{mq}, C_{np}, C_{lp}	Aerodynamic moment velocity derivatives (–)
C_{nr}, C_{lr}	Aerodynamic moment velocity derivatives (–)
$C_{z\delta si}, C_{y\delta si}$	Normal and side force aeroderivatives due to control surface deflections δ_{csi} (1/deg)
$C_{m\delta i}, C_{l\delta i}, C_{n\delta i}$	Aero moment derivatives due to control surface δ_{si} (1/deg)
S_{ref}	Vehicle reference area (ft^2)
Q-bar	Dynamic pressure (lb/ft^2)
l_{ch}, l_{sp}	Mean aerodynamic chord and span vehicle reference lengths (ft)
ϕ_{hs}, λ_{hs}	Control surface hinge line orientation angles (deg)
η_j	Generalized modal displacement of the jth mode in feet
$M_g(j)$	Generalized modal mass for the jth mode (lb-$\sec^2$/ft)
$\varphi_{yek}(j), \varphi_{zek}(j)$	Mode (j) shapes along y- and z-axes at the gimbal of the kth engine (ft/ft)
$\sigma_{yek}(j), \sigma_{zek}(j)$	Mode (j) slopes about y- and z-axes at the gimbal of the kth engine (rad/ft)

$\varphi_{ysi}(j)$, $\varphi_{zsi}(j)$	Mode (j) shapes along y and z at the ith slosh mass location (ft/ft)
$\varphi_{yai}(j)$, $\varphi_{zai}(j)$	Mode (j) shapes along y and z at the ith accelerometer (ft/ft)
$\sigma_{hsk}(j)$	Modal slope for mode (j) at the kth surface about the hinge vector (rad/ft)
$\sigma_{ygi}(j)$, $\sigma_{zgi}(j)$	Modal slope for mode (j) about the y- and z-axes, at gyro (i) location (rad/ft)
$M_{hs}(k)$	Hinge moment about the hinge vector of the kth aerosurface (ft-lb)
$C_{nj\alpha}$, $C_{nj\beta}$	Generalized modal force stability derivatives of the jth mode with respect to (α, β) (1/rad)
C_{njp}, C_{njq}, C_{njr}	Generalized modal force stability derivatives of the jth mode with respect to (p, q, and r) (1/rad/sec)
$C_{nj\delta k}$	Generalized modal force stability derivatives of the jth mode with respect to the kth control surface deflection (1/rad)
$C_{nj\eta i}$	Generalized modal force stability derivatives of the jth mode with respect to the generalized modal displacement η_i of the ith mode (1/ft)
$C_{hi\alpha}$, $C_{hi\beta}$	Hinge moment derivative at the ith control surface due to α and β variations, respectively (1/rad)
$C_{hi\delta k}$	Hinge moment derivative at the ith control surface due to the deflection at the kth control surface (1/rad)
C_{hinj}	Hinge moment derivative at control surface (i) due to generalized modal displacement η_j of mode (j) (1/ft)
C_{lnj}, C_{mnj}, $C_{n\,nj}$	Roll, pitch, and yaw moment derivatives with respect to modal displacement η_j of the jth mode (1/ft)
C_{Ynj}, C_{Znj}	Force derivatives along the Y- and Z-axes with respect to modal displacement η_j of the jth mode (1/ft)
N_{aer}	Number of aerosurfaces
N_{mod}	Number of bending modes
N_{eng}	Number of engines
N_{sl}	Number of slosh masses

2

NONLINEAR RIGID-BODY EQUATIONS USED IN 6-DOF SIMULATIONS

We begin with the nonlinear earth-based, simultaneous equations of motion used in atmospheric 6-DOF, large-angle simulations.

2.1 FORCE AND ACCELERATION EQUATIONS

The combined external forces applied on the vehicle consist of the following types: aerodynamic, engines (TVC), propellant sloshing, wind-gust disturbance, and gravity forces. Although slosh is an internal force we treat it as external in this case because it is modeled as a separate interacting body.

$$F_X = F_{Xaero} + F_{Xeng} + F_{Xslosh} + F_{Xgrav} + F_{Xgust}$$

$$F_Y = F_{Yaero} + F_{Yeng} + F_{Yslosh} + F_{Ygrav} + F_{Ygust}$$

$$F_Z = F_{Zaero} + F_{Zeng} + F_{Zslosh} + F_{Zgrav} + F_{Zgust} \tag{2.1}$$

The vehicle acceleration measured at the CG is the result of all forces, either external or internal, with the exception of gravity and centripetal forces. Accelerometers do not "feel" the gravity force. The accelerations measured by an accelerometer are described in equation 2.2 and they are derived from the "$F=ma$" equation. Note that, we are using M_T to represent the total vehicle mass including the slosh masses, engines, and aerosurfaces. We use M_V to represent the vehicle mass that excludes the

---

Performance Evaluation and Design of Flight Vehicle Control Systems, First Edition. Eric T. Falangas.
© 2016 by Eric T. Falangas. Published 2016 by John Wiley & Sons, Inc.

moving masses that interact through forces.

$$\ddot{x}_{CG} = \left(F_{Xeng} + F_{Xaero} + F_{Xslosh} + F_{Xgust} \right) / M_{T}$$

$$\ddot{y}_{CG} = \left(F_{Yeng} + F_{Yaero} + F_{Yslosh} + F_{Ygust} \right) / M_{V} \qquad (2.2)$$

$$\ddot{z}_{CG} = \left(F_{Zeng} + F_{Zaero} + F_{Zslosh} + F_{Zgust} \right) / M_{V}$$

The relationship between the accelerations (A_X, A_Y, A_Z) measured by an accelerometer at the CG and the vehicle inertial accelerations (A_{XI}, A_{YI}, A_{ZI}) is as follows:

$$A_X = A_{XI} + g \sin \Theta$$

$$A_Y = A_{YI} - g \cos \Theta \sin \Phi$$

$$A_Z = A_{ZI} - g \cos \Theta \cos \Phi \qquad (2.3)$$

Note that, when the tail-wags-dog (TWD) dynamics are included in the model, the masses of the engines and the control surfaces should not be included in the calculation of vehicle mass properties, CG, or moments of inertia calculations. The TWD generates reaction forces and moments on the vehicle due to the rotational accelerations of the engines or the control surfaces about the hinge. The effects of the swiveling masses are included by the forces in the equations of motion. Otherwise, if the TWD dynamics is not used, the vehicle mass (M_T), the CG, and the moments of inertia should include the masses of the engines and the control surfaces. Also, if an engine is defined as "fixed," such as, a reaction jet that is throttling and not gimbaling, then there is no TWD involved and the engine masses should be included in the vehicle mass property calculation.

Accelerations due to External Forces

The acceleration of the vehicle relative to the body frame that has a rotational rate ω_b is obtained from equation 2.4.

$$\dot{v} + \omega_b \times v = \frac{\sum F_i}{M_V} \qquad (2.4)$$

where M_V is the vehicle mass and F_i is the external force. This equation is resolved along the three body axes to calculate accelerations along x, y, z, as shown in equation 2.5, where P, Q, and R are the body rates.

$$\dot{U} + QW - RV = \sum \frac{F_X}{M_V}$$

$$\dot{V} + RU - PW = \sum \frac{F_Y}{M_V} \qquad (2.5)$$

$$\dot{W} + PV - QU = \sum \frac{F_Z}{M_V}$$

The left side of the equations is the rate of change of linear momentum, ΣF_x, ΣF_y, ΣF_z are the summations of all external forces including gravity. The vehicle velocities U, V, W, along the body axes are obtained by integrating equation 2.5. The total vehicle velocity along the flight path is V_0. The velocity components along the body x-, y-, and z-axes are related to the aerodynamic angles of attack and sideslip as shown in equation 2.6.

$$U = V_0 \cos \alpha \cos \beta \qquad \alpha = \sin^{-1} \left(\frac{W}{V_0 \cos \beta} \right)$$
$$V = V_0 \sin \beta$$
$$W = V_0 \sin \alpha \cos \beta \qquad \beta = \sin^{-1} \left(\frac{V}{V_0} \right)$$
$$V_0 = \sqrt{U^2 + V^2 + W^2}$$

(2.6)

where α and β are the angles of attack and sideslip. After substituting (U, V, and W) from equations 2.5 in 2.6, the acceleration equations can be written in the following form which helps solving for the aerodynamic angles.

$$
\begin{aligned}
\dot{U} = \dot{V}_0 \cos \alpha \cos \beta \quad &= V_0 \sin \alpha \cos \beta \, (\dot{\alpha} - Q) + V_0 \cos \alpha \, \sin \beta \, \dot{\beta} \\
&+ V_0 R \sin \beta + \sum {}^{F_X}\!/\!_{M_V} \\
\dot{V} = \dot{V}_0 \sin \beta \quad &= -V_0 \cos \beta \, \dot{\beta} + V_0 \cos \beta \, (P \sin \alpha - R \cos \alpha) \\
&+ \sum {}^{F_Y}\!/\!_{M_V} \\
\dot{W} = V_0 \cos \alpha \cos \beta \dot{\alpha} \quad &= -\dot{V}_0 \sin \alpha \cos \beta + V_0 \sin \alpha \, \sin \beta \, \dot{\beta} - V_0 P \sin \beta \\
&+ V_0 Q \, \cos \alpha \, \cos \beta + \sum {}^{F_Z}\!/\!_{M_V}
\end{aligned}
$$

(2.7)

2.2   MOMENT AND ANGULAR ACCELERATION EQUATIONS

The vehicle roll, pitch, and yaw body rates (P, Q, R) about its center of mass are obtained by integrating the nonlinear rate of change of momentum equation 2.8.

$$I_{xx}\dot{P} = I_{xy} \left(\dot{Q} - PR \right) + I_{xz} \left(\dot{R} + PQ \right) + I_{yz} \left(Q^2 - R^2 \right) + \left(I_{yy} - I_{zz} \right) QR + \sum L_X$$
$$I_{yy}\dot{Q} = I_{xy} \left(\dot{P} + QR \right) + I_{yz} \left(\dot{R} - PQ \right) + I_{xz} \left(R^2 - P^2 \right) + \left(I_{zz} - I_{xx} \right) PR + \sum M_Y$$
$$I_{zz}\dot{R} = I_{xz} \left(\dot{P} - QR \right) + I_{yz} \left(\dot{Q} + PR \right) + I_{xy} \left(P^2 - Q^2 \right) + \left(I_{xx} - I_{yy} \right) PQ + \sum N_Z$$

(2.8)

where ΣL_X, ΣM_Y, and ΣN_Z are the total roll, pitch, and yaw external moments which are applied to the vehicle about the x-, y-, and z-axes. They are due to aerodynamics, TVC, propellant sloshing, wind gusts, and other external disturbances. The moments and products of inertia and the vehicle CG are calculated without the slosh masses and the gimbaling engines, because their presence in the vehicle equations is captured by the slosh and the TWD forces which are treated as external.

2.3 GRAVITATIONAL FORCES

Equation 2.9 calculates the earth's gravity force acting on the vehicle CG and it is resolved along the body x-, y-, and z-axes as a function of the pitch and roll Euler angles relative to the local horizontal frame.

$$
\begin{pmatrix} F_{Xgrav} \\ F_{Ygrav} \\ F_{Zgrav} \end{pmatrix} = M_V g \begin{pmatrix} -\sin\Theta \\ \cos\Theta \, \sin\Phi \\ \cos\Theta \, \cos\Phi \end{pmatrix} \tag{2.9}
$$

where the gravity constant (g) is:

$$
g = g_0 \left(\frac{R_e}{R_e + h_0} \right)^2 - \left(\frac{V_0^2}{R_e + h_0} \right)
$$

The acceleration due to gravity constant g is a function of altitude h_0. The equation also takes into consideration the orbital effect on g due to the velocity V_0 which is causing a centripetal force, where

g_0 is the gravity acceleration at the earth's surface,
M_v is the vehicle mass, and
R_e is the earth's radius.

2.4 ENGINE TVC FORCES

The attitude and direction of a rocket vehicle is controlled either by gimbaling the TVC engines, or by varying the thrust of the engines (throttling), or both, Figure 2.1. In our models the TVC nozzles are attached to the vehicle by means of spherical pivots that allow them to gimbal in two directions. By deflecting the engine nozzles the TVC generates forces at the gimbals perpendicular to the velocity vector which are used to guide and stabilize the vehicle. Typically, two orthogonal actuators per engine are used to provide the forces needed to rotate the nozzle in pitch (δ_y) and in yaw (δ_z) directions relative to the trim angles. The maximum gimbal angles in typical launch vehicles vary between $\pm 5°$ and $\pm 10°$. A typical engine nozzle (k) with a nominal thrust (T_{ek}) is not always aligned with the vehicle x-axis, but it is tilted in pitch and yaw at angles Δ_E and Δ_Z respectively, see Figure 2.2.

This nonzero position is either due to intentional mounting of the nozzle or due to trimming, as the engines rotate to balance the aero moments. Δ_E is the elevation angle, which is positive down from the x–y plane. It is the angle between the nozzle thrust direction and the vehicle x–y plane. Δ_Z is the yaw rotation angle of the nozzle and

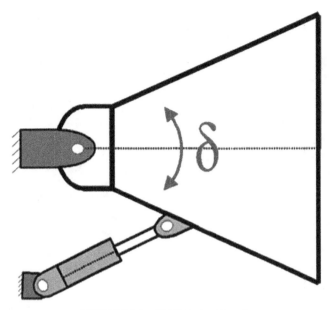

FIGURE 2.1   TVC Engine Nozzle

the thrust vector at the gimbal. It is the rotation angle of the thrust vector projection in the x–y plane about the z-axis, measured from to the $-x$ direction, see Figure 2.2.

The force components at the gimbal which are perpendicular to the flight path are used to control the vehicle attitude and direction. The vehicle velocity can be controlled by varying the engine thrust T_{ek}. In some cases the engines do not gimbal and the vehicle attitude and velocity are both controlled by varying the engine thrusts

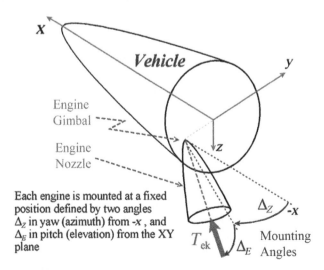

FIGURE 2.2   Engine Nozzle Orientation Angles with Respect to the Vehicle Axes

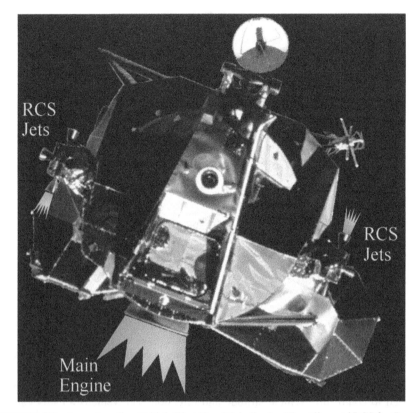

FIGURE 2.3   Lunar Landing Vehicle That Uses Main Engine Throttling and RCS for Control

(throttling) such as in the case of the lunar landing vehicle shown in Figure 2.3. Each engine at its trimmed position generates a force vector on the vehicle that can be resolved into three components $F_{XE}(k)$, $F_{YE}(k)$, and $F_{ZE}(k)$ as a function of thrust and the trim angles along the vehicle body axes, as shown in equation 2.10.

$$
\begin{aligned}
F_{XE}(k) &= T_{ek} \cos \Delta_E(k) \cos \Delta_Z(k) \\
F_{YE}(k) &= T_{ek} \cos \Delta_E(k) \sin \Delta_Z(k) \\
F_{ZE}(k) &= -T_{ek} \sin \Delta_E(k)
\end{aligned}
\tag{2.10}
$$

2.5   AERODYNAMIC FORCES AND MOMENTS

The aerodynamic forces and moments which are applied to the flight vehicle as it flies through the atmosphere are based on an aerodynamic model that calculates the forces along the x, y, and z body axes and also the moments about the body axes, as a function of the Mach number, the angles of attack and sideslip, the body rates, and the surface deflections. The aerodynamic model consists of a set of coefficients for the base vehicle, plus increment coefficients that provide additional forces for each

TABLE 2.1   Basic Aerodynamic Force and Moment Coefficients Which Are Functions of Mach Number, Alpha, and Beta; Each Coefficient Is a Three-Dimensional Matrix

```
Basic Aero Data (Pitch and Lateral) for the F-16 Fighter Aircraft
S_reference, Aero Chord, Span   (ft)         300.0       11.32     30.0
Moment Reference Center, (MRC) (ft)          -17.29       0.000     1.2
Numbers of: Mach, Beta, Alpha=               5,   5,   51
Aero Force Coefficients,     Aero Moment Coefficients
    Mach#      Beta,      Alpha (deg), Ca (-Cx),    Cy,        Cz (-CN),     Cl,           Cm,          Cn
   0.2000    -12.00      -20.00     0.2600E-01    0.2400      1.748      -0.1220E-01   -0.9800E-01   -0.3960E-01
   0.2000    -12.00      -19.00     0.2560E-01    0.2400      1.647      -0.1076E-01   -0.9280E-01   -0.4016E-01
   0.2000    -12.00      -18.00     0.2520E-01    0.2400      1.546      -0.9320E-02   -0.8760E-01   -0.4072E-01
   0.2000    -12.00      -17.00     0.2480E-01    0.2400      1.444      -0.7880E-02   -0.8240E-01   -0.4128E-01
   0.2000    -12.00      -16.00     0.2440E-01    0.2400      1.343      -0.6440E-02   -0.7720E-01   -0.4184E-01
   0.2000    -12.00      -15.00     0.2400E-01    0.2400      1.242      -0.5000E-02   -0.7200E-01   -0.4240E-01
   0.2000    -12.00      -14.00     0.2360E-01    0.2400      1.141      -0.3560E-02   -0.6680E-01   -0.4296E-01
   ...
   0.2000     -5.000     -20.00     0.2600E-01    0.1000      1.814      -0.5000E-02   -0.9800E-01   -0.1600E-01
   0.2000     -5.000     -19.00     0.2560E-01    0.1000      1.709      -0.4400E-02   -0.9280E-01   -0.1620E-01
   0.2000     -5.000     -18.00     0.2520E-01    0.1000      1.604      -0.3800E-02   -0.8760E-01   -0.1640E-01
   0.2000     -5.000     -17.00     0.2480E-01    0.1000      1.499      -0.3200E-02   -0.8240E-01   -0.1660E-01
   0.2000     -5.000     -16.00     0.2440E-01    0.1000      1.394      -0.2600E-02   -0.7720E-01   -0.1680E-01
   0.2000     -5.000     -15.00     0.2400E-01    0.1000      1.289      -0.2000E-02   -0.7200E-01   -0.1700E-01
   0.2000     -5.000     -14.00     0.2360E-01    0.1000      1.184      -0.1400E-02   -0.6680E-01   -0.1720E-01
   ...
   0.2000      0.000     -20.00     0.2600E-01    0.000       1.828       0.000       -0.9800E-01    0.000
   0.2000      0.000     -19.00     0.2560E-01    0.000       1.722       0.000       -0.9280E-01    0.000
   0.2000      0.000     -18.00     0.2520E-01    0.000       1.616       0.000       -0.8760E-01    0.000
   0.2000      0.000     -17.00     0.2480E-01    0.000       1.511       0.000       -0.8240E-01    0.000
   0.2000      0.000     -16.00     0.2440E-01    0.000       1.405       0.000       -0.7720E-01    0.000
   0.2000      0.000     -15.00     0.2400E-01    0.000       1.299       0.000       -0.7200E-01    0.000
   0.2000      0.000     -14.00     0.2360E-01    0.000       1.193       0.000       -0.6680E-01    0.000
   ...
```

aerosurface deflection. Table 2.1 shows a sample of the six basic force and moment aerodynamic coefficients (C_A, C_Y, C_Z, C_l, C_m, and C_n) for an F16 aircraft which vary as a function of the Mach number and the angles of attack and sideslip. Table 2.2 shows a sample of the force and moment increment coefficients for the same aircraft as a function of the Mach number, the angles of attack and sideslip in (deg), and also

TABLE 2.2   Force and Moment Increment Coefficients for Three Control Surfaces (Elevon, Aileron, and Rudder), Which Are Functions of Mach Number, Alpha, Beta, and Surface Deflection (degree); Each of the Six Increments Is a Four-Dimensional Matrix for Each Aerosurface

```
Aero-Surface Coefficient Increments for the F-16 Falcon Aircraft
Numbers of: Mach=5, Beta=5, Alpha=51
-----------------------------------------------------------------------------------------
Elevon          Delta=18
Surface Area,   Chord,   Span    Hinge Line Angles,Location,  Bias Deflect,  Deflect Range Min, Max,  Surface Mass,Inert,Mom Arm,
(ft^2)          (ft)     (ft)    Lambda,Phi (deg)  x,y,z (ft)  (degrees)      (degrees)              Mas,Ias,Las, (slugs), (ft)
  80.0           5.0      5.5    0.0  0.0  -42.0   0.0  0.0    0.0            -45.0, +45.0,           5.0, 1.5, 0.3,
Angles in (degree)                Aero Force Coefficients ............        Aero Moment Coefficients ..........
Mach#           Beta        Alpha       Delta,  Ca (-Cx),    Cy,        Cz (-CN),     Cl,           Cm,          Cn
  0.2000       -12.00      -20.00     -40.00   0.1983      0.000       0.3040      0.000       0.6303       0.000
  0.2000       -12.00      -19.00     -40.00   0.1930      0.000       0.3040      0.000       0.6089       0.000
  0.2000       -12.00      -18.00     -40.00   0.1877      0.000       0.3040      0.000       0.5875       0.000
  0.2000       -12.00      -17.00     -40.00   0.1823      0.000       0.3040      0.000       0.5661       0.000
  0.2000       -12.00      -16.00     -40.00   0.1770      0.000       0.3040      0.000       0.5447       0.000
  0.2000       -12.00      -15.00     -40.00   0.1717      0.000       0.3040      0.000       0.5233       0.000
-----------------------------------------------------------------------------------------
Aileron         Delta=18
Surface Area,   Chord,   Span    Hinge Line Angles,Location,  Bias Deflect,  Deflect Range Min, Max,  Surface Mass,Inert,Mom Arm,
(ft^2)          (ft)     (ft)    Lambda,Phi (deg)  x,y,z (ft)  (degrees)      (degrees)              Mas,Ias,Las, (slugs), (ft)
  35.0           3.0      7.0    0.0  0.0  -34.0,  0.0, 0.0,   0.0            -45.0, +45.0,           2.5, 0.7, 0.3,
Angles in (degrees)               Aero Force Coefficients ............        Aero Moment Coefficients ..........
Mach#           Beta        Alpha       Delta,  Ca (-Cx),    Cy,        Cz (-CN),     Cl,           Cm,          Cn
  0.2000       -12.00      -20.00     -42.50   0.000      -0.4462E-01   0.000       0.8245      0.000       -0.2210
  0.2000       -12.00      -19.00     -42.50   0.000      -0.4462E-01   0.000       0.9197      0.000       -0.1802
  0.2000       -12.00      -18.00     -42.50   0.000      -0.4462E-01   0.000       1.015       0.000       -0.1394
  0.2000       -12.00      -17.00     -42.50   0.000      -0.4462E-01   0.000       1.110       0.000       -0.9860E-01
  0.2000       -12.00      -16.00     -42.50   0.000      -0.4462E-01   0.000       1.205       0.000       -0.5780E-01
  0.2000       -12.00      -15.00     -42.50   0.000      -0.4462E-01   0.000       1.301       0.000       -0.1700E-01
-----------------------------------------------------------------------------------------
Rudder          Delta=18
Surface Area,   Chord,   Span    Hinge Line Angles,Location,  Bias Deflect,  Deflect Range Min, Max,  Surface Mass,Inert,Mom Arm,
(ft^2)          (ft)     (ft)    Lambda,Phi (deg)  x,y,z (ft)  (degrees)      (degrees)              Mas,Ias,Las, (slugs), (ft)
  20.0           2.0      8.0    20.0 90.0  -47.0  0.0  -14.0  0.0            -45.0, +45.0,           2.0, 0.5, 0.2,
Angles in (degrees)               Aero Force Coefficients ............        Aero Moment Coefficients ..........
Mach#           Beta        Alpha       Delta,  Ca (-Cx),    Cy,        Cz (-CN),     Cl,           Cm,          Cn
  0.2000       -12.00      -20.00     -42.50   0.000      -0.1218      0.000       -0.3485      0.000       0.8330
  0.2000       -12.00      -19.00     -42.50   0.000      -0.1218      0.000       -0.3638      0.000       0.8993
  0.2000       -12.00      -18.00     -42.50   0.000      -0.1218      0.000       -0.3791      0.000       0.9656
  0.2000       -12.00      -17.00     -42.50   0.000      -0.1218      0.000       -0.3944      0.000       1.032
  0.2000       -12.00      -16.00     -42.50   0.000      -0.1218      0.000       -0.4097      0.000       1.098
  0.2000       -12.00      -15.00     -42.50   0.000      -0.1218      0.000       -0.4250      0.000       1.164
  0.2000       -12.00      -14.00     -42.50   0.000      -0.1218      0.000       -0.4403      0.000       1.231
-----------------------------------------------------------------------------------------
```

the aerosurface deflection in (deg). In 6-DOF simulations the coefficients are stored in four-dimensional lookup matrix tables and interpolated.

The aerodynamic forces along the x-, y-, and z-axes are described by equation 2.11, and the aerodynamic moments by equation 2.12. They are used in 6-DOF simulations and calculate the aero forces and moments applied on the vehicle base by the airflow as a function of Mach number and the angles of attack and sideslip. The equations also include terms that represent the forces and moments generated by the control surfaces combined. The vehicle "feels" the presence of a wind shear or a wind gust by the changes it produces in the angles of attack and sideslip and also in the velocity relative to the wind. The angles of attack and sideslip relative to the wind (α_w, β_w) and the velocity V_w relative to the wind capture the effects due to the wind gust and they are used in the calculations of the aerodynamic forces, see Figures 3.8–3.10.

Aerodynamic Forces

$$F_{Xaero} = -\overline{Q}S_{ref}C_A\left(M,\ \alpha_w,\ \beta_w\right) + \sum_{k=1}^{Ncs} F_{Xcs}(k)$$

$$F_{Yaero} = \overline{Q}S_{ref}\left(C_Y\left(M,\ \alpha_w,\ \beta_w\right) + \left(\frac{l_{sp}}{2V_0}\right)\left(C_{Yp}\,p + C_{Yr}\,r + C_{Y\beta}\,\dot{\beta}_w\right)\right)$$

$$+ \sum_{k=1}^{Ncs} F_{Ycs}(k) \tag{2.11}$$

$$F_{Zaero} = \overline{Q}S_{ref}\left(C_Z\left(M,\ \alpha_w,\ \beta_w\right) + \left(\frac{l_{ch}}{2V_0}\right)\left(C_{Zq}\,q + C_{Y\dot{\alpha}}\,\dot{\alpha}_w\right)\right) + \sum_{k=1}^{Ncs} F_{Zcs}(k)$$

where $\alpha_w = \alpha + \dfrac{w_{gust}}{V_0 \cos\alpha_0}$; $\beta_w = \beta + \dfrac{v_{gust}}{V_0}$

Aerodynamic Moments

$$L_{Xaero} = \overline{Q}S_{ref}l_{sp}\left(C_l\left(M,\ \alpha_w,\ \beta_w\right) + \frac{l_{sp}}{2V_0}\left(C_{lp}\,p + C_{lr}\,r + C_{l\beta}\,\dot{\beta}_w\right)\right) + \sum_{k=1}^{Nsurf} L_{Xs}(k)$$

$$M_{Yaero} = \overline{Q}S_{ref}l_{ch}\left(C_m\left(M,\ \alpha_w,\ \beta_w\right) + \frac{l_{ch}}{2V_0}\left(C_{mq}\,q + C_{m\dot{\alpha}}\,\dot{\alpha}_w\right)\right) + \sum_{k=1}^{Nsurf} M_{Ys}(k)$$

$$\tag{2.12}$$

$$N_{Zaero} = \overline{Q}S_{ref}l_{sp}\left(C_n\left(M,\ \alpha_w,\ \beta_w\right) + \frac{l_{sp}}{2V_0}\left(C_{np}\,p + C_{nr}\,r + C_{n\beta}\,\dot{\beta}_w\right)\right) + \sum_{k=1}^{Nsurf} N_{Zs}(k)$$

The aerosurface aerodynamic coefficients in Table 2.2 calculate increments in forces and moments due to the deflection of each surface that must be added to the basic moments and forces. The forces on the vehicle $F_{Xcs}(k)$, $F_{Ycs}(k)$, $F_{Zcs}(k)$, and the moments $L_{Xcs}(k)$, $M_{Ycs}(k)$, $N_{Zcs}(k)$ which are generated by the deflection of the kth control surface relative to its zero position are defined in equation 2.13. They are functions of four variables: Mach number, angles of attack and sideslip relative to the wind, and control surface increment from zero position. Damping coefficients are often included which are also functions of Mach, α_w, and β_w. The equations also include the "TWD" forces and moments at the aerosurface hinges generated by the acceleration of the control surfaces.

Aerosurface Forces

$$F_{Xcs}(k) = -\overline{Q}S_{ref}\left[C_{A\delta(k)}\left(M, \alpha_w, \beta_w, \delta_{s(k)}\right) + \left(\frac{\overline{c}_{sk}}{2V_0}\right) C_{A\dot{\delta}(k)}\, \dot{\delta}_{s(k)} \right] + F_{XS_{TWD}}(k)$$

$$F_{Ycs}(k) = \overline{Q}S_{ref}\left[C_{Y\delta(k)}\left(M, \alpha_w, \beta_w, \delta_{s(k)}\right) + \left(\frac{\overline{c}_{sk}}{2V_0}\right) C_{Y\dot{\delta}(k)}\, \dot{\delta}_{s(k)} \right] + F_{YS_{TWD}}(k)$$

$$F_{Zcs}(k) = \overline{Q}S_{ref}\left[C_{Z\delta(k)}\left(M, \alpha_w, \beta_w, \delta_{s(k)}\right) + \left(\frac{\overline{c}_{sk}}{2V_0}\right) C_{Z\dot{\delta}(k)}\, \dot{\delta}_{s(k)} \right] + F_{ZS_{TWD}}(k)$$

Aerosurface Moments

$$L_{Xs}(k) = \overline{Q}\,S_{ref}\,l_{sp}\left[C_{l\delta(k)}\left(M, \alpha_w, \beta_w, \delta_{s(k)}\right) + \left(\frac{\overline{c}_{sk}}{2V_0}\right) C_{l\dot{\delta}(k)}\, \dot{\delta}_{s(k)} \right]$$
$$+ F_{ZS_{TWD}(k)}l_{Ysk} - F_{YS_{TWD}(k)}l_{Zsk}$$

$$M_{Ys}(k) = \overline{Q}\,S_{ref}\,l_{ch}\left[C_{m\delta(k)}\left(M, \alpha_w, \beta_w, \delta_{s(k)}\right) + \left(\frac{\overline{c}_{sk}}{2V_0}\right) C_{m\dot{\delta}(k)}\, \dot{\delta}_{s(k)} \right]$$
$$+ F_{XS_{TWD}(k)}l_{Zsk} - F_{ZS_{TWD}(k)}l_{Xsk} - I_{hs(k)}\cos\varphi_{h(k)}\,\ddot{\delta}_{s(k)}$$

$$N_{Zs}(k) = \overline{Q}\,S_{ref}\,l_{sp}\left[C_{n\delta(k)}\left(M, \alpha_w, \beta_w, \delta_{s(k)}\right) + \left(\frac{\overline{c}_{sk}}{2V_0}\right) C_{n\dot{\delta}(k)}\, \dot{\delta}_{s(k)} \right]$$
$$+ F_{YS_{TWD}(k)}l_{Xsk} - F_{XS_{TWD}(k)}l_{Ysk} - I_{hs(k)}\sin\varphi_{h(k)}\,\ddot{\delta}_{s(k)} \qquad (2.13)$$

where $F_{XS_{TWD}}(k)$, etc., are the TWD forces
$F_{XS_{TWD}}(k) = -m_{sk}L_{hs(k)}\cos\lambda_{hs(k)}\sin\Delta_{s(k)}\,\ddot{\delta}_{s(k)}$
$F_{YS_{TWD}}(k) = +m_{sk}L_{hs(k)}\sin\varphi_{hs(k)}\,\ddot{\delta}_{s(k)}$
$F_{ZS_{TWD}}(k) = -m_{sk}L_{hs(k)}\cos\varphi_{hs(k)}\cos\Delta_{s(k)}\,\ddot{\delta}_{s(k)}$
$l_{Xsk} = X_{sk} - X_{CG} \qquad l_{Ysk} = Y_{sk} - Y_{CG} \qquad l_{Zsk} = Z_{sk} - Z_{CG}$

2.6   PROPELLANT SLOSHING USING THE PENDULUM MODEL

The forces acting on a vehicle from a sloshing propellant in a partially filled tank can be characterized by the nonlinear model shown in Figure 2.4. The model is a pendulum that is suspended from a pivot point in the tank centerline, a little below the liquid surface. The pendulum mass represents the sloshing part of the propellant near the surface. The vehicle is accelerating and it is assumed that the x-acceleration is considerably larger than in z. The remaining of the propellant that is not sloshing is assumed to be rigid and its mass is located at a fixed point further down in the centerline of the tank. This point mass is included in the calculation of the vehicle mass properties and CG. The oscillations are excited by variations in the vehicle acceleration, mainly in A_z. The figure shows the sloshing motion in the longitudinal plane but the same principle also applies in lateral. The nonlinear nature of the pendulum constraints the amplitude of the oscillation, where the pendulum angle θ does not exceed $\pm 90°$, so this model is suitable for nonlinear, limit-cycle analysis. The tension of the string T_s is a function of total acceleration A_T, but it is also affected by the centripetal force due to the angular rate. The string tension creates a disturbance force that is resolved in two components along z and x and applied as an external force on the vehicle. The slosh frequency is adjusted by the length of the pendulum (l), and at low amplitudes it is $\omega_s = \sqrt{A_T/l}$ in (rad/sec). The accelerations of the slosh mass relative to the tank in the z and x directions are functions of the vehicle accelerations, string tension, and pendulum angle θ. The equations are integrated twice to calculate the slosh mass position relative to the tank and the pendulum angle.

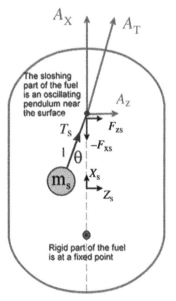

A_X and A_Z are the sensed vehicle accelerations along the body x and z directions. A_Y is assumed zero

Total acceleration: $A_T = \sqrt{A_X^2 + A_Z^2}$

String tension: $T_s = m_s\left(A_T\cos\theta + l|\dot\theta|^2\right)$

Forces on vehicle due to slosh mass:

$F_{ZS} = -T_s\sin\theta$; $F_{XS} = -T_s\cos\theta$

Slosh mass displacement relative to Tank:

$$\ddot z_s = -D_{zs}\dot z_s - A_z + \frac{T_s}{m_s}\sin\theta$$

$$\ddot x_s = -D_{xs}\dot x_s - A_x + \frac{T_s}{m_s}\cos\theta$$

Where: $\tan\theta = \dfrac{-z_s}{l-x_s}$; $\sin\theta = \dfrac{-z_s}{l}$

D_{xs} and D_{zs} are small damping terms

FIGURE 2.4   Nonlinear Pendulum Slosh Model

The slosh moments applied to the vehicle are calculated from the forces at the pivot and by the moment arms' distances between the pivot point and the vehicle CG.

2.7 EULER ANGLES

The Euler angles (Φ, Θ, Ψ) are used to relate the vehicle body axes attitude with respect to an earth-bonded reference axis or with respect to a local-vertical–local-horizontal (LVLH) frame. They are obtained by integrating a set of kinematic differential equations which are functions of the body rates. The Euler angles, however, are not suitable for large-angle maneuver simulations because they are vulnerable to singularities. Quaternions are a better choice for representing attitude because they do not have singularities. There are 12 sets of kinematic equations that can be used to solve for the Euler angles by integrating the body rates. They depend on the rotational sequence chosen. The analyst should choose a set of equations that avoids singularities of a particular application in the angles' range of operation. Equation 2.14 includes three commonly used sets of differential equation for propagating the Euler angles as a function of body rates. They must be initialized at some attitude values. The angle Θ is measured between the vehicle x-axis and the local horizontal plane. The angle Φ is the angle between the vehicle y-axis and the horizontal plane. The angle Ψ is between an arbitrary reference line in the horizontal plane and the projection of the vehicle x-axis in that plane.

1-2-3 Rotation
$$\dot{\Phi} = P + \tan\Theta \ (Q\sin\Phi + R\cos\Phi)$$
$$\dot{\Theta} = Q\cos\Phi - R\sin\Phi \qquad\qquad \dot{\Psi} = \frac{Q\sin\Phi + R\cos\Phi}{\cos\Theta}$$

3-2-1 Rotation
$$\dot{\Phi} = P\cos\Theta + R\sin\Theta$$
$$\dot{\Theta} = Q + \tan\Phi \ (P\sin\Theta - R\cos\Theta) \quad \dot{\Psi} = \frac{R\cos\Theta - P\sin\Theta}{\cos\Phi}$$

1-3-2 Rotation
$$\dot{\Phi} = P + \tan\Psi \ (R\sin\Phi - Q\cos\Phi)$$
$$\dot{\Theta} = \frac{Q\cos\Phi - R\sin\Phi}{\cos\Psi} \qquad\qquad \dot{\Psi} = Q\sin\Phi + R\cos\Phi$$

$$(2.14)$$

The position of the vehicle in the earth frame is obtained from the velocities (U, V, W) by integrating equation 2.15 using the 1-2-3 rotation.

$$\begin{pmatrix} \dot{X} \\ \dot{Y} \\ \dot{Z} \end{pmatrix}_{earth} = \begin{bmatrix} c\psi & -s\psi & 0 \\ s\psi & c\psi & 0 \\ 0 & 0 & 1 \end{bmatrix} \begin{bmatrix} c\Theta & 0 & s\Theta \\ 0 & 1 & 0 \\ -s\Theta & 0 & c\Theta \end{bmatrix} \begin{bmatrix} 1 & 0 & 0 \\ 0 & c\Phi & -s\Phi \\ 0 & s\Phi & c\Phi \end{bmatrix} \begin{pmatrix} U \\ V \\ W \end{pmatrix} \qquad (2.15)$$

2.8 VEHICLE ALTITUDE AND CROSS-RANGE VELOCITY CALCULATION

The rate of change in vehicle altitude (h) is related to the flight path angle by the following equation:

$$\dot{h} = V_0 \sin \gamma \qquad (2.16)$$

We can also prove that the rate of change in vehicle altitude with respect to the ground is expressed by the following equation.

$$\dot{h} = V_0 \left(\sin \Theta \cos \Phi \cos \alpha \cos \beta - \sin \Phi \sin \beta - \cos \Theta \cos \Phi \sin \alpha \cos \beta \right) \qquad (2.17)$$

The cross-range velocity, perpendicular to the flight path, in the lateral direction is:

$$V_{CR} = V_0 \cos(\gamma) \sin(\beta + \Psi) \qquad (2.18)$$

where

γ is the flight path angle.
V_0 is the nominal vehicle speed along the velocity vector.

2.9 RATES WITH RESPECT TO THE STABILITY AXES

When a flight vehicle is operating at high angles of attack, roll maneuvering is usually performed by rotating about the velocity vector V_0 and not about the body x-axis. This minimizes the sideslip transients and the undesirable side-loads due to beta. A rotation about V_0 requires the vehicle to be commanded to rotate about both roll and yaw (P_b and R_b) body axes simultaneously in order to minimize beta. It is useful, therefore, to create vehicle models that have roll and yaw output rates measured in stability axes, rather than the body axis which is more commonly used in simulations. In stability axes the roll rate P_s is measured about the velocity vector V_0, assuming zero beta. The pitch rate is the same as in the body axes. The stability axes yaw rate R_s is orthogonal to the roll and pitch rates. Since the rate gyro measurements are in body axis they must be converted to stability axis rates by using the transformation in equation 2.19.

$$\begin{aligned} P_s &= P_b \cos \alpha + R_b \sin \alpha \\ R_s &= R_b \cos \alpha - P_b \sin \alpha \end{aligned} \qquad (2.19)$$

2.10   TURN COORDINATION

During a coordinated turn an aircraft is experiencing zero lateral acceleration. It banks at an angle Φ and allows a small gravity component to offset the centripetal force due to turning. It maintains the same pitch and roll attitude but its heading is changing at a constant rate. Turn coordination is desirable for passenger comfort and also the pilot functions more efficiently. It minimizes sideslip and undesirable aeroloading by rolling about the velocity vector V_0, and it maximizes aerodynamic efficiency. To perform coordinated stability axis rolls, both roll and yaw systems are used to provide roll control about the velocity vector and to maintain zero sideslip. At low angles of attack there is usually adequate rudder power to obtain the desired motion. However, as the angle of attack increases, the demand on rudder authority increases rapidly.

In level flight with small sideslip, the yaw rate R that is required to produce a coordinated turn at a constant bank angle Φ and an airspeed V_0 is:

$$R = \frac{g \cos \Theta \tan \Phi}{V_0} \approx \frac{g \sin \Phi}{V_0} \qquad (2.20)$$

The circle radius (r_c) of the coordinated turn is:

$$r_c = \frac{V_0^2}{g \cos \gamma \tan \Phi} \qquad (2.21)$$

When the flight control system receives a roll rate command in stability axes, the turn coordination system applies a feed-forward signal from bank angle Φ and the velocity V_0 to the yaw rate R as defined in equation 2.20. This loop allows the vehicle to achieve its commanded roll rate, otherwise, without the turn coordination feedback it would result in a nonzero steady-state roll rate error.

2.11   ACCELERATION SENSED BY AN ACCELEROMETER

Normal, lateral, and axial accelerometers are used to measure the vehicle accelerations along the vehicle z-, y-, and x-axes, respectively. Accelerometer feedback is used by the flight control system to provide load relief and to protect the vehicle structure from excessive normal and lateral aerodynamic loading. It is also used to control the angles of attack and sideslip and the rate of descent. Excessive aeroloading due to α and β may be catastrophic especially at high dynamic pressures. The aeroloads are measured in terms of $Q\alpha = (\overline{Q} \times \alpha)$. Typically, normal and lateral loads should not exceed 3000–3500 (psf-deg). In launch vehicles during high dynamic pressures, a load-relief feedback loop is included in parallel with the attitude control loop in order to alleviate normal and lateral loads. When the load-relief system is operating, the flight control system steers the vehicle in a direction that trades-off a certain amount of directional controllability in order to gain some reduction in lateral loading. The control gains in the load-relief feedback loop are typically phased in-and-out proportionally with the dynamic pressure.

The acceleration measured by normal and lateral accelerometers in a nonlinear simulation in equation 2.22, consists of three components:

(a) a term due to rigid-body acceleration at the CG,

(b) terms due to angular acceleration times the distance from the CG, and

(c) a term due to structural bending at the accelerometer.

$$A_{Yaccel} = A_{YCG} + l_{Xacc}\left(\dot{R} + PQ\right) - l_{Zacc}\left(\dot{P} - QR\right) - l_{Yacc}\left(P^2 + R^2\right)$$

$$+ \sum_{j=1}^{N\,mod} \phi_{Yacc}(j)\,\ddot{\eta}(j) \tag{2.22}$$

$$A_{Zaccel} = A_{ZCG} - l_{Xacc}\left(\dot{Q} - PR\right) - l_{Yacc}\left(\dot{P} + QR\right) - l_{Zacc}\left(P^2 + Q^2\right)$$

$$+ \sum_{j=1}^{N\,mod} \phi_{Zacc}(j)\,\ddot{\eta}(j)$$

2.12   VEHICLE CONTROLLED WITH A SYSTEM OF MOMENTUM EXCHANGE DEVICES

Space vehicles are usually controlled by momentum exchange devices (MED), such as a cluster of control moment gyros (CMG) or a reaction wheels assembly (RWA). The CMGs are flywheels spinning at constant rate and the direction of the spinning rotor relative to the spacecraft can be varied by a servomotor. The reaction wheels (RW) have a fixed spin axis but the wheel speed can be varied by a torque motor. The MEDs generate torque through angular momentum transfer to and from the main vehicle body. This is achieved either by changing the angular momentum vector direction of the CMG spin axes, or by changing the angular momentum magnitude of the RW. The advantage of MEDs is that, contrary to the on–off jets, they achieve precise control because they operate in a continuous manner, plus they do not consume propellant-only electric power. Their disadvantage is that they have a limited amount of momentum, that is torque times time, and they require desaturation using thrusters or other means. The MED system generates an equal and opposite internal torque between the vehicle and the devices. In the absence of external torques the spacecraft/MED system momentum remains constant.

Let us first consider a space vehicle controlled with a cluster of CMGs. Equation 2.23 is a simple dynamic model that describes the exchange of angular momentum between the space vehicle and the CMGs. It calculates the combined CMG momentum and torque generated from the momentum exchange system in the body axes. It bypasses the detailed modeling of the individual CMGs geometry and the steering logic that converts the torque commands to gimbal rates or to RW accelerations when an RWA is used instead of CMG.

$$J_v \dot{\omega} = -(\omega \times J_v \omega) + T_c + \sum T_{ext}$$
$$\dot{H}_{cmg} = -(\omega \times H_{cmg}) - T_c$$

(2.23)

where

H_{cmg} is the combined CMG momentum in body axes,
ω is the body rate,
T_c is the CMG control torque applied to the vehicle,
T_{ext} is the combined external torque applied to the vehicle,
J_v is the moment of inertia dyadic of the vehicle.

The first equation states that the rate of change in vehicle momentum is equal to the external torques plus the CMG control torque. The second equation states that the rate of change of CMG momentum is equal to the torque generated by the CMG cluster, which is equal and opposite to the vehicle control torque. T_c is the torque in the body frame generated from the combined CMG array that controls the vehicle. The gimbal angles and rates of each CMG must be manipulated by a steering logic in order to produce this torque T_c, as we shall see. If you add the terms together in the two equations given in equation 2.23 and set the external torques to zero, the gimbal torques cancel and you end up with the conservation of system momentum equation, which means that the system momentum is not affected by the internal torque T_c.

$$J_v \dot{\omega} + \dot{H}_{cmg} + \omega \times (J_v \omega + H_{cmg}) = 0$$

(2.24)

2.13   SPACECRAFT CONTROLLED WITH REACTION WHEELS ARRAY

Reaction wheels consist of a spinning rotor whose spin rate is nominally zero. Its spin axis is fixed to the spacecraft and its speed is increased or decreased to generate a reaction torque about the spin axis. They are typically used to control small 3-axes stabilized satellites. They are simple and least expensive of all MEDs but their control torque capability is small. Figure 2.5 shows a typical RW with the control electronics. Its angular rate can be varied by applying a torque to the motor that rotates it about its spin axis. As the wheel accelerates it applies an equal and opposite reaction torque to the spacecraft that is used to control attitude. In general, a 3-axes stabilized spacecraft requires an RW array consisting of at least three reaction wheels.

Figure 2.6 represents a rigid-body spacecraft with a reaction wheel (RW) and an externally applied force F. The spacecraft origin is at the CG. The unit vector $\underline{a}_i$ defines the spinning direction for the ith wheel. The force vector $\underline{F}_j$ defines an external force j, for example, due to a jet firing. The displacement $\underline{d}_j$ represents the distance between the force F_j application point, and the vehicle CG. By combining

FIGURE 2.5 Typical Reaction Wheel with Control Electronics

the RW assembly spin column vectors $\underline{a}_i$ together we can create a transformation matrix C_w^b that transforms the wheel momentum from individual wheel axis to the spacecraft body axes (x, y, z). For example, equation 2.25 shows the transformation from four reaction wheel spin directions to body. The transformation from body to wheel C_b^w is the pseudo-inverse of C_w^b.

$$C_w^b = \begin{bmatrix} \underline{a}_1 & \underline{a}_2 & \underline{a}_3 & \underline{a}_4 \end{bmatrix} \tag{2.25}$$

The following equations describe the nonlinear dynamics of a spacecraft with reaction wheels and reaction control jets for large-angle 6-DOF simulations. The rate of change of the total system momentum (spacecraft plus wheels) is not affected by

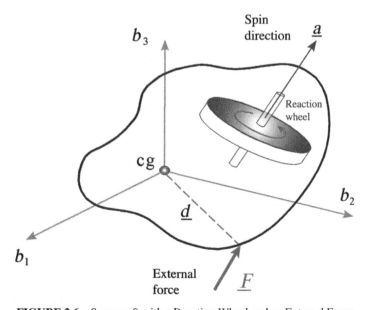

FIGURE 2.6 Spacecraft with a Reaction Wheel and an External Force

the internal wheel control torques. It is affected only by the external disturbances or the RCS jets torques.

$$\dot{H}_{sys} = -\underline{\omega}_b \times \underline{H}_{sys} + \underline{T}_{ext} \tag{2.26}$$

$\underline{T}_{ext}$ is the sum of all external moments applied to the spacecraft. It consists of three terms: the aerodynamic disturbance torques T_D, the gravity gradient torques T_{GG}, and the torques generated by the reaction control jets $\underline{F}_{RCS(j)}$.

$$\underline{T}_{ext} = \underline{T}_D + T_{GG} + \sum_{j}^{Nj} \underline{d}_j \times \underline{F}_{RCS(j)} \tag{2.27}$$

When the spacecraft has an orbital rate ω_0 (rad/sec) the gravity gradient torque is a function of the LVLH Euler angles (ϕ, θ, ψ), as described in equation 2.28.

$$T_{GG} = 3\omega_0^2 \left(c \times J_v c \right), \qquad c = \begin{bmatrix} -\sin\theta \cos\psi \\ \cos\varphi \sin\theta \sin\psi + \sin\varphi \cos\theta \\ -\sin\varphi \sin\theta \sin\psi + \cos\varphi \cos\theta \end{bmatrix} \tag{2.28}$$

Equation 2.29 calculates the rate of change of spacecraft momentum and the spacecraft rotational acceleration $\dot{\omega}$ as a function of internal and external torques. The second equation calculates the rate of change in RW momentum as a function of the torque generated by the wheels, which is opposite to spacecraft torque.

$$\begin{aligned} J_v\dot{\omega} &= -\omega \times J_v\omega + \underline{T}_{RW} + \underline{T}_{ext} \\ \dot{h}_b + \underline{\omega} \times \underline{h}_b &= -\underline{T}_{RW} \end{aligned} \tag{2.29}$$

where

$\underline{h}_b$ is the RW array momentum in the spacecraft body frame,
T_{RW} is a vector of RW torques in spacecraft body axes,
T_{ext} is the external torques applied to the spacecraft in body axes,
J_v is the spacecraft moment of inertial dyadic.

The total system momentum H_{sys} consists of the spacecraft momentum plus the combined wheel momentum. During maneuvering the RW momentum varies significantly but the system momentum is maintained small by frequent momentum dumps. In the absence of external torques the system momentum is constant and close to zero.

$$\underline{H}_{sys} = J_{sc}\underline{\omega} + \underline{h}_b \quad ; \quad \underline{h}_b = C_w^b \underline{h}_w \tag{2.30}$$

The torque applied to the spacecraft body axes by the reaction wheels array is

$$\underline{T}_{RW} = C_w^b \underline{T}_{wi} - \omega \times C_w^b \underline{h}_{wi} \tag{2.31}$$

where the first term is due to the torque motors that accelerate the RWs and the second term is a gyroscopic torque due to coupling between the ith wheel momentum with the spacecraft rate. Resolving this equation for each individual wheel, the torque applied to the spacecraft body axis $\underline{T}_{rwi}$ by the ith RW is a combination of the motor torque T_{wi} applied about the wheel spin vector $\underline{a}_{wi}$ plus the gyroscopic torque due to the coupling between the wheel momentum and the spacecraft rate $\underline{\omega}$. The torque at each individual wheel can be used to excite structural flexibility.

$$\underline{T}_{rwi} = \underline{a}_{wi} T_{wi} - \underline{\omega} \times \underline{a}_{wi} h_{wi} \tag{2.32}$$

The spacecraft rate of change of momentum under the influence of an array of RW torques and the external torques is

$$J_v \underline{\dot{\omega}} = -\underline{\omega} \times J_v \underline{\omega} + C_w^b \underline{T}_{wi} - \underline{\omega} \times C_w^b \underline{h}_{wi} + \underline{T}_{ext} \tag{2.33}$$

Assuming that the system momentum is zero

$$\begin{aligned} \underline{H}_{sys} &= J_v \underline{\omega} + C_w^b \underline{h}_{wi} = 0 \\ J_v \underline{\dot{\omega}} &= C_w^b \underline{T}_{wi} + \underline{T}_{ext} \end{aligned} \tag{2.34}$$

The inertial rate of change of the ith RW momentum about its spin axis is

$$\dot{h}_{wi} = -T_{wi} \tag{2.35}$$

The measured angular rate of the ith reaction wheel relative to the spacecraft is equal to its inertial rate minus the spacecraft rate resolved in the wheel spin axis, that is,

$$w_i = \frac{h_{wi}}{J_{wi}} - \underline{a}_{wi}^T \underline{\omega} \tag{2.36}$$

The Euler angles are measured with respect to the rotating LVLH frame,

$$\begin{pmatrix} \dot{\varphi} \\ \dot{\theta} \\ \dot{\psi} \end{pmatrix} = \frac{1}{\cos\psi} \begin{bmatrix} \cos\psi & -\cos\varphi\,\sin\psi & \sin\varphi\,\sin\psi \\ 0 & \cos\varphi & -\sin\varphi \\ 0 & \sin\varphi\,\cos\psi & \cos\varphi\,\cos\psi \end{bmatrix} \omega + \begin{pmatrix} 0 \\ \omega_o \\ 0 \end{pmatrix} \tag{2.37}$$

where

ω_o is the orbital rate,
J_w is the individual wheel moment of inertia about their spin axis,
J_v is the spacecraft inertia and it includes the weights of the reaction wheels.

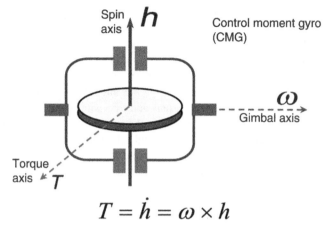

$$T = \dot{h} = \omega \times h$$

FIGURE 2.7   Single-Gimbal Control Moment Gyro

2.14   SPACECRAFT CONTROLLED WITH AN ARRAY OF SINGLE-GIMBAL CMGS

A single-gimbal control moment gyroscope (SGCMG) is shown in Figures 2.7 and 2.8. It consists of a spinning rotor that is mounted on a structure that can be gimbaled perpendicular to the rotor axis. The rotor spin rate is maintained at a constant speed by a small motor that produces a constant angular momentum h. It is the precession of this vector that produces a useful output torque that is substantially greater than a reaction wheel torque, and for this reason it is very attractive in high torque and

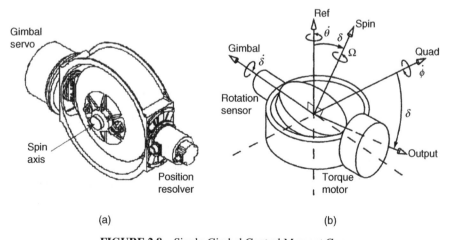

(a)   (b)

FIGURE 2.8   Single-Gimbal Control Moment Gyro

fast maneuvering applications. The direction of the spinning rotor and hence the flywheel momentum can be rotated with respect to the spacecraft by a stronger motor that is mounted at the gimbal. The gimbal motor controls the gimbaling rate, and hence the output torque. By commanding the gimbal to rotate, by means of a servo system that receives rate command from the steering logic, high precession torques are generated by changing the orientation of the angular momentum vector. The reaction torque on the spacecraft T is equal and opposite to the rate of change in momentum vector $\dot{h}$, which is orthogonal to the momentum vector h and also to the gimbaling vector according to the right-hand rule. However, the torque direction at any instant is a function of the gimbal position. The SGCMG essentially acts as a torque amplification device because the output torque magnitude is equal to the CMG momentum multiplied by the gimbal rate. In addition, this does not require much power, because, a small input torque from the gimbal actuator produces a much greater torque in a plane formed by the rotating momentum vector. This large torque amplification plus their capability of storing large amounts of angular momentum over long periods of time makes them especially advantageous as attitude control actuators for agile spacecraft that require fast maneuvering, and high precision. They are also used in large space structures, such as a space station. Single-gimbal and double-gimbal CMGs have been used for attitude control of the Skylab, the MIR, and the International Space Station (ISS).

Another attractive feature of CMGs compared with reaction wheels is that the rotor in a CMG spins at a constant rate which places the vibrations at known frequencies while in a RW, the rotor speed changes, thus, exciting the spacecraft structure in multiple frequencies which may not be desirable in precision applications. CMGs, however, are complex systems, expensive, and require complex controls with singularity avoidance algorithms.

2.14.1 Math Model of a SGCMG Array

The momentum of one CMG about the Gimbal, Output, and Spin axes can be calculated in CMG axes, as a function of the spinning rotor momentum and the spacecraft rotation rate.

$$\begin{pmatrix} h_g \\ h_o \\ h_s \end{pmatrix} = \begin{pmatrix} J_g \dot{\delta}_i \\ J_o \left(\dot{\varphi} \cos \delta - \dot{\theta} \sin \delta \right) \\ h_0 + J_s \left(\dot{\theta} \cos \delta + \dot{\varphi} \sin \delta \right) \end{pmatrix} \tag{2.38}$$

Figure 2.8 shows the local coordinates of a SGCMG vectors which are defined relative to the spacecraft axes. They are the Gimbal, Ref, and Quad axes ($\underline{m}$, $\underline{r}$, $\underline{q}$) which are fixed relative to the spacecraft. The Gimbal ($\underline{m}$) is the axis about which the rotor is gimbaled. Ref ($\underline{r}$) is the direction of the spin vector when the gimbal angle δ_i is zero. Quad ($\underline{q}$) is the quadrature formed by the cross product of the Gimbal and Ref axes, see Figure 2.9. We also have the Spin and the Output torque axes which vary as a function of the gimbal angle. Equation 2.39 denotes the rate of change in SGCMG momentum which is the moment generated by each SGCMG in the Gimbal, Output, and Spin axes respectively, as a result of gimbaling.

$$
\begin{bmatrix} M_G \\ M_O \\ M_S \end{bmatrix} =
$$

$$
\begin{bmatrix} T_{gi} \\ J_o(\ddot{\varphi}\cos\delta - \dot{\varphi}\dot{\delta}_i\sin\delta - \dot{\theta}\dot{\delta}_i\cos\delta - \ddot{\theta}\sin\delta) + h_0\dot{\delta}_i + \dot{\delta}_i(J_s - J_g)(\dot{\theta}\cos\delta + \dot{\varphi}\sin\delta) \\ J_s(\dot{\Omega} + \ddot{\theta}\cos\delta + \ddot{\varphi}\sin\delta + \dot{\varphi}\dot{\delta}_i\cos\delta - \dot{\theta}\dot{\delta}_i\sin\delta) + \dot{\delta}_i(J_g - J_o)(\dot{\varphi}\cos\delta - \dot{\theta}\sin\delta) \end{bmatrix}
$$

$$(2.39)$$

The reaction torque which is applied to the spacecraft is the opposite: $-[M_G, M_O, M_S]$. Note that the angular rotation of the spacecraft couples with the stored CMG momentum vector producing additional torque on the interface. The rotation rates of the CMG can be related to the spacecraft body rate. If $\underline{\omega}$ is the spacecraft roll, pitch and yaw body rate vector (ω_X, ω_Y, ω_Z), equation 2.40 resolves the spacecraft rates about the CMG axes: ($\underline{r}$, $\underline{q}$, $\underline{m}$). $\dot{\theta}$ and $\dot{\varphi}$ are the spacecraft rates resolved about the CMG reference and quad axes. The gimbal rate $\dot{\delta}_i$ in addition to the gimbal rate relative to spacecraft $\dot{\delta}$ also includes the spacecraft rotational rate about the CMG gimbal.

$$
\begin{aligned}
\dot{\theta} &= \omega_X \cos\gamma + \omega_Y \sin\gamma \\
\dot{\varphi} &= -\omega_X \cos\beta \sin\gamma + \omega_Y \cos\beta \cos\gamma + \omega_Z \sin\beta \\
\dot{\xi} &= \omega_X \sin\beta \sin\gamma - \omega_Y \sin\beta \cos\gamma + \omega_Z \cos\beta \\
\dot{\delta}_i &= \dot{\delta} + \dot{\xi}
\end{aligned}
$$

$$(2.40)$$

where

J_g is the CMG moment of inertia about its gimbal axis,
J_o is the CMG moment of inertia about its output axis,
J_s is the CMG moment of inertia about its spin axis,
T_{gi} is the torque applied by the torque motor at the gimbal,
$\ddot{\delta}_i$ is the inertial angular acceleration of the rotor about the gimbal including spacecraft,
δ is the CMG gimbal rotation about the $\underline{m}$ axis measured from the Ref axis,
h_0 is the constant CMG momentum about its spin axis ($I_s\Omega$),
$\dot{\Omega}$ is the rotor spin acceleration,
$\dot{\theta}$ is the vehicle rate about the CMG reference $\underline{r}$ axis,
$\dot{\varphi}$ is the vehicle rate about the CMG quad $\underline{q}$ axis,
$\dot{\xi}$ is the vehicle rate about the CMG gimbal $\underline{m}$ axis.

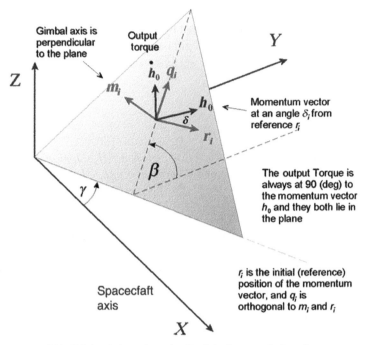

FIGURE 2.9   Orientation of a CMG in Spacecraft Coordinates

The projection matrix P in equation 2.41 transforms the CMG torques from CMG reference frame to the spacecraft (x, y, z) axes.

$$
\begin{pmatrix} M_X \\ M_Y \\ M_Z \end{pmatrix} =
$$

$$
\begin{bmatrix} \sin\beta\sin\gamma & -\sin\delta\cos\gamma - \cos\delta\cos\beta\sin\gamma & \cos\delta\cos\gamma - \sin\delta\cos\beta\sin\gamma \\ -\sin\beta\cos\gamma & -\sin\delta\sin\gamma + \cos\delta\cos\beta\cos\gamma & \cos\delta\sin\gamma + \sin\delta\cos\beta\cos\gamma \\ \cos\beta & \cos\delta\sin\beta & \sin\delta\sin\beta \end{bmatrix} \begin{pmatrix} M_G \\ M_O \\ M_S \end{pmatrix}
$$

$$(2.41)$$

When a spacecraft is controlled by CMGs at least three CMGs are needed to provide 3-axes control. Let us consider an array of SGCMG mounted on the surfaces of a pyramid with their gimbal axes directions $(\underline{m}_i)$ perpendicular to one of the surfaces and the momentum direction (h_i) always aligned with the surface of the pyramid as the gimbal σ_i rotates. The output torque from each CMG is equal to the rate of change of angular momentum which is in the $(\underline{m}_i \times \underline{h}_i)$ direction and proportional to the gimbal rate $\dot{\delta}_i$. From the pyramid surface orientations we can calculate three important matrices that will be used in the equations of motion.

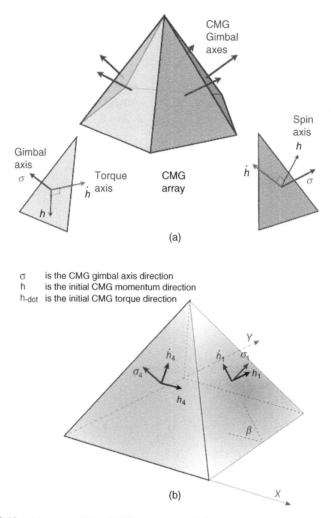

FIGURE 2.10   (a) Array of Five CMGs in a Pyramid Configuration; (b) Array of Four CMGs in a Pyramid Configuration

Let us consider an SGCMG pyramid arrangement shown in Figure 2.10b, where the spacecraft has four SGCMGs mounted onto the four faces of the four-sided pyramid. All CMGs have the same constant angular momentum $h_0 = 1200$ (ft-lb-sec). Their CMG momentum vectors $\underline{h}_i$ are initially (at zero gimbal, $\sigma_{0i} = 0$) parallel to the spacecraft X–Y plane producing zero total momentum. Their momentum vectors are constrained to lay parallel to the surface of the pyramid and they can be continuously rotated about the gimbal vectors $\underline{\sigma}_i$, which are perpendicular to each surface. The orientation of each CMG relative to the spacecraft is shown in Figure 2.10b. The pyramid angle β is 68°, and the γ_i angles of the four surfaces are: 90°, 180°, 270°, and $0°$. We can create the 3×4 gimbal to body transformation matrix M_g^b by stacking together in columns the four gimbal direction unit vectors $\underline{m}_i$ as shown in equation 2.42.

$$M_g^b = \begin{bmatrix} \underline{m}_1 & \underline{m}_2 & \underline{m}_3 & \underline{m}_4 \end{bmatrix} \text{ where, } \underline{m}_i = \begin{bmatrix} \sin \beta_i \sin \gamma_i \\ -\sin \beta_i \cos \gamma_i \\ \cos \beta_i \end{bmatrix}$$

(2.42)

$$M_g^b = \begin{bmatrix} \sin \beta & 0 & -\sin \beta & 0 \\ 0 & \sin \beta & 0 & -\sin \beta \\ \cos \beta & \cos \beta & \cos \beta & \cos \beta \end{bmatrix} = \begin{bmatrix} 0.927 & 0 & -0.927 & 0 \\ 0 & 0.927 & 0 & -0.927 \\ 0.375 & 0.375 & 0.375 & 0.375 \end{bmatrix}$$

Similarly, we create the 3×4 reference directions matrix R by stacking together the initial directions $\underline{r}_i$ of the momentum vectors $\underline{h}_i$ when the gimbal angles are zero, $\sigma_{0i}=0$. In this case the initial gimbal angles produce zero momentum bias.

$$R = \begin{bmatrix} \underline{r}_1 & r_2 & r_3 & \underline{r}_4 \end{bmatrix}, \quad \text{where } \underline{r}_i = \begin{bmatrix} \cos \gamma_i \\ \sin \gamma_i \\ 0 \end{bmatrix}; \; R = \begin{bmatrix} 0 & -1 & 0 & 1 \\ 1 & 0 & -1 & 0 \\ 0 & 0 & 0 & 0 \end{bmatrix} \quad (2.43)$$

We must also define a third 3×4 quad-matrix Q that contains column vectors of the cross product direction unit vectors $\underline{q}_i$.

$$Q = \begin{bmatrix} \underline{q}_1 & \underline{q}_2 & \underline{q}_3 & \underline{q}_4 \end{bmatrix}; \quad \underline{q}_i = \left(\underline{m}_i \times \underline{r}_i \right);$$
$$Q = \begin{bmatrix} -0.375 & 0 & 0.375 & 0 \\ 0 & -0.375 & 0 & 0.375 \\ 0.927 & 0.927 & 0.927 & 0.927 \end{bmatrix}$$

(2.44)

Notice, that the pyramid structure is only used for visualization. The CMGs do not have to be physically mounted on the four surfaces of an actual pyramid as shown in Figure 2.10, but they can be translated anywhere on the spacecraft as long as their gimbal axes $\underline{m}_i$ and their reference momentum vectors $\underline{r}_i$ are parallel to the directions shown in the pyramid. See, for example, the CMG cluster in Figure 2.11. The CMGs are typically mounted on a structure that is attached to the spacecraft by means of vibration isolation struts.

2.14.2 Steering Logic for a Spacecraft with SGCMGs

In order to control the spacecraft by generating the required control torques from an array of SGCMGs it is necessary to develop a steering logic. An optimal steering logic is one that maneuvers the SGCMG gimbal rates to generate torques that are

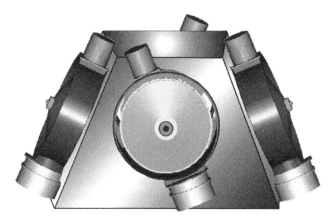

FIGURE 2.11   Cluster of Four Single-Gimbal CMGs Mounted on a Pyramid Structure

equal to the commanded spacecraft control torques. However, one of the principal difficulties in using SGCMGs for spacecraft attitude control is the geometric singularity problem in which no control torque is generated for the commanded gimbal rates. The development of a SGCMG steering logic should also consider the avoidance of the singularities.

Similar to the RWA model, the combined spacecraft plus CMG rate of change of momentum is not affected by the internal CMG torque but it is related only to the external torques.

$$\dot{\underline{H}}_{sys} + \underline{\omega} \times \underline{H}_{sys} = \underline{T}_{ext} \tag{2.45}$$

where

$\underline{H}_{sys}$ is the combined CMG plus spacecraft system momentum,
$\underline{T}_{ext}$ is the external torque vector, and
$\underline{\omega}$ is the spacecraft angular rate.

By introducing the internal CMG torque we separate the spacecraft and CMG rate of change of momentum equations and solve for the spacecraft rate $\underline{\omega}$ as a function of internal plus external torques.

$$\begin{aligned} J_v\dot{\underline{\omega}} &= -\underline{\omega} \times J_v\underline{\omega} + \underline{T}_{cmg} + \underline{T}_{ext} \\ \dot{\underline{H}}_{cmg} + \underline{\omega} \times \underline{H}_{cmg} &= -\underline{T}_{cmg} \end{aligned} \tag{2.46}$$

where the total system momentum consists of spacecraft plus CMG momentum.

$$\underline{H}_{sys} = J_v\underline{\omega} + \underline{H}_{cmg} \tag{2.47}$$

The internal CMG torque $\underline{T}_{cmg}$ applied to the spacecraft is equal and opposite to the torque applied to the CMG array. It consists of two terms: the control torque

$\underline{T}_{con}$ intended to control the spacecraft which is also the rate of change in the CMG momentum, plus a component to cancel the gyroscopic torque $\omega \times H_{cmg}$.

$$T_{cmg} = T_{con} - \omega \times H_{cmg} \tag{2.48}$$

The control torque $\underline{T}_{con}$ is a nonlinear function of the gimbal angles and gimbal rates.

$$T_{con} = -[A(\delta)]\,\underline{\dot{\delta}} = -\dot{H}_{cmg} \tag{2.49}$$

The angular momentum vector of the entire CMG array in body axes $\underline{H}_{cmg}$ is obtained by combining the individual CMG momentum defined in CMG axes: Gimbal, Output, and Spin axes from equation 2.38, and transform them into spacecraft axes using the transformation matrix P_i, as follows:

$$\underline{H}_{cmg} = \sum_{i=1}^{N_{cmg}} P_i \begin{pmatrix} h_G \\ h_O \\ h_S \end{pmatrix}_i \tag{2.50}$$

where matrix P_i transforms the ith CMG momentum from (Gimbal, Output, Spin) axes to spacecraft axes.

$$P_i = \begin{bmatrix} \sin\beta\sin\gamma & -\sin\delta\cos\gamma - \cos\delta\cos\beta\sin\gamma & \cos\delta\cos\gamma - \sin\delta\cos\beta\sin\gamma \\ -\sin\beta\cos\gamma & -\sin\delta\sin\gamma + \cos\delta\cos\beta\cos\gamma & \cos\delta\sin\gamma + \sin\delta\cos\beta\cos\gamma \\ \cos\beta & \cos\delta\sin\beta & \sin\delta\sin\beta \end{bmatrix} \tag{2.51}$$

Similarly, the control torque from each CMG is transformed from CMG axes to spacecraft axes, and they are combined to form the total CMG torque in body axes.

$$\underline{T}_{cmg} = \sum_{i=1}^{N_{cmg}} P_i \begin{pmatrix} M_G \\ M_O \\ M_S \end{pmatrix}_i \tag{2.52}$$

where

$$\begin{bmatrix} M_G \\ M_O \\ M_S \end{bmatrix}$$

$$= \begin{bmatrix} T_{gi} \\ J_o(\ddot{\varphi}\cos\delta - \dot{\varphi}\dot{\delta}_i\sin\delta - \dot{\theta}\dot{\delta}_i\cos\delta - \ddot{\theta}\sin\delta) + h_0\dot{\delta}_i + \dot{\delta}_i(J_s - J_g)(\dot{\theta}\cos\delta + \dot{\varphi}\sin\delta) \\ J_s(\dot{\Omega} + \ddot{\theta}\cos\delta + \ddot{\varphi}\sin\delta + \dot{\varphi}\dot{\delta}_i\cos\delta - \dot{\theta}\dot{\delta}_i\sin\delta) + \dot{\delta}_i(J_g - J_o)(\dot{\varphi}\cos\delta - \dot{\theta}\sin\delta) \end{bmatrix} \tag{2.53}$$

A simplified model can be obtained when we ignore the momentum about the gimbal and output axes, because they are small, and consider only the CMG momentum about the spin axes. The combined CMG angular momentum vector is calculated by equation 2.54, as a function of the individual CMG momentums h_{0i} which depend on their spin axis orientations.

$$\underline{H}_{cmg} = \sum_{i=1}^{N_{cmg}} (\cos \delta_i \, \underline{r}_i + \sin \delta_i \, \underline{q}_i) \, h_{cmg(i)} \qquad (2.54)$$

Also, by combining equations 2.46 and 2.49 we can rewrite equation 2.46 in terms of only the control torque instead of the total CMG torque $\underline{T}_{cmg}$ as shown in 2.55. T_{con} is the steering torque designed to shape the spacecraft rate as commanded by the attitude control system. Equation 2.54 can be used instead of 2.46 in simple 6-DOF simulations that do not require gimbal torque dynamics and coupling with structural flexibility.

$$J_{sc}\underline{\dot{\omega}} + (\underline{\omega} \times \underline{H}_{sys}) = T_{con} + \underline{T}_{ext}$$
$$\underline{\dot{H}}_{cmg} = -\underline{T}_{con} \qquad (2.55)$$

For a more accurate calculation of the CMG momentum $\underline{H}_{cmg}$ that also includes the effects due to the gimbal torques, we may integrate equation 2.46 using T_{cmg} from 2.52. The matrix A used in equation 2.49 relates the gimbal rates to rate of change in CMG momentum which is the control torque. It is a time-varying $3 \times N_{cmg}$ matrix consisting of N_{cmg} column vectors $\underline{a}_i$, Its elements $\underline{a}_i$ are related to the Ref and Quad vectors (r_i and q_i) of each CMG and vary with the positions of the gimbal angles δ_i.

$$A(\delta) = \begin{pmatrix} \underline{a}_1 \ \underline{a}_2 \ \underline{a}_3 \ \underline{a}_4 \end{pmatrix}, \text{ where } a_i = (\cos \delta_i \underline{q}_i - \sin \delta_i \underline{r}_i)h_{0i} \qquad (2.56)$$

δ_i	is the gimbal angle for CMG (i),
$\underline{r}_i$	is a unit vector of the initial momentum direction for CMG (i),
$\underline{m}_i$	is a unit vector of the gimbal direction for CMG (i),
$\underline{q}_i$	is the orthogonal direction ($m_i \times r_i$) for CMG (i),
T_{con}	is the control torque applied to the gimbal by the motor,
T_{gi}	is the gyroscopic torque $-\omega \times H_{cmg}$ resolved in gimbal (i) direction,
M_g^b	is the (3×4) transformation matrix from gimbal axis to body axis,
h_{0i}	is the constant momentum of the ith CMG about its spin axis,
N_{CMG}	is the number of CMGs used.

The CMG gimbal rates are controlled by a servo system that generates gimbal torques T_{gi}. The servo torque at the gimbal of each CMG is also counteracting the gyroscopic disturbance torque created by the spacecraft rates $\dot{\theta}$ and $\dot{\varphi}$ resolved about the CMG reference and quad axes, as defined in equation 2.40. Ignoring friction, the inertial acceleration of each CMG gimbal is obtained from the gimbal moment equation 2.57, where T_{gi} is the motor torque applied at each gimbal. Even though the CMG moment of inertia about the gimbal J_g is relatively small, the $\omega \times h_0$ gyroscopic moment produced by the CMG momentum coupling with spacecraft rate is a big torque that requires a powerful gimbal servomotor in order to be able to achieve the required gimbal rate. In simple models we can assume that the gimbal rates are equal to the commanded rates $\dot{\delta}_i = \dot{\delta}_{comd(i)}$.

$$J_g \ddot{\delta}_i + h_{0i}\left(\dot{\theta}_i \sin\delta - \dot{\varphi}_i \cos\delta\right) = T_{gi} \tag{2.57}$$

The relative gimbal angle δ for each CMG is obtained by integrating $\dot{\delta} = \dot{\delta}_i - \dot{\xi}$ The spacecraft attitude error is used by the control law to calculate the acceleration command $\dot{\omega}_{comd}$. The attitude quaternion is updated by integrating the quaternion rate which is a function of the body rate and the previous quaternion, as shown in equation 2.58.

$$\dot{\underline{Q}} = 0.5 \begin{bmatrix} 0 & \omega_3 & -\omega_2 & \omega_1 \\ -\omega_3 & 0 & \omega_1 & \omega_2 \\ \omega_2 & -\omega_1 & 0 & \omega_3 \\ -\omega_1 & -\omega_2 & -\omega_3 & 0 \end{bmatrix} \underline{Q} \tag{2.58}$$

The pseudo-inverse of matrix A is used at each iteration for calculating the gimbal rate commands in the SGCMG steering law. The steering law makes the vehicle acceleration to be equal to the commanded acceleration.

$$\dot{\delta}_{com} = -A^+(\delta)\left(J_v \dot{\omega}_{com} + \left(\omega \times H_{sys}\right)_{estim}\right), \quad \text{where } A^+ = A^T\left(A A^T\right)^{-1},$$

$$\dot{\omega} = \dot{\omega}_{com} \tag{2.59}$$

The development of a CMG steering logic, however, should also consider the avoidance of the singularities. In Section 10.3 we demonstrate the design of a space station attitude control system using CMGs.

3

LINEAR PERTURBATION EQUATIONS USED IN CONTROL ANALYSIS

The following differential equations are linearized about an operating point or a trim condition. They are known as "perturbation equations" because the states represent variations of vehicle variables from the trim variables, which are denoted with the subscript "0." They are used for analyzing vehicle stability in the frequency domain, robustness to parameter uncertainties, and for designing control laws at fixed flight conditions. They are also used in linear simulations to appraise the vehicle performance by examining its response to commands and transient disturbances. The inputs and outputs to the dynamic models, however, represent variations from the trim variables and not the actual values. They are implemented in the Flixan$^®$ program that converts them into state-space form that can be loaded into, and processed by MATLAB$^®$ for control analysis, as shown in the examples in Chapter 10.

3.1 FORCE AND ACCELERATION EQUATIONS

The total variation in the external forces on a vehicle consists of force variations due to aerodynamics, engines (TVC and throttling), propellant sloshing, wind-gust disturbance, and gravity. Although slosh is an internal force we treat it as external in this case because it is modeled as a separate body.

$$
\begin{aligned}
\delta F_X &= F_{Xaero} + F_{Xeng} + F_{Xslosh} + F_{Xgrav} + F_{Xgust} \\
\delta F_Y &= F_{Yaero} + F_{Yeng} + F_{Yslosh} + F_{Ygrav} + F_{Ygust} \\
\delta F_Z &= F_{Zaero} + F_{Zeng} + F_{Zslosh} + F_{Zgrav} + F_{Zgust}
\end{aligned}
\tag{3.1}
$$

Performance Evaluation and Design of Flight Vehicle Control Systems, First Edition. Eric T. Falangas.
© 2016 by Eric T. Falangas. Published 2016 by John Wiley & Sons, Inc.

Equation 3.2 calculates the variations in measured acceleration at the CG, as a function of the external force variations, with the exception of the gravity forces, since the accelerometers do not "feel" the gravity force. They are derived from the "F = ma" equation. Note, M_T represents the vehicle total mass including the slosh masses, engines, and aerosurfaces and it is used in the x-acceleration. M_V is the vehicle mass excluding the moving masses (slosh masses, engines, and aerosurfaces). M_V is used instead of M_T in the normal and lateral acceleration equations because the effects of the moving masses are included in the external forces.

$$\ddot{x}_{CG} = \left(F_{Xeng} + F_{Xaero} + F_{Xslosh} + F_{Xgust}\right) / M_T$$
$$\ddot{y}_{CG} = \left(F_{Yeng} + F_{Yaero} + F_{Yslosh} + F_{Ygust}\right) / M_V \qquad (3.2)$$
$$\ddot{z}_{CG} = \left(F_{Zeng} + F_{Zaero} + F_{Zslosh} + F_{Zgust}\right) / M_V$$

3.2   LINEAR ACCELERATIONS

Changes in the external forces δF_x, δF_y, δF_z along the vehicle x-, y-, and z–axes, respectively, produce variations in vehicle acceleration $(\dot{u}, \dot{v}, \dot{w})$, as shown in equations 3.3. Equations 3.3 denotes changes in vehicle acceleration due to external force variations. It also calculates changes in the angles of attack and sideslip and the change in vehicle velocity δV as a function of the external forces. Note that δV represents the change in velocity along the velocity vector V_0. Equations 3.3 are obtained by perturbing the nonlinear equation 2.7, and substituting $(w$ and $v)$ with $(\alpha$ and $\beta)$, respectively, from equation 3.4.

$$\dot{u} = \delta\dot{V}\cos\alpha_0 = V_0\sin\alpha_0\,(\dot{\alpha} - q) + V_0\left(\frac{(\dot{\alpha}_0 - Q_0)}{\cos\alpha_0} + R_0\beta_0\tan\alpha_0\right)\alpha + V_0\beta_0 r$$

$$+ V_0\left(R_0 + \dot{\beta}_0\cos\alpha_0\right)\beta + V_0\beta_0\cos\alpha_0\,\dot{\beta}$$

$$+ \left[\beta_0\dot{\beta}_0\cos\alpha_0 + R_0\beta_0 + (\dot{\alpha}_0 - Q_0)\sin\alpha_0\right]\delta V + \frac{\delta F_X}{M_V} \qquad (3.3)$$

$$\dot{v} = V_0\dot{\beta} = V_0\sin\alpha_0 p - V_0\cos\alpha_0 r + V_0\left(R_0\sin\alpha_0 + P_0\cos\alpha_0\right)\alpha - \beta_0\delta\dot{V} - \dot{V}_0\beta$$

$$+ \left[\frac{\dot{V}_0\beta_0}{V_0} - R_0\cos\alpha_0\right]\delta V + \frac{\delta F_Y}{M_V}$$

$$\dot{w} = V_0\cos\alpha_0\,\dot{\alpha}$$

$$= V_0\cos\alpha_0 q + V_0\left(\dot{\beta}_0\sin\alpha_0 - P_0\right)\beta - \sin\alpha_0\,\delta\dot{V} + V_0\beta_0\left(\sin\alpha_0\,\dot{\beta} - p\right)$$

$$+ \left[\frac{(V_0\beta_0\dot{\beta}_0 - \dot{V}_0)}{\cos\alpha_0} - P_0\beta_0 V_0\tan\alpha_0\right]\alpha + \left[\frac{\dot{V}_0\sin\alpha_0}{V_0} + Q_0\cos\alpha_0\right]\delta V + \frac{\delta F_Z}{M_V}$$

Small changes in the velocities (w and v) along z- and y-axes create variations in the angles of attack and sideslip α and β relative to the trim aerodynamic angles α_0 and β_0, respectively, as shown in equation 3.4.

$$\alpha = \frac{w}{V_0 \cos \alpha_0} \qquad \beta = \frac{v}{V_0} \tag{3.4}$$

Linear Accelerations at the Engine Gimbals

The translational accelerations at the TVC engine gimbals are used in the load-torque calculations. They include the effects of the vehicle rotations plus bending. The accelerations at the kth gimbal consist of the following three terms:

(a) linear acceleration at the CG,

(b) contributions due to the vehicle angular acceleration, and

(c) translational acceleration components due to structural flexibility at the gimbals.

$$a_{xe}(k) = \ddot{x}_{CG} + l_{Zek}\dot{q} - l_{Yek}\dot{r} + \sum_{j=1}^{Nm} \varphi_{xe}(k,j)\ddot{\eta}(j)$$

$$a_{ye}(k) = \ddot{y}_{CG} + l_{Xek}\dot{r} - l_{Zek}\dot{p} + \sum_{j=1}^{Nm} \varphi_{ye}(k,j)\ddot{\eta}(j) \tag{3.5}$$

$$a_{ze}(k) = \ddot{z}_{CG} + l_{Yek}\dot{p} - l_{Xek}\dot{q} + \sum_{j=1}^{Nm} \varphi_{ze}(k,j)\ddot{\eta}(j)$$

where

$\phi_{xe}(k,j)$ is the modal shape along x of the jth mode at the kth engine location,

$\eta(j)$ is the generalized modal displacement for the jth mode,

$l_{Xek}, l_{Yek}, l_{Zek}$ are the moment arms between the kth engine and the vehicle CG in the x, y, and z directions, respectively, $l_{Zek} = Z_{e(k)} - Z_{CG}$.

Linear Accelerations at the Control Surface Hinges

Similarly, the acceleration along x, y, and z, at the midpoint of the kth control surface hinge line is calculated from equation 3.6. The accelerations at the control surfaces are used in the hinge moment equations.

$$a_{xcs}(k) = \ddot{x}_{CG} + l_{Zsk}\dot{q} - l_{Ysk}\dot{r} + \sum_{j=1}^{Nm} \varphi_{xcs}(k,j)\ddot{\eta}(j)$$

$$a_{ycs}(k) = \ddot{y}_{CG} + l_{Xsk}\dot{r} - l_{Zsk}\dot{p} + \sum_{j=1}^{Nm} \varphi_{ycs}(k,j)\ddot{\eta}(j) \tag{3.6}$$

$$a_{zcs}(k) = \ddot{z}_{CG} + l_{Ysk}\dot{p} - l_{Xsk}\dot{q} + \sum_{j=1}^{Nm} \varphi_{zcs}(k,j)\ddot{\eta}(j)$$

The moment arm distances between the center of the kth control surface hinge line and the vehicle CG are defined in the following equations:

$$1_{XSk} = X_{Sk} - X_{CG} \quad 1_{YSk} = Y_{Sk} - Y_{CG} \quad 1_{ZSk} = Z_{Sk} - Z_{CG} \tag{3.7}$$

3.3   MOMENT AND ANGULAR ACCELERATION EQUATIONS

Equation 3.8 is the moment equations linearized at a fixed flight condition and they calculate the variations in roll, pitch, and yaw body rates (p, q, r), relative to the nominal trajectory rates (P_0, Q_0, R_0), as a function of variations in the total external torques: ΣL_x, ΣM_y, and ΣN_z (roll, pitch, and yaw), which are applied to the vehicle. The moments and products of inertia values should not include the slosh masses and the gimbaling engine nozzles because their presence in the equations is captured by the slosh and the tail-wags-dog (TWD) forces.

$$
\begin{aligned}
I_{xx}\dot{p} &= I_{xy}\left(\dot{q} - P_0 r - R_0 p\right) + I_{xz}\left(\dot{r} + Q_0 p + P_0 q\right) + 2I_{yz}\left(Q_0 q - R_0 r\right) \\
&\quad + \left(I_{yy} - I_{zz}\right)\left(Q_0 r + R_0 q\right) + \sum L_x \\
I_{yy}\dot{q} &= I_{xy}\left(\dot{p} + Q_0 r + R_0 q\right) + I_{yz}\left(\dot{r} - Q_0 p - P_0 q\right) + 2I_{xz}\left(R_0 r - P_0 p\right) \\
&\quad + \left(I_{zz} - I_{xx}\right)\left(P_0 r + R_0 p\right) + \sum M_y \\
I_{zz}\dot{r} &= +I_{yz}\left(\dot{q} + P_0 r + R_0 p\right) + I_{xz}\left(\dot{p} - Q_0 r - R_0 q\right) + 2I_{xy}\left(P_0 p - Q_0 q\right) \\
&\quad + \left(I_{xx} - I_{yy}\right)\left(Q_0 p + P_0 q\right) + \sum N_z
\end{aligned}
\tag{3.8}
$$

The sum of external torques in equation 3.8 are due to the engine TVC forces, jet firing, aerodynamic forces, wind-gust disturbances, and sloshing, and they are calculated in equation 3.9.

$$
\begin{aligned}
\sum L_X &= L_{Xaero} + L_{X_{eng}} + L_{Xslosh} + L_{Xdist} \\
\sum M_Y &= M_{Yaero} + M_{Y_{eng}} + M_{Yslosh} + M_{Ydist} \\
\sum N_Z &= N_{Zaero} + N_{Z_{eng}} + N_{Zslosh} + N_{Zdist}
\end{aligned}
\tag{3.9}
$$

The dynamic coupling between axes is almost negligible in cylindrical boosters. Aircraft and rocket-plane type of vehicles, such as the Space Shuttle and most atmospheric vehicles have a significant amount of cross-coupling between the roll and yaw axes due to the I_{xz} product of inertia term and also due to the aerodynamic coefficients $C_{l\beta}$, $C_{n\beta}$, etc. Coupling between all three: roll, pitch, and yaw moment axes can also occur, mainly due to the products of inertia, steady vehicle rates (P_0, Q_0, R_0), CG offsets, lack of symmetry in the aerodynamic coefficients, and asymmetric TVC forces. It can also happen when an aircraft is flying at constant bank angle Φ_0 and sideslip β_0, or when a rocket vehicle is experiencing a steady sideslip angle β_0 due to cross-wind or because of the loss of thrust in an off-centered engine. Therefore, in the equations of motion we are not ignoring the cross-coupling terms between axes.

3.4   GRAVITATIONAL FORCES

Equations 3.10 calculate the variations in the earth's gravity force acting on the vehicle CG as a function of variations in $(\alpha, \delta h, \theta, \phi, \delta V)$. They are obtained by linearizing Equations 2.9 along the vehicle x, y, and z axes, as a function of the Euler angles and other variables. The gravitational constant g in equation 3.10 is a function of both altitude and velocity because we include the centripetal force due to V_0.

$$\frac{F_{Xgrav}}{M_V} = -g\cos\Theta_0\,\theta - g\sin\Theta_0\tan\alpha_0\,\alpha$$

$$+ \frac{\sin\Theta_0}{\left(R_e + h_0\right)^2}\left[\frac{2gR_e^2}{R_e + h_0} - V_0^2\right]\delta h + \left[\frac{2V_0}{R_e + h_0}\right]\sin\Theta_0\,\delta V$$

$$\frac{F_{Ygrav}}{M_V} = g\left(\cos\Theta_0\cos\Phi_0\varphi - \sin\Theta_0\sin\Phi_0\,\theta\right) \tag{3.10}$$

$$- \frac{\cos\Theta_0\sin\Phi_0}{\left(R_e + h_0\right)^2}\left[\frac{2gR_e^2}{R_e + h_0} - V_0^2\right]\delta h - \left[\frac{g}{V_0} + \frac{2V_0}{R_e + h_0}\right]\cos\Theta_0\sin\Phi_0\,\delta V$$

$$\frac{F_{Zgrav}}{M_V} = -g\left(\sin\Theta_0\cos\Phi_0\theta + \cos\Theta_0\sin\Phi_0\,\varphi\right) + g\left(\cos\Theta_0\cos\Phi_0\tan\alpha_0\right)\alpha$$

$$- \frac{\cos\Theta_0\cos\Phi_0}{\left(R_e + h_0\right)^2}\left[\frac{2gR_e^2}{R_e + h_0} - V_0^2\right]\delta h - \left[\frac{g}{V_0} + \frac{2V_0}{R_e + h_0}\right]\cos\Theta_0\cos\Phi_0\,\delta V$$

R_e is the radius of the earth,
h_0 is the vehicle altitude from sea level,
V_0 is the vehicle horizontal velocity,
α_0, α are the nominal angles of attack and its variation from nominal,
Φ_0, Θ_0 are the nominal vehicle roll and pitch Euler angles,
ϕ, θ are the attitude variation angles in roll and pitch,
g, g_0 is the gravity acceleration at the flight condition, and at zero altitude.

The acceleration constant g is calculated from the following equation:

$$g = g_0 \left(\frac{R_e}{R_e + h_0} \right)^2 - \left(\frac{V_0^2}{R_e + h_0} \right) \tag{3.11}$$

3.5   FORCES AND MOMENTS DUE TO AN ENGINE PIVOTING AND THROTTLING

A typical engine nozzle k with a nominal thrust T_{ek} is not necessarily aligned with the vehicle x-axis, but it is tilted at angles Δ_E and Δ_Z in pitch and yaw, respectively, see Figure 3.1 and Figure 3.3. This is due to hard mounting or for trimming the steady moments. The angle Δ_E is the elevation or pitch between the thrust direction and the vehicle x–y plane, positive down from the x–y plane. The angle Δ_Z is the azimuth or yaw rotation about the gimbal z-axis of the thrust vector projection in the x–y plane, measured relative to the -x direction. Figure 3.2 shows the reaction control thrusters or RCS. They are fixed and do not gimbal, and they generate moments by varying the thrust.

Total Deflection at an Engine Gimbal

The nozzles of gimbaling engines further deflect from their trim (Δ_E and Δ_Z) positions in order to guide and stabilize the vehicle against disturbances. Sometimes they even vary their thrust in order to control the vehicle. These additional deflections generate small components of thrust force variations perpendicular to the velocity vector that affect the vehicle attitude and direction. The actuators provide the forces required to rotate the nozzles in pitch δ_{ya} and in yaw δ_{za} directions, relative to their trim positions (Δ_E, Δ_Z). The intended deflections, however, are corrupted by

FIGURE 3.1   Shuttle TVC Engine Nozzles and RCS Thrusters

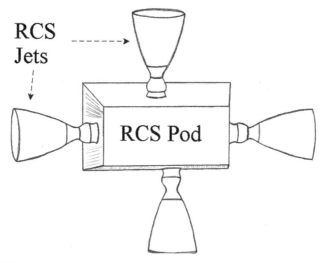

FIGURE 3.2   Reaction Control Thrusters Used for Attitude Control

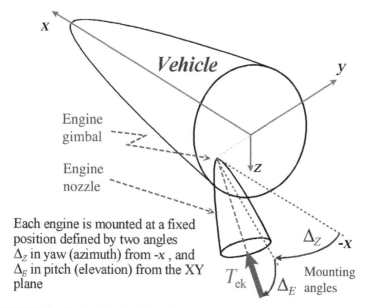

Each engine is mounted at a fixed position defined by two angles Δ_Z in yaw (azimuth) from -x , and Δ_E in pitch (elevation) from the XY plane

FIGURE 3.3   Engine Nozzle Orientation Angles with Respect to the Vehicle Axes

structural flexibility at the engine gimbal. In equation 3.12 the deflections $\delta_{ye}(i)$ and $\delta_{ze}(i)$ represent the total deflections of the nozzle in pitch and yaw, relative to the trim angles. It consists of two components as illustrated in Figure 3.4

   (a) the rotation of the nozzle about its gimbal as a result of the actuator displacement, and

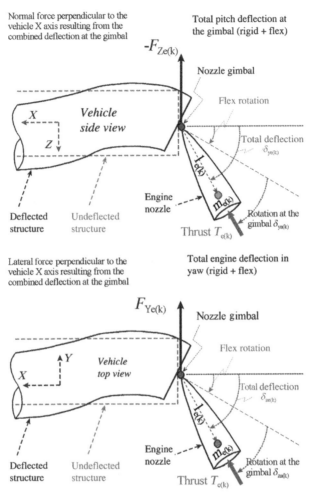

FIGURE 3.4   Total Nozzle deflection consists of two components: a rotation at the gimbal, plus structure deformation

(b) an additional deflection due the elastic deformation of the vehicle structure at the gimbal, where σ_{yek} and σ_{zek} are the pitch and yaw modal slopes at the kth engine gimbal, respectively.

The pitch deflection δ_{ye} is measured with respect to the trim position (Δ_E) and along the same direction. When $\Delta_z = 0$, the pitch deflection δ_y is exactly in the pitch axis.

$$
\delta_{ye}(k) = \delta_{ya}(k) + \sum_{j=1}^{N\,\mathrm{mod}} \sigma_{ye}(k,j)\,\eta(j)
$$

$$
\delta_{ze}(k) = \delta_{za}(k) + \sum_{j=1}^{N\,\mathrm{mod}} \sigma_{ze}(k,j)\,\eta(j)
$$

(3.12)

Force Variations at the Engine Gimbals

Equation 2.10 calculates the trim forces at the gimbal of an engine k. These are the steady-state forces that accelerate the vehicle and balance the steady aerodynamic moments. The engines are also gimbaling in pitch and yaw with respect to their trim positions. The vehicle attitude and direction is controlled by force variations at the engine gimbals which are perpendicular to the flight path. They are generated by small deflection commands in pitch δ_{yak} and in yaw δ_{zak}, and also by thrust variations δT_{ek} relative to the nominal engine thrust T_{ek}, which are commanded by the flight control system. The vehicle velocity can be controlled by varying the engine thrust closed loop. However, in most launch vehicles throttling is scheduled open loop and not included in the flight control system.

Each engine k generates a force variation vector at its gimbal due to the combined pivoting and throttling. It can be resolved into three components $F_{xe}(k)$, $F_{ye}(k)$, and $F_{ze}(k)$ at the pivot along the vehicle axes. Equation 3.13 defines the force variations at the kth gimbal, derived from equation 2.10 by replacing the engine trim angles Δ_E and Δ_Z by $(\Delta_E + \delta_y)$ and $(\Delta_Z + \delta_z)$, and the engine thrust T_{ek} with $(T_{ek} + \delta T_{ek})$ and solving for the variations. The gimbal force variation equations consist of four parts:

(a) forces due to small pitch engine deflections δ_{yek}, from trim
(b) forces due to small yaw engine deflections δ_{zek} from trim, about the local z-axis
(c) force variation due to engine thrust variation $\pm \delta T_{ek}$ relative to nominal thrust T_{ek}, and
(d) TWD forces caused by angular accelerations of the engine nozzles.

The last term in equation 3.13 represents the reaction forces at the gimbal due to the angular acceleration of the engine nozzle as it rotates about its pivot. This term creates the "TWD" dynamics. We assume that the engine nozzles are rigid and when they rotate they generate reaction forces against the vehicle at the gimbals. The TWD terms create a pair of zeros in the transfer function. The location of the zeros in the complex s-plane depends on the sign of the nozzle moment arm l_{ek}, which is the distance between the nozzle's center of mass from its pivot. A typical engine nozzle has its center of mass further back from the pivot point ($l_{ek} > 0$), and the TWD zeros are a complex pair along the ($j\omega$) axis. If the moment arm is negative (nozzle center of mass is above gimbal) the zeros are along the real axis. The TWD terms consist of two components:

- the angular acceleration of the nozzle relative to the gimbal and
- the acceleration of the structure around the gimbal.

Equation 3.13 denotes thrust force variations at engine gimbal due to gimbaling and throttling. The relative rotation, rate, and acceleration at the gimbals are calculated from separate actuator models for each TVC rotational DOF. The total deflection, rate, and acceleration at the gimbal, however, must also include the structural deformation component from the flex mode equations, as described in equation 3.12. The gimbal forces, therefore, contain undesirable low-damped oscillatory components due to structural flexibility at the gimbals.

$$F_{xe}(k) = T_{ek} \left(-\cos \Delta_{Z_k} \sin \Delta_{E_k} \delta_{yek} - \cos \Delta_{E_k} \sin \Delta_{Z_k} \delta_{zek} + \cos \Delta_{Z_k} \cos \Delta_{E_k} \delta_{Thr_k} \right)$$

$$- m_{ek} l_{ek} \left(\cos \Delta_{Z_k} \sin \Delta_{E_k} \ddot{\delta}_{yek} + \cos \Delta_{E_k} \sin \Delta_{Z_k} \ddot{\delta}_{zek} \right)$$

$$F_{ye}(k) = T_{ek} \left(-\sin \Delta_{Z_k} \sin \Delta_{E_k} \delta_{yek} + \cos \Delta_{E_k} \cos \Delta_{Z_k} \delta_{zek} + \sin \Delta_{Z_k} \cos \Delta_{E_k} \delta_{Thr_k} \right)$$

$$+ m_{ek} l_{ek} \cos \Delta_{Z_k} \ddot{\delta}_{zek} \tag{3.13}$$

$$F_{ze}(k) = -T_{ek} \left(\cos \Delta_{E_k} \delta_{yek} + \sin \Delta_{E_k} \delta_{Thr_k} \right) - m_{ek} l_{ek} \cos \Delta_{E_k} \ddot{\delta}_{yek}$$

where the throttle control $\delta_{Thr_k} = \left(\dfrac{\delta T_{ek}}{T_{ek}} \right)$

For a thrust varying engine, the ratio of thrust variation divided by the nominal engine thrust ($\delta T_{ek}/T_{ek}$) is defined to be the throttle control input δ_{Thr_k}. The maximum throttle control represents the maximum amount of thrust variation that the thruster is capable of producing above and below its nominal thrust. A positive value for δ_{Thr_k} indicates a thrust increase from its nominal thrust. A negative value indicates a thrust reduction from nominal thrust. For example, if the thrust of an engine can be made to vary from 80 to 120 lb by means of a throttle control valve, the thrust variation in the equations should be represented by having a nominal thrust $T_{ek} = 100$ lb and a throttle control input $\delta_{Thr_k} = \pm 0.2$. In this situation the throttle control input is expected to assume values that vary between +0.2 and −0.2. A throttle input of +0.2 corresponds to the maximum thrust of 120 (lb), zero corresponds to nominal thrust of 100 (lb), and a throttle input −0.2 corresponds to the minimum thrust of 80 (lb). Throttle inputs of magnitudes greater than 0.2 would violate the engine throttling specifications. A pair of back-to-back RCS jets can be modelled as having zero nominal thrust with a thrust variation equal to $\pm T_{max}$ (maximum thrust).

Total Engine Forces

The forces from the individual engines are combined together along the vehicle x-, y-, and z-axes to calculate the total force and moment variations due to the combined TVC that will maneuver the vehicle. Note that when the inertial coupling coefficients (h_e) are not included in the flex equations, the pitch and yaw gimbal accelerations should include the flexibility components similar to the equation 3.12. Otherwise, when the h-parameters are included, the gimbal accelerations in equation 3.13 should not include the flex acceleration terms because the coupling of the vehicle structure with the gimbal accelerations is captured by the inertial coupling coefficients which are provided by the Finite Elements Model (FEM). Equation 3.14 combines the TVC forces from all engines along the x-, y-, and z-axes, consisting of the summation of the components from each individual engine as described in equation 3.13.

$$F_{X_{eng}} = \sum_{k=1}^{Neng} F_{xe}(k) \qquad F_{Y_{eng}} = \sum_{k=1}^{Neng} F_{ye}(k) \qquad F_{Z_{eng}} = \sum_{k=1}^{Neng} F_{ze}(k) \qquad (3.14)$$

Moments due to the Combined TVC Forces

Equation 3.15 denotes vehicle moments due to combined TVC forces. It calculates the roll, pitch, and yaw moments on the vehicle generated by the force variations at the gimbals of the kth engine, about the x-, y-, and z-axes. The equations also include the reaction forces and moments generated by the nozzle angular accelerations.

$$L_{X_{eng}} = \sum_{k=1}^{Neng} \left(F_{ze(k)} l_{yek} - F_{ye(k)} l_{zek} \right)$$

$$M_{Y_{eng}} = \sum_{k=1}^{Neng} \left\{ \begin{array}{l} F_{xe(k)} l_{zek} - F_{ze(k)} l_{xek} - I_{ek} \ddot{\delta}_{ye(k)} - m_{ek} l_{ek} A_X \cos \Delta_{E(k)} \delta_{ye(k)} \\ + \left(T_{ek} \cos \Delta_{E(k)} \cos \Delta_{Z(k)} - m_{ek} A_X \right) \sum_{j=1}^{N\,mod} \varphi_{ze}(k,j) \eta(j) \end{array} \right\} \qquad (3.15)$$

$$N_{Z_{eng}} = \sum_{k=1}^{Neng} \left\{ \begin{array}{l} F_{ye(k)} l_{xek} - F_{xe(k)} l_{yek} - I_{ek} \ddot{\delta}_{ze(k)} - m_{ek} l_{ek} A_X \cos \Delta_{E(k)} \cos \Delta_{Z(k)} \delta_{ze(k)} \\ - \left(T_{ek} \cos \Delta_{E(k)} \cos \Delta_{Z(k)} - m_{ek} A_X \right) \sum_{j=1}^{N\,mod} \varphi_{ye}(k,j) \eta(j) \end{array} \right\}$$

The engine forces $F_{xe(k)}$, $F_{ye(k)}$, $F_{ze(k)}$ are calculated from equation 3.13 and as we have already mentioned, the calculation of those forces depends on whether the inertial coupling coefficients (h_e) are included in the structural equations. When the h-parameters are not included, the pitch and yaw gimbal accelerations in equation 3.13 are calculated with flexibility included as shown in equation 3.12. Otherwise, when the h-parameters are present in the flex modes, only the rigid component of the gimbal acceleration is included in equation 3.12 because flexibility is introduced via the coupling coefficients (h_e) in the bending equation. The equation 3.15 consists of the following terms:

(a) The first two terms in each axis consist of the cross products between x, y, z engine forces times the moment arms (vehicle CG to engine gimbal equation 3.16).

(b) The third term in the pitch and yaw equations are TWD torques generated by the nozzle's rotational ac-celeration.

(c) The fourth term in the pitch and yaw equations represents the moment generated when an engine nozzle has its center of mass displaced due to gimbal rotations (δ_{ye}, δ_{ze}) and is coupling with the vehicle axial acceleration A_x.

(d) The fifth term in the pitch and yaw equations represents the moment variation generated by the engine thrust that has its application point being displaced from nominal rigid position due to normal or lateral flex displacements at the gimbal.

(e) The sixth term in the pitch and yaw equations is the additional moment produced due to the flex displacement at the gimbal coupling with the vehicle acceleration A_x.

The relationship between inertial and the sensed accelerations is defined in equation 2.3. The moment arms between the kth engine gimbal and the vehicle CG are defined by equation 3.16.

$$l_{xek} = X_{ek} - X_{CG} \qquad l_{yek} = Y_{ek} - Y_{CG} \qquad l_{zek} = Z_{ek} - Z_{CG} \qquad (3.16)$$

3.6   AERODYNAMIC FORCES AND MOMENTS

Equation 3.17 calculates the variations in aerodynamic forces along x, y, and z as a function of variations in vehicle variables, structural elasticity, and forces from the aerosurfaces. They are derived by linearizing equation 2.11, and they represent variations from the steady-state trim forces. The angles of attack and sideslip α_w and β_w describe variations from the nominal trim α_0 and β_0 values and they include the effects due to the wind-gust disturbance as it is described in Section 2.9. The change in vehicle velocity δV_w includes the effects due to the wind-gust velocity also. The variables w_{gust} and v_{gust} are the wind-gust velocity components relative to the vehicle in the z and y directions, respectively.

The coefficients C_{Znj}, C_{Ynj}, etc. are aeroelastic coefficients that define variations in the aerodynamic forces along the z- and y-axes due to the generalized modal displacements η_j and their derivatives. They define, for example, the normal force variations on a flexible wing as a result of the wing oscillations. The forces $F_{Xs}(k)$, $F_{Ys}(k)$, and $F_{Zs}(k)$ represent variations from steady-state forces due to the deflection of the kth aerosurface and they are defined in equation 3.18.

$$F_{Xaero} = -\bar{Q}S_{ref} \left(\begin{array}{l} C_{A0}\tan\alpha_0\,\alpha + C_{A\alpha}\alpha_w + C_{A\beta}\beta_w + \left(\dfrac{l_{ch}}{2V_0}C_{Aq} \right) q \\[2mm] + \left(\dfrac{2C_{A0}}{V_0} + \dfrac{\partial C_A}{\partial V} \right) \delta V_w + \left(\dfrac{\partial C_A}{\partial h} + \dfrac{\partial \rho}{\partial h}\left(\dfrac{V_0^2}{2\bar{Q}} \right)C_{A0} \right) \delta h \end{array} \right) + \sum_{k=1}^{Ncs} F_{Xs}(k)$$

$$
F_{Yaero} = \overline{Q}S_{ref}
\begin{pmatrix}
C_{Y\alpha}\alpha_w + C_{Y\beta}\beta_w + \left(\dfrac{l_{sp}}{2V_0}\right)(C_{Yp}p + C_{Yr}r + C_{Y\dot{\beta}}\dot{\beta}_w) + \left(\dfrac{2C_{Y0}}{V_0}\right)\delta V_w \\[2mm]
+\dfrac{\partial\rho}{\partial h}\left(\dfrac{V_0^2}{2\overline{Q}}\right)C_{Y0}\delta h + \displaystyle\sum_{j=1}^{N\,mod}\left(C_{Y\eta_j}\eta_j + \dfrac{C_{Y\dot{\eta}_j}}{V_0}\dot{\eta}_j\right)
\end{pmatrix}
$$

$$
+ \sum_{k=1}^{Ncs} F_{Ys}(k) \tag{3.17}
$$

$$
F_{Zaero} = \overline{Q}S_{ref}
\begin{pmatrix}
C_{Z0}\tan\alpha_0\,\alpha + C_{Z\alpha}\alpha_w + C_{Z\beta}\beta_w + \left(\dfrac{l_{ch}}{2V_0}\right)(C_{Zq}q + C_{Z\dot{\alpha}}\dot{\alpha}_w) \\[2mm]
+\left(\dfrac{2C_{Z0}}{V_0} + \dfrac{\partial C_Z}{\partial V}\right)\delta V_w + \left(\dfrac{\partial C_Z}{\partial h} + \dfrac{\partial\rho}{\partial h}\left(\dfrac{V_0^2}{2\overline{Q}}\right)C_{Z0}\right)\delta h \\[2mm]
+\displaystyle\sum_{j=1}^{N\,mod}\left(C_{Z\eta_j}\eta_j + \dfrac{C_{Z\dot{\eta}_j}}{V_0}\dot{\eta}_j\right)
\end{pmatrix}
+ \sum_{k=1}^{Ncs} F_{Zs}(k)
$$

where $\alpha_w = \alpha + \dfrac{w_{gust}}{V_0\cos\alpha_0}$; $\beta_w = \beta + \dfrac{v_{gust}}{V_0}$

Forces Generated by a Control Surface

In this book we attempt to model each aerosurface as a separate control input to the vehicle model because it is easier to include flexibility and TWD forces at the hinges of each control surface. It is a lot more complex to model the TWD when the surfaces are combined together as a single control input. For example, an elevon is often defined as the average deflection of two flaps and an aileron is the differential deflection of two flaps. Each rotating panel should be defined as a separate control surface. The reason is because each control surface creates a separate force and torque on the vehicle at a specific location and direction. The forcing functions on the vehicle are not only from aerodynamic forces, which are defined by the aero coefficients, but they are also affected by TWD forces and structural deformations which are specific to each surface. This information may be lost or become corrupted when you combine aerosurfaces together. Figure 3.5 shows a typical control surface (k) rotating about a hinge vector. A positive control surface deflection $+\delta_s$ is defined to be a clockwise rotation of the surface about its hinge vector. The orientation of the hinge vector with respect to the vehicle axes is defined in terms of two angles:

1. The back-sweep angle $\lambda_{hs}(k)$ which is the angle between the hinge vector and the y–z plane that slices the vehicle perpendicular to the x-axis. In most airplanes the hinge line is perpendicular to the vehicle x-axis and $\lambda_{hs} = 0°$, or it may be slightly positive. It would be negative on a forward-sweep aircraft.

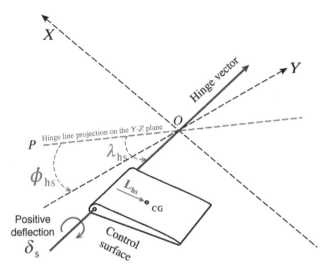

FIGURE 3.5   Definition of the Hinge Vector Orientation Angles

2. The aerosurface bank angle $\phi_{hs}(k)$ is defined between the projection line OP of the hinge line on the y–z plane and the vehicle x–y plane. For example, a left elevator or flap has $\phi_{hs} = 0°$, a right elevator has $\phi_{hs} = 180°$, and a rudder has $\phi_{hs} = 90°$.

The aerodynamic increment coefficients are important for the calculation of the aerodynamic forces. The midpoint location of the hinge, the hinge line orientation angles, and the mass properties for each control surface are also needed for the calculation of the TWD forces. The inputs to the flight vehicle dynamic model are control surface deflections in (rad), rates, and accelerations for each individual rotating panel and not a combination of panels. An aileron, for example, should not be defined as a single control surface, unless the TWD, load-torque feedback, and bending dynamics are not included, in which case it does not matter if surfaces are combined. The transformation from roll, pitch, and yaw flight control demands to the individual surface deflections is defined by the mixing logic matrix that connects between the flight control system output and the vehicle inputs. The derivation of the mixing logic matrix is discussed in Chapter 5.

Figure 3.6 is an example of a vehicle that uses five control surfaces: a body flap, two flaps, and two rudders in a V-tail configuration. The hinge vectors of the body flap and the two flaps are pointing in the y direction. The hinge lines of the two flaps are slightly tilted from the y-axis because the wing has a small dihedral. The surface bank angle φ_{hs} is measured between the hinge vector and the vehicle x–y plane. The left flap has a small positive bank angle ($\varphi_{hs} = +2°$) and the right flap has a small negative angle ($\varphi_{hs} = -2°$). The two flaps also have a small back-sweep angle $\lambda_{hs} = 5°$. The two rudders are tilted more, forming a 45° V-tail. The hinge vector of the left rudder is pointing toward the vehicle and has a bank angle ($\varphi_{hs} = +45°$) relative to

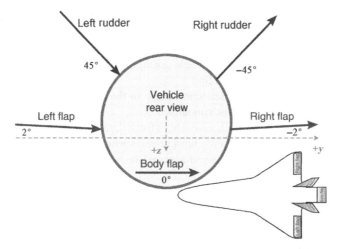

FIGURE 3.6   Lifting-Body Aircraft Showing the Directions of the Hinge Orientation Angles (λ_{hs} and ϕ_{hs})

the x–y plane. The hinge vector of the right rudder is pointing away from the vehicle and has a $\varphi_{hs} = -45°$. Both rudders have a positive back-sweep angle $\lambda_{hs} = 10°$.

Equation 3.18 calculates the force variations at the hinge of the kth control surface, along the vehicle x-, y-, and z-axes, as a function of surface deflection, deflection rate, and angular acceleration. It consists of two parts:

1. An aerodynamics force part that consists of two terms: a term due the deflection δ_s obtained from increment coefficients which is a function of: Mach, α, β, and surface deflection, as shown in equation 2.13, and a damping term due to the surface rate $\dot{\delta}_s$ and the damping coefficients which are typically a function of Mach and α.

2. A reaction "TWD" force caused by the angular acceleration of the control surface and the structural deformation at the hinge.

$$F_{Xcs}(k) = -\overline{Q}S_{ref}\left[C_{A\delta(k)}\,\delta_{s(k)} + \left(\frac{\overline{c}_{sk}}{2V_0}\right)C_{A\dot{\delta}(k)}\,\dot{\delta}_{s(k)}\right] + F_{XS_{TWD}}(k)$$

$$F_{Ycs}(k) = \overline{Q}S_{ref}\left[C_{Y\delta(k)}\,\delta_{s(k)} + \left(\frac{\overline{c}_{sk}}{2V_0}\right)C_{Y\dot{\delta}(k)}\,\dot{\delta}_{s(k)}\right] + F_{YS_{TWD}}(k)$$

$$F_{Zcs}(k) = \overline{Q}S_{ref}\left[C_{Z\delta(k)}\,\delta_{s(k)} + \left(\frac{\overline{c}_{sk}}{2V_0}\right)C_{Z\dot{\delta}(k)}\,\dot{\delta}_{s(k)}\right] + F_{ZS_{TWD}}(k)$$

where $F_{XS_{TWD}}(k)$, etc., are the TWD forces   (3.18)

$$F_{XS_{TWD}}(k) = -m_{sk}\,L_{hs(k)}\,\cos\lambda_{hs(k)}\,\sin\Delta_{s(k)}\,\ddot{\delta}_{s(k)}$$

$$F_{YS_{TWD}}(k) = +m_{sk}\,L_{hs(k)}\,\sin\varphi_{hs(k)}\,\ddot{\delta}_{s(k)}$$

$$F_{ZS_{TWD}}(k) = -m_{sk}\,L_{hs(k)}\,\cos\varphi_{hs(k)}\,\cos\Delta_{s(k)}\,\ddot{\delta}_{s(k)}$$

$L_{hs(k)} = X_{\text{hinge}} - X_{\text{cscg}}$ is the moment arm distance between the hinge line and the control surface CG. It is positive when the surface CG is behind the hinge.

$\Delta_s(k)$ is the trim angle position for the kth control surface, about its hinge line.

$\delta_s(k)$ is the deflection angle of the kth control surface measured from its trim angle.

$C_{Y\delta k}, C_{Z\delta k}$ are the aerodynamic force derivatives per surface deflection.

c_{sk} is the aero chord of aerosurface k.

m_{sk} is the mass of aerosurface k.

The aerosurface deflections, rates, and accelerations: $\delta_s(k), \dot{\delta}_s(k), \ddot{\delta}_s(k)$ in equation 3.18 have different interpretations depending on whether the inertial coupling coefficients $h_s(k)$ for the aerosurfaces are included or not in the flex mode equations. In the absence of the inertial coupling coefficients the flexibility at the surfaces is introduced by mode shapes at the surface support structure. We assume that the surfaces themselves are rigid, but the hinges where the surfaces are attached have a torsional stiffness representing the combined backup and load flexibility. In equation 3.18, the inertial surface rotations $\delta_{s(k)}$ are replaced with $\delta_{fs(k)}$ consisting of rigid deflections due to actuator rotation plus the deformation of the supporting structure at the hinges, as shown in equation 3.19 which calculates the total aerosurface deflection at the hinge. The same applies for the surface rate and also for the acceleration $\ddot{\delta}_s(k)$ that excites the TWD forces at the hinges. It should include the flex accelerations caused by the deformation of the structure around the hinge in addition to the relative acceleration from the actuator model, similar to equation 3.19 which written for the deflection:

$$\delta_{fs}(k) = \delta_s(k) + \sum_{j=1}^{N \bmod} \sigma_{hs}(k,j)\,\eta(j), \text{ where} \tag{3.19}$$
$$\sigma_{hs}(k) = \left(\sigma_{Ys}(k)\cos\varphi_{h(k)} + \sigma_{Zs}(k)\sin\varphi_{h(k)}\right)\cos\lambda_{h(k)} + \sigma_{Xs}(k)\sin\lambda_{h(k)}$$

where

$\sigma_{hs}(k,j)$ is the modal slope for the jth mode at the kth hinge line, resolved about the hinge vector as shown in Figure 3.5.

However, when the inertial coupling coefficients are included in the flexible model the dynamic coupling between surface rotation and structural excitation is incorporated in the finite elements model by the h-parameters. In this case, the surface deflections, rates, and acceleration inputs in 3.18, should include only the rigid components due to the actuator rotation alone: $\delta_s(k), \dot{\delta}_s(k), \ddot{\delta}_s(k)$ without the structural deformation term.

Aerodynamic Moments

Equation 3.20 calculates the variations in the aerodynamic moments about roll, pitch, and yaw as a result of variations in the angles of attack α_w, sideslip β_w, variations in velocity δV_w, changes in altitude δh, vehicle rates (p, q, r), and control surface deflections δ_{sk}. Notice that the changes in the angles of attack and sideslip, and also in velocity are relative to the airflow, including the effects due to the wind gust as described in Section 3.7.

$$
L_{Xaero} = \overline{Q} S_{ref} l_{sp} \left(\begin{array}{l} C_{l\alpha}\alpha_w + C_{l\beta}\beta_w + \dfrac{l_{sp}}{2V_0}\left(C_{lp}\,p + C_{lr}\,r + C_{l\dot\beta}\,\dot\beta_w\right) + \left(\dfrac{2C_{l0}}{V_0}\right)\delta V_w \\[2ex] + \displaystyle\sum_{j=1}^{N\,mod}\left(C_{l\eta_j}\eta_j + \dfrac{C_{l\dot\eta_j}}{V_0}\dot\eta_j\right) + C_{l0}\left(\dfrac{V_0^2}{2\overline{Q}}\right)\dfrac{\partial\rho}{\partial h}\delta h \end{array} \right)
$$

$$
+ \sum_{k=1}^{N surf} L_{Xs}(k)
$$

$$
M_{Yaero} = \overline{Q} S_{ref} l_{ch} \left(\begin{array}{l} C_{m\alpha}\alpha_w + C_{m\beta}\beta_w + \dfrac{l_{ch}}{2V_0}\left(C_{mq}\,q + C_{m\dot\alpha}\,\dot\alpha_w\right) + \left(\dfrac{2C_{m0}}{V_0} + \dfrac{\partial C_m}{\partial V}\right)\delta V_w \\[2ex] + \displaystyle\sum_{j=1}^{N\,mod}\left(C_{m\eta_j}\eta_j + \dfrac{C_{m\dot\eta_j}}{V_0}\dot\eta_j\right) + \left(\dfrac{\partial C_m}{\partial h} + C_{m0}\right)\left(\dfrac{V_0^2}{2\overline{Q}}\right)\dfrac{\partial\rho}{\partial h}\delta h \end{array} \right)
$$

$$
+ \sum_{k=1}^{N surf} M_{Ys}(k) \tag{3.20}
$$

$$
N_{Zaero} = \overline{Q} S_{ref} l_{sp} \left(\begin{array}{l} C_{n\alpha}\alpha_w + C_{n\beta}\beta_w + \dfrac{l_{sp}}{2V_0}\left(C_{np}\,p + C_{nr}\,r + C_{n\dot\beta}\,\dot\beta_w\right) + \left(\dfrac{2C_{n0}}{V_0}\right)\delta V_w \\[2ex] + \displaystyle\sum_{j=1}^{N\,mod}\left(C_{n\eta_j}\eta_j + \dfrac{C_{n\dot\eta_j}}{V_0}\dot\eta_j\right) + C_{n0}\left(\dfrac{V_0^2}{2\overline{Q}}\right)\dfrac{\partial\rho}{\partial h}\delta h \end{array} \right)
$$

$$
+ \sum_{k=1}^{N surf} N_{Zs}(k)
$$

$\overline{Q}, S_{ref}$	are the dynamic pressure in (lb/ft^2), and reference area in (ft^2).
$C_{l\beta}, C_{m\alpha}, C_{n\beta}$	are the aerodynamic moment derivatives due to the angles of attack and sideslip, in (1/deg).
l_{sp} and l_{ch}	are the wing span and mean aerodynamic chord, respectively, in feet.
$L_{Xs}, M_{Ys}, N_{Zs}(k)$	are the moments about x-, y-, and z-axes generated by the kth control surface deflections δ_{sk}.

The terms $(\Sigma L_{xs(k)}, \Sigma M_{ys(k)},$ and $\Sigma N_{zs(k)})$ are the combined moments on the vehicle generated by the deflections of N_{surf} control surfaces ($k = 1$ to N_{surf}). The moments generated by a single control surface are shown in equation 3.21. The aerodynamic coupling coefficients $C_{l\alpha}$, $C_{m\beta}$, and $C_{n\alpha}$ are zero when the vehicle is aerodynamically symmetric. They are included only when there is significant aerodynamic coupling between pitch and lateral. The coefficients $C_{l\eta j}$, C_{mnj}, C_{nnj} are aeroelastic moment derivatives. They define variations in vehicle moments due to a jth bending mode excitation η_j. For example, the flexing of a wing mode causes oscillatory moments on the vehicle. The aeroelastic coefficients are generated by combining Computational Fluid Dynamics (CFD) aero models with finite element models and it will be discussed in Section 3.9.

Vehicle Moments due to Control Surface Deflections

Equation 3.21 calculates the roll, pitch, and yaw moments on the vehicle generated by the deflection $\delta_{s(k)}$ of a control surface k from trim position $\Delta_{s(k)}$. The orientation of the control surface hinge line is defined by the angles $\varphi_{hs(k)}$ and $\lambda_{hs(k)}$ with respect to the vehicle axes, as shown in Figure 3.5. The coefficients $C_{l\delta(k)}$, $C_{m\delta(k)}$, $C_{n\delta(k)}$ in equation 3.21 are the aerodynamic moment derivatives that generate the aero moment variations due to surface deflection and rate. They are originally calculated relative to the moments reference center (MRC) and are converted to the vehicle CG using the transformations in equation 3.23. The equation also includes the TWD forces which are calculated in equation 3.18. There are also reaction terms generated by the angular accelerations of the aerosurfaces, and terms generated by the coupling of the vehicle axial and normal accelerations A_x and A_z with the displacements of the surface center of mass.

$$L_{Xs}(k) = \overline{Q}\, S_{ref}\, l_{sp} \left[C_{l\delta(k)}\, \delta_{s(k)} + \left(\frac{\overline{c}_{sk}}{2V_0} \right) C_{l\dot\delta(k)}\, \dot\delta_{s(k)} \right] + F_{ZS_{TWD}(k)} l_{Ysk} - F_{YS_{TWD}(k)} l_{Zsk}$$

$$M_{Ys}(k) = \overline{Q}\, S_{ref}\, l_{ch} \left[C_{m\delta(k)}\, \delta_{s(k)} + \left(\frac{\overline{c}_{sk}}{2V_0} \right) C_{m\dot\delta(k)}\, \dot\delta_{s(k)} \right] + F_{XS_{TWD}(k)} l_{Zsk}$$

$$- F_{ZS_{TWD}(k)} l_{Xsk} - I_{hs(k)} \cos\varphi_{h(k)}\, \ddot\delta_{s(k)} + m_{sk} \left(A_Z d_{Xsk} - A_X d_{Zsk} \right) \tag{3.21}$$

$$N_{Zs}(k) = \overline{Q}\, S_{ref}\, l_{sp} \left[C_{n\delta(k)}\, \delta_{s(k)} + \left(\frac{\overline{c}_{sk}}{2V_0} \right) C_{n\dot\delta(k)}\, \dot\delta_{s(k)} \right] + F_{YS_{TWD}(k)} l_{Xsk}$$

$$- F_{XS_{TWD}(k)} l_{Ysk} - I_{hs(k)} \sin\varphi_{h(k)}\, \ddot\delta_{s(k)} + m_{sk} A_X d_{Ysk}$$

where $l_{Xsk} = X_{sk} - X_{CG}$ $l_{Ysk} = Y_{sk} - Y_{CG}$ $l_{Zsk} = Z_{sk} - Z_{CG}$

Equation 3.22 calculates the displacements d_{Xsk}, d_{Ysk}, d_{Zsk} along x, y, and z of the aerosurface center of mass from the hinge line. They consist of two terms:

1. a term due to the rotation δ_{sk} of the kth surface, and
2. a term due to the structural deformation at the hinge of the kth surface.

When these displacements couple with the vehicle accelerations A_X and A_Z they create additional moments. The moment arm l_{Xsk} of a control surface k is the distance between the hinge line midpoint and the vehicle CG.

$$d_{Zsk} = L_{hs(k)} \cos \Delta_s \cos \varphi_{h(k)} \delta_{s(k)} + \sum_{j=1}^{N \text{ mod}} \varphi_{Zs}(k,j) \eta(j)$$

$$d_{Ysk} = -L_{hs(k)} \cos \Delta_s \sin \varphi_{h(k)} \delta_{s(k)} + \sum_{j=1}^{N \text{ mod}} \varphi_{Ys}(k,j) \eta(j) \qquad (3.22)$$

$$d_{Xsk} = L_{hs(k)} \sin \Delta_s \cos \lambda_{h(k)} \delta_{s(k)} + \sum_{j=1}^{N \text{ mod}} \varphi_{Xs}(k,j) \eta(j)$$

where

$C_{l\delta(k)}, C_{m\delta(k)}, C_{n\delta(k)}$	are the roll, pitch, and yaw aerodynamic moment derivatives for the control aerosurfaces (δ_{sk}) in (1/deg), where $k = 1$ to N_{surf},
$F_{YS_{TWD}}, F_{ZS_{TWD}}$	are the TWD forces at the hinges of the control surfaces, obtained from equation 3.18,
$m_{s(k)}, I_{hs(k)}$	are the mass and the moment of inertia of the kth aerosurface about its hinge,
Δ_s	is the surface trim position with respect to the vehicle x-axis.

Equation 3.21 in addition to the aero moment terms generated by the deflections of the kth control surface, also includes reaction moments generated by the TWD forces due to surface accelerations, as calculated in equation 3.18. Note that, when the inertial coupling coefficients h_s of the control surfaces are not included in the flex equations, we assume that flexibility at the hinges is introduced by the actuators and the mode shapes at the effector backup structure, as already described. The TWD forces are excited by the inertial surface accelerations, defined as $\ddot{\delta}_{fs}(k)$ in equation 3.19, which includes rotation at the hinge plus rotational deformation about the hinge line. However, when the inertial coupling coefficients are included in the structural model and in the hinge moment equations, only the first term in equation 3.19 due to the actuator rotation alone $\delta_s(k), \dot{\delta}_s(k), \ddot{\delta}_s(k)$ should be used, without the structural deformation term because the dynamic coupling between the surface acceleration with flexibility is introduced by means of the h-parameters.

Transforming the Aero Moment Coefficients

The aerodynamic moment coefficients C_L, C_M, and C_N, and their derivatives with respect to α, β, p, q, r, η, etc. in equations 3.20 and 3.21 are assumed to be defined about the vehicle center of mass. However, they are originally calculated relative to a fixed location on the vehicle, called the MRC. This is because the CG usually varies during flight and the aero moment coefficients, therefore, must be transformed from the MRC to the vehicle CG before they can be used in the equations. The location of the MRC $(X_{MRC}, Y_{MRC}, Z_{MRC})$ with respect to the vehicle coordinates is defined in the aerodynamic data. The following aerodynamic moment transformation equations are used to convert the aero moment coefficients and the derivatives from the MRC to the vehicle center of mass, where (X_{CG}, Y_{CG}, Z_{CG}) are the CG coordinates. Figure 3.7 illustrates this transformation in pitch

$$C_{L_{CG}} = C_{L_{MRC}} + C_Y \frac{Z_{CG} - Z_{MRC}}{l_{sp}} - C_Z \frac{Y_{CG} - Y_{MRC}}{l_{sp}}$$

$$C_{M_{CG}} = C_{M_{MRC}} + C_Z \frac{X_{CG} - X_{MRC}}{l_{ch}} + C_A \frac{Z_{CG} - Z_{MRC}}{l_{ch}} \qquad (3.23)$$

$$C_{N_{CG}} = C_{N_{MRC}} - C_Y \frac{X_{CG} - X_{MRC}}{l_{sp}} - C_A \frac{Y_{CG} - Y_{MRC}}{l_{sp}}$$

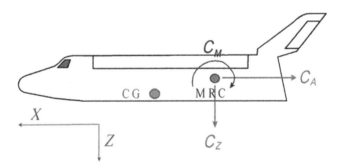

FIGURE 3.7 Pitch Moment Transformation from MRC to CG

Definition of the Aero Coefficients

The coefficients $C_{z\alpha}$ and $C_{y\beta}$ are the aerodynamic force derivatives in units of (1/deg). They are used to calculate changes in aerodynamic forces due to small variations in the angles of attack and sideslip. The coefficients $C_{z\delta si}$ and $C_{y\delta si}$ are the aerodynamic force deflection derivatives in units of (1/deg) and they calculate changes in aerodynamic forces due to small deflections of the control surfaces. The axial force coefficients C_{Au} and $C_{A\alpha}$ calculate the force variation along the x-axis due to variations in the velocity along the x-axis and in the angle of attack, respectively. They are defined as positive along the negative vehicle x-axis because they generate drag forces on the vehicle. The cross-coupling derivatives $C_{z\beta}$, $C_{Y\alpha}$, and $C_{A\beta}$ are usually very small and in most cases they are set to zero.

Force Derivatives: The force derivatives $C_{A\alpha}$, $C_{Z\alpha}$, and $C_{Y\beta}$ define variations in vehicle forces along the $-x$, z, and y–axes, respectively, due to small changes of angle of attack and sideslip in (rad). They are defined as follows:

$$\frac{\partial F_Z}{\partial \alpha} = \bar{Q} S_r C_{Z\alpha}, \qquad \frac{\partial F_Y}{\partial \beta} = \bar{Q} S_r C_{Y\beta}, \quad \text{etc.}$$

Moment Derivatives: The moment derivatives $C_{m\alpha}$, $C_{n\beta}$, $C_{l\beta}$, etc. define variations in vehicle moments about roll, pitch, and yaw axes, due to small changes in the angles of attack or sideslip. They are defined by the following equations:

$$\frac{\partial M_Y}{\partial \alpha} = \bar{Q} S_r l_{ch} C_{m\alpha} \qquad \frac{\partial L_X}{\partial \beta} = \bar{Q} S_r l_{sp} C_{l\beta}, \quad \text{etc.}$$

Force Rate Derivatives: The force rate derivatives C_{Zq}, C_{Yr}, C_{Yp}, etc. define force variations along the z- and y-axes due to variations in the vehicle body rates p, q, r in (rad/sec). They are defined as follows:

$$\frac{\partial F_Z}{\partial q} = \left(\frac{\bar{Q} S_r l_{ch}}{2V_0} \right) C_{Zq} \qquad \frac{\partial F_Y}{\partial p} = \left(\frac{\bar{Q} S_r l_{sp}}{2V_0} \right) C_{Yp}$$

Moment Rate Derivatives: The moment rate derivatives C_{mq}, C_{nr}, C_{lp}, etc. define variations in roll, pitch, and yaw vehicle moments due to variations in vehicle body rates p, q, r in (rad/sec). They are defined as follows:

$$\frac{\partial M_Y}{\partial q} = \left(\frac{\bar{Q} S_r l_{ch}^2}{2V_0} \right) C_{mq} \qquad \frac{\partial N_Z}{\partial r} = \left(\frac{\bar{Q} S_r l_{sp}^2}{2V_0} \right) C_{nr}$$

Velocity Derivatives: The velocity derivatives C_{Av}, and C_{Zv}, etc. define the change in vehicle force along the negative x-axis, and along the z-axis, respectively, due to changes in vehicle velocity in (ft/sec). They are defined as follows:

$$\frac{\partial F_X}{\partial V} = - \left(\frac{\bar{Q} S_r}{V_0} \right) C_{AV} \qquad \frac{\partial F_Z}{\partial V} = \left(\frac{\bar{Q} S_r}{V_0} \right) C_{ZV}$$

Steady state Force and Moment Coefficients: The coefficients C_{m0}, C_{A0}, and C_{Z0}, etc. are not derivatives. They are aerodynamic force and moment coefficients at the trim condition. They are dimensionless, functions of a, b, and Mach and they define the nominal forces (F_{X0}, F_{Z0}) and moments (M_{y0}, N_{z0}) on the vehicle (not variations), at the trim condition. They are defined by the following equations:

$$M_{Y0} = \bar{Q} S_r l_{ch} C_{m0} \qquad L_{X0} = \bar{Q} S_r l_{sp} C_{l0} \qquad F_{X0} = \bar{Q} S_r C_{A0} \qquad F_{Z0} = \bar{Q} S_r C_{Z0}$$

Variations in Dynamic Pressure due to δV and δh: The change in dynamic pressure due to changes in vehicle velocity (δV), and the change in dynamic pressure due to changes in altitude (δh) can be obtained from the following equations, where h_0 is the vehicle altitude and b_0 is a constant.

$$\frac{\partial \bar{Q}}{\partial V} = \rho_o V_0 e^{-b_o h_o} \qquad \frac{\partial \bar{Q}}{\partial h} = \rho_o \left(\frac{b_o V_0^2}{2} \right) e^{-b_o h_o}$$

Velocity Derivatives: The velocity derivatives define the effect of velocity variations (δV) on the vehicle forces and moments. They are used in equations (x) and are defined as follows:

$$\frac{\partial F_X}{\partial V} = -\bar{Q}S_r \left(\frac{2}{V_0} \right) (C_{AV} + C_{A0}), \text{ where } C_{AV} = \frac{V_0}{2} \left(\frac{\partial C_A}{\partial V} \right)$$

$$\frac{\partial F_Z}{\partial V} = \bar{Q}S_r \left(\frac{2}{V_0} \right) (C_{ZV} + C_{Z0}), \text{ where } C_{ZV} = \frac{V_0}{2} \left(\frac{\partial C_Z}{\partial V} \right)$$

$$\frac{\partial F_Y}{\partial V} = \bar{Q}S_r \left(\frac{2}{V_0} \right) C_{Y0}$$

$$\frac{\partial M_y}{\partial V} = \bar{Q}S_r l_{ch} \left(\frac{2}{V_0} \right) (C_{mV} + C_{m0}), \text{ where } C_{mV} = \frac{V_0}{2} \left(\frac{\partial C_m}{\partial V} \right)$$

Altitude Derivatives: They define the effects of altitude variations from nominal on the vehicle forces and moments.

$$\frac{\partial F_X}{\partial h} = -\bar{Q}S_r \left[\frac{\partial C_A}{\partial h} + \frac{\partial \rho}{\partial h} \left(\frac{V_0^2}{2\bar{Q}} \right) C_{A0} \right] \qquad \frac{\partial F_Z}{\partial h} = \bar{Q}S_r \left[\frac{\partial C_Z}{\partial h} + \frac{\partial \rho}{\partial h} \left(\frac{V_0^2}{2\bar{Q}} \right) C_{Z0} \right]$$

$$\frac{\partial F_Y}{\partial h} = \bar{Q}S_r \frac{\partial \rho}{\partial h} \left(\frac{V_0^2}{2\bar{Q}} \right) C_{Y0}$$

$$\frac{\partial M_y}{\partial h} = \bar{Q}S_r l_{ch} \left[\frac{\partial C_m}{\partial h} + \frac{\partial \rho}{\partial h} \left(\frac{V_0^2}{2\bar{Q}} \right) C_{m0} \right] \qquad \frac{\partial N_z}{\partial h} = \bar{Q}S_r l_{sp} \left[\frac{\partial \rho}{\partial h} \left(\frac{V_0^2}{2\bar{Q}} \right) C_{n0} \right]$$

Angle of Attack Rate Derivatives: The derivatives $C_{Z\alpha\_dot}, C_{Y\beta\_dot}, C_{m\alpha\_dot}$ define the changes in vehicle forces and moments due to variations in the rates of the angles of attack and sideslip. They are used in the force and moment equations 3.17 and 3.20, and are defined as follows:

$$\frac{\partial F_Z}{\partial \dot{\alpha}} = \bar{Q}S_r \left(\frac{l_{ch}}{2V_0} \right) C_{Z\dot{\alpha}} \qquad \frac{\partial M_y}{\partial \dot{\alpha}} = \bar{Q}S_r l_{ch} \left(\frac{l_{ch}}{2V_0} \right) C_{m\dot{\alpha}}$$

Generalized Aerodynamic Force Derivatives (GAFD): The aeroelastic coefficients are not always available. They are included only when the coupling between the aerodynamics and structure is significant. They are defined by the following equations. The generalized aero moment derivatives $(C_{l\eta j}\ C_{m\eta j}\ C_{n\eta j})$ define the change in vehicle roll, pitch, and yaw moments due to variations in the generalized modal displacement $\eta(j)$ of mode (j). Similarly, the generalized aero moment derivatives $(C_{l\dot\eta j}\ C_{m\dot\eta j}\ C_{n\dot\eta j})$ define the change in vehicle roll, pitch, and yaw moments due to variations of the generalized modal rates of a mode (j). Where l_{ch} and l_{sp} are the vehicle mean aero chord and span reference lengths, and S_{ref} is the vehicle reference area.

$$
\begin{aligned}
\frac{\partial M_y}{\partial \eta_{(j)}} &= \overline{Q}\, S_r\, l_{ch}\, C_{m\,\eta(j)} &
\frac{\partial M_y}{\partial \dot\eta_{(j)}} &= \overline{Q}\, S_r\, l_{ch}\left(\frac{C_{m\,\dot\eta(j)}}{V_0}\right) \\
\frac{\partial N_z}{\partial \eta_{(j)}} &= \overline{Q}\, S_r\, l_{sp}\, C_{n\,\eta(j)} &
\frac{\partial N_z}{\partial \dot\eta_{(j)}} &= \overline{Q}\, S_r\, l_{sp}\left(\frac{C_{n\,\dot\eta(j)}}{V_0}\right)
\end{aligned}
$$

The generalized aero force derivatives also define the variations in vehicle forces along the y- and z-axes due to variations in the modal displacement $\eta(j)$ and also the displacement rate $\dot\eta(j)$ of mode (j).

$$
\begin{aligned}
\frac{\partial F_Y}{\partial \eta_{(j)}} &= \overline{Q}\, S_r\, C_{Y\,\eta(j)} &
\frac{\partial F_Y}{\partial \dot\eta_{(j)}} &= \overline{Q}\, S_r\left(\frac{C_{Y\,\dot\eta(j)}}{V_0}\right) \\
\frac{\partial F_Z}{\partial \eta_{(j)}} &= \overline{Q}\, S_r\, C_{Z\,\eta(j)} &
\frac{\partial F_Z}{\partial \dot\eta_{(j)}} &= \overline{Q}\, S_r\left(\frac{C_{Z\,\dot\eta(j)}}{V_0}\right)
\end{aligned}
$$

The GAFD derivatives $(C_{\eta j\alpha},\ C_{\eta j\beta},\ C_{\eta jp},\ C_{\eta jq},\ C_{\eta j\delta s}$, etc.) define the amount of modal excitation $\eta(j)$ of a mode (j) caused by the variations in $(\alpha,\ \beta,\ p,\ q,\ r,\ \delta_s)$ etc. and also their rates. The reference length (l_{chg}) is used to normalize the data and it may not be the same as the vehicle reference length (l_{ch}).

$$
\begin{aligned}
\frac{\partial \eta_j}{\partial \alpha} &= \bar{Q}S_{ref}C_{\eta j\alpha} &
\frac{\partial \eta_j}{\partial \delta_{sk}} &= \bar{Q}S_{ref}C_{\eta_j\delta_{sk}} \\
\frac{\partial \eta_j}{\partial \dot\alpha} &= \bar{Q}S_{ref}\left(\frac{l_{chg}}{2V_0}\right)C_{\eta j\dot\alpha} &
\frac{\partial \eta_j}{\partial \dot\delta_{sk}} &= \bar{Q}S_{ref}\left(\frac{\bar{c}_k}{2V_0}\right)C_{\eta_j\dot\delta_{sk}} \\
\frac{\partial \eta_j}{\partial p} &= \bar{Q}S_{ref}\left(\frac{l_{chg}}{2V_0}\right)C_{\eta jp} &
\frac{\partial \eta_j}{\partial q} &= \bar{Q}S_{ref}\left(\frac{l_{chg}}{2V_0}\right)C_{\eta jq} \\
\frac{\partial \eta_j}{\partial \dot p} &= \bar{Q}S_{ref}\left(\frac{l_{chg}}{2V_0}\right)^2 C_{\eta j\dot p} &
\frac{\partial \eta_j}{\partial \dot q} &= \bar{Q}S_{ref}\left(\frac{l_{chg}}{2V_0}\right)^2 C_{\eta j\dot q} \\
\frac{\partial \eta_j}{\partial \eta_i} &= \bar{Q}S_{ref}C_{\eta_j\eta_i} &
\frac{\partial \eta_j}{\partial \dot\eta_i} &= \bar{Q}S_{ref}\left(\frac{1}{V_0}\right)C_{\eta_j\dot\eta_i}
\end{aligned}
$$

The hinge moment derivatives $\{C_{h\alpha(k)},\ C_{h\beta(k)},\ C_{hq(k)},\ C_{hp(k)},\ C_{hr(k)}\}$ define how the moments (HM_k) at the hinge of a control surface (k) vary as a result of variations in vehicle states $(\alpha,\ \beta,\ q,\ p,\ r)$ and the derivatives of these states. The hinge moment derivatives $C_{h\delta(k,i)}$ define how the moments at the hinge of a control surface (k) vary as a result of surface (i) deflections (δ_{si}), and the hinge moment derivatives $C_{h\eta(k,j)}$ define how the moments at the hinge vary as a result of mode (j) generalized displacement (η_j), similarly for the rates of surface deflections and the modal displacements. The parameters (c_k) and (s_k) are the chord reference length and the surface area, respectively, for each control surface (k). The hinge moments are defined as follows:

$$\frac{\partial HM_k}{\partial \alpha} = \overline{Q}\, s_k \overline{c}_k\, C_{h\alpha}(k) \qquad\qquad \frac{\partial HM_k}{\partial \dot{\alpha}} = \overline{Q}\, s_k \overline{c}_k \left(\frac{l_{ch}}{2V_0}\right) C_{h\dot{\alpha}}(k)$$

$$\frac{\partial HM_k}{\partial q} = \overline{Q}\, s_k \overline{c}_k \left(\frac{l_{ch}}{2V_0}\right) C_{hq}(k) \qquad \frac{\partial HM_k}{\partial \dot{q}} = \overline{Q}\, s_k \overline{c}_k \left(\frac{l_{ch}}{2V_0}\right)^2 C_{h\dot{q}}(k)$$

$$\frac{\partial HM_k}{\partial \delta_{si}} = \overline{Q}\, s_k \overline{c}_k\, C_{h\delta}(k,i) \qquad\quad \frac{\partial HM_k}{\partial \dot{\delta}_{si}} = \overline{Q}\, s_k \overline{c}_k \left(\frac{\overline{c}_i}{2V_0}\right) C_{h\dot{\delta}}(k,i)$$

$$\frac{\partial HM_k}{\partial \eta_j} = \overline{Q}\, s_k \overline{c}_k\, C_{h\eta}(k,j) \qquad\quad \frac{\partial HM_k}{\partial \dot{\eta}_j} = \overline{Q}\, s_k \overline{c}_k \left(\frac{1}{V_0}\right) C_{h\dot{\eta}}(k,j)$$

3.7   MODELING A WIND-GUST DISTURBANCE

The air through which an aircraft flies is never still, and as a consequence, the wind causes disturbances on the flight vehicle, and its motion is always erratic. The total velocity of the vehicle relative to the air mass is obtained by the combination of two component vectors: a component of the vehicle velocity relative to the earth, plus the velocity of the wind relative to the ground. The nature of those air disturbances is influenced by many factors, but it is usually turbulence which occurs in and around clouds, wind shears which are violent atmospheric phenomena caused due to severe downburst of air, and sudden changes in the direction and velocity of the wind as the altitude changes, and gusts which are random, short duration sharp impulses of variations in wind velocity relative to the vehicle. In simulations, turbulence can be represented by a white noise generator passing through a low-pass second-order filter. A wind shear is a steady and persistent disturbance and it can be represented with a step in velocity filtered by a low-pass filter. A short wind gust (W_G) represents an unexpected change in the air-mass velocity with respect to the vehicle. It can be defined by the function.

$$w_G(t) = w_{\text{Gust}} \left(1 - \cos\left(\omega_G\, t\right)\right)$$

The wind disturbances cause changes in the angles of attack and sideslip of the vehicle relative to the wind α_w and β_w in comparison to the steady α_0 and β_0 angles without the wind. It also changes the magnitude and direction of the vehicle velocity relative to the air mass. The direction of the wind gust with respect to the vehicle can be defined in terms of two angles. The gust vector W_{GUST} can be resolved in three velocity components along the vehicle body axes, as shown in Figure 3.8.

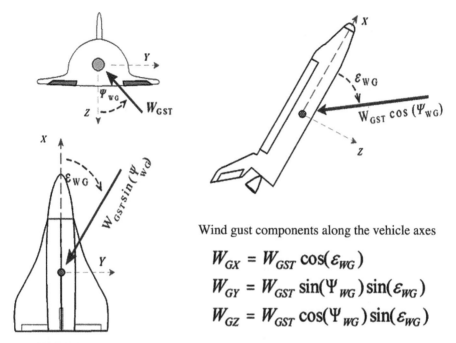

Wind gust components along the vehicle axes

$$W_{GX} = W_{GST}\cos(\varepsilon_{WG})$$
$$W_{GY} = W_{GST}\sin(\Psi_{WG})\sin(\varepsilon_{WG})$$
$$W_{GZ} = W_{GST}\cos(\Psi_{WG})\sin(\varepsilon_{WG})$$

FIGURE 3.8   Wind-Gust Velocity Vector W_{GST} is Resolved Along the Vehicle Body Axes

Prior to launching a vehicle the average wind velocity and direction near and above the launch site is measured at different altitudes by releasing weather balloons. The velocity and direction of the winds vary considerably with seasons and altitude, and the wind data received are used to adjust the vehicle trajectory in order to maximize performance.

In the equations the wind-gust input W_{GUST} is a cosine pulse scalar function, representing the wind-gust velocity relative to the ground in (ft/sec). Its direction relative to the vehicle body axes is fixed and defined by two angles, an elevation angle ε_{WG} of the wind vector with respect to the vehicle x-axis, and an azimuth angle Ψ_{WG} which is the angle between the vehicle z-axis and the projection of W_{GST} in the y–z plane. In a typical airplane, for example, an elevation gust angle $\varepsilon_{WG} = 0$ means that the wind-gust direction is coming head-on toward the pilot, along the $-x$-axis. When the azimuth and elevation angles are: Ψ_{WG}=90° and ε_{WG}=90°, it means that the wind-gust direction is coming toward the pilot from the right side along the $-y$-axis. When Ψ_{WG}=0° and ε_{WG}=90°, it means that the gust is coming from the bottom of the aircraft along the $-z$-axis, toward the pilot's feet.

In Figure 3.8 the wind-gust velocity is resolved in three components along the vehicle body axes (W_{GX}, W_{GY}, and W_{GZ}). The gust causes the angles of attack, sideslip, and also the vehicle velocity to change relative to the air mass. The angles of attack and sideslip (α, β) are defined in equation 2.6 and for variations in equation 3.4. They are generated by the vehicle motion relative to earth alone, assuming that the air

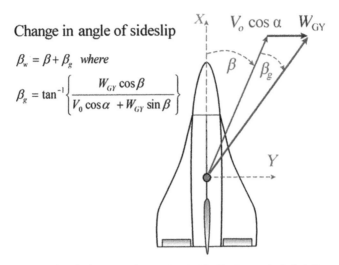

Change in angle of sideslip

$$\beta_w = \beta + \beta_g \quad \text{where}$$

$$\beta_g = \tan^{-1}\left\{\frac{W_{GY}\cos\beta}{V_0\cos\alpha + W_{GY}\sin\beta}\right\}$$

FIGURE 3.9   Change in the Angle of Sideslip due to the Wind Gust

mass is not moving. In the presence of the wind motion, however, the angles of attack and sideslip are modified to (α_w, β_w) to include the effects of the wind disturbance on the aerodynamic angles. The equations in Figures 3.9 and 3.10 calculate the total angles of attack and sideslip relative to the air mass, including the effects of the wind velocity. Instead of $(\alpha, \beta, \text{and } \delta V)$, due to the vehicle motion alone, the variables $(\alpha_w, \beta_w, \text{and } \delta V_w)$ are used in equations 3.17 and 3.20, which include the effects due to the wind. They are used to calculate the total aerodynamic forces and moments acting on the vehicle, which includes also the motion of the air mass.

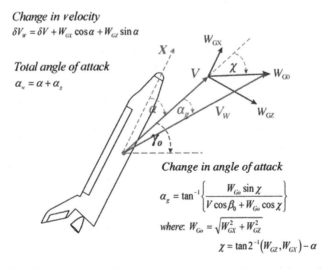

Change in velocity
$$\delta V_w = \delta V + W_{GX}\cos\alpha + W_{GZ}\sin\alpha$$

Total angle of attack
$$\alpha_w = \alpha + \alpha_g$$

Change in angle of attack

$$\alpha_g = \tan^{-1}\left\{\frac{W_{Go}\sin\chi}{V\cos\beta_0 + W_{Go}\cos\chi}\right\}$$

$$\text{where: } W_{Go} = \sqrt{W_{GX}^2 + W_{GZ}^2}$$

$$\chi = \tan 2^{-1}(W_{GZ}, W_{GX}) - \alpha$$

FIGURE 3.10   Changes in the Vehicle Angle of Attack and Air Speed due to the Wind Gust

3.8   PROPELLANT SLOSHING (SPRING–MASS ANALOGY)

Sloshing is defined as the periodic motion of the free surface of a liquid in a partially filled tank or container. Sloshing can be excited by the vehicle motion as a result of control system commands or from changes in vehicle acceleration caused by thrust variations, jet firing, or gust disturbances. If the liquid is allowed to slosh freely, the uncontrolled oscillations can produce disturbance forces that cause additional accelerations on the vehicle. These accelerations are then sensed and responded to by the flight control system, forming a closed loop that can lead to an oscillatory instability. A slosh-induced instability may lead to a structural failure, premature engine shutdown, or inability of the spacecraft to achieve upper-stage engine start through loss of propellant head at the drain port, loss of propellant through the vent system, and even loss of the vehicle itself. Even in a low or near zero-gravity environment, where the slosh frequencies and the torques exerted are low, instability can occur (e.g., due to coupling with the RCS) and the liquid motion can build up to amplitudes that may cause a failure. In general, the slosh frequencies should be greater than the control system bandwidth and below the fundamental bending frequency.

When a vehicle is accelerating the motion of the liquid inside the tank can be approximated with a spring–mass model. We may assume that the propellant can be separated in two components: (a) a solid mass that does not move relative to the tank and it is located near the bottom of the tank, and (b) a sloshing mass that oscillates slightly below the surface perpendicular to the acceleration A_T and the oscillations are excited by the vehicle lateral motion. The solid propellant mass is rigidly attached to the vehicle mass properties at a fixed point in the tank centerline located at the center of the nonsloshing propellant mass. This point mass is used in the calculations of the vehicle CG and moments of inertia, because, the moment of inertia of the nonsloshing propellant mass about its center of mass is zero, since it does not rotate with the vehicle.

The sloshing part of the liquid near the surface is approximated with a low damped, oscillating spring–mass system whose weight and location depends on the properties of the liquid, the tank shape and size, and the liquid fill level. One side of the spring is attached at a fixed point in the tank centerline a little below the surface, and the other end of the spring is attached to a mass representing the sloshing part of the propellant. The mass is free to oscillate in two directions perpendicular to the steady-state acceleration vector A_T, which is mostly in the x direction, but there may also be a small acceleration component in the z direction. We ignore the y component. The frequency of the slosh oscillation is proportional to the square root of the acceleration A_T, and the mass displacements are perpendicular to A_T. The slosh model can also be coupled with structural flexibility because the attachment point of the spring at the tank centerline is not necessarily rigid but it may oscillate due to structural deformation. A structural node must be selected for that attachment point in order to introduce slosh/flex excitation in both directions, that is, slosh exciting flexibility and flexibility exciting slosh.

The slosh motion is excited by variations in vehicle normal and lateral accelerations at the point where the spring is attached to the tank centerline. The slosh mass cannot

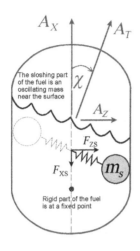

A_X and A_Z are the sensed vehicle accelerations along the body x and z directions. A_Y is assumed zero

Total acceleration: $A_T = \sqrt{A_X^2 + A_Z^2}$

$$\sin \chi = \frac{A_Z}{A_T}$$

The slosh frequency:

$$\omega_s = \omega_{s0} \sqrt{\frac{A_T}{g}} \; (rad \, / \, sec)$$

Forces due to slosh mass:

$$F_{ZS} = m_s \left(\omega_s^2 z_s + 2\zeta_s \omega_s \dot{z}_s \right) \cos \chi$$

$$F_{XS} = -m_s \left(\omega_s^2 z_s + 2\zeta_s \omega_s \dot{z}_s \right) \sin \chi$$

$$F_{YS} = m_s \left(\omega_s^2 y_s + 2\zeta_s \omega_s \dot{y}_s \right)$$

FIGURE 3.11   Slosh Mass Oscillating in a Plane That Is Perpendicular to the Total Acceleration Vector A_T and Generating Oscillatory Forces on the Vehicle

move along the acceleration vector A_T, which is the sensed acceleration: $(F_{\text{thrust}} - F_{\text{aero}})/M_T$, but the displacements are perpendicular to A_T. The displacements of the ith slosh mass, z_{si} and y_{si}, are measured relative to the spring attachment point at the tank centerline. If we assume that the steady-state A_Y acceleration is zero, the y_{si} slosh mass displacement is along the vehicle y-axis. The z_{si} slosh displacement, however, is not necessarily along the vehicle z-axis but it is slightly tilted by an angle χ in the x–z plane, perpendicular to A_T, see Figure 3.11.

The slosh mass displacement in two directions perpendicular to A_T is described by two low-damped second-order transfer functions 3.24, whose inputs are changes in vehicle acceleration at the tank centerline, perpendicular to A_T.

$$z_{si}(s) \left(s^2 + 2\zeta_i \omega_{si} s + \omega_{si}^2 \right) = a_{xsi} \sin \chi - a_{zsi} \cos \chi - A_T \sum_{j=1}^{N \, \text{mod}} \sigma_{ysi}(j) \, \eta(j)$$

$$y_{si}(s) \left(s^2 + 2\zeta_i \omega_{si} s + \omega_{si}^2 \right) = -a_{ysi} + A_T \sum_{j=1}^{N \, \text{mod}} \sigma_{zsi}(j) \, \eta(j) \qquad (3.24)$$

where

ω_{si}, ζ_{si} are the frequencies of the ith slosh mass in (rad/sec) and the damping coefficient.

$a_{xsi}, a_{zsi}, a_{ysi}$ are the translational accelerations of the vehicle at the ith slosh mass location along the x, z and y directions, respectively, as defined in equation 3.25.

z_{si} and y_{si} are the displacements of the slosh mass relative to the tank centerline attachment point perpendicular to A_T.

χ is the angle between the acceleration vector A_T and the vehicle x-axis.

The slosh mass is excited to oscillations by variations in vehicle acceleration at the spring attachment point in the tank centerline. The vehicle accelerations along x, y, and z at that point consist of three components (see equation 3.25).

1. Linear accelerations at the vehicle CG,

2. rotational acceleration components due to the vehicle angular accelerations, and

3. structural acceleration components because the tank wall structure is flexing around the slosh mass and it is approximated by a single point at the tank centerline.

$$a_{xsi} = \ddot{x}_{CG} + l_{Zsi}\dot{q} - l_{Ysi}\dot{r} + \sum_{j=1}^{N \text{ mod}} \varphi_{xs}(i,j)\,\ddot{\eta}\,(j)$$

$$a_{ysi} = \ddot{y}_{CG} + l_{Xsi}\dot{r} - l_{Zsi}\dot{p} + \sum_{j=1}^{N \text{ mod}} \varphi_{ys}(i,j)\,\ddot{\eta}\,(j) \qquad (3.25)$$

$$a_{zsi} = \ddot{z}_{CG} + l_{Ysi}\dot{p} - l_{Xsi}\dot{q} + \sum_{j=1}^{N \text{ mod}} \varphi_{zs}(i,j)\,\ddot{\eta}\,(j)$$

where

$\phi_{ys}(i,j),\ \phi_{zs}(i,j)$ are the modal shapes along y and z of the jth mode at the ith slosh mass location,

$\eta(j)$ is the generalized modal displacement for the jth mode,

$l_{Xsi},\ l_{Ysi},\ l_{Zsi}$ are the moment arms between the ith slosh mass and the vehicle CG in the x, y, and z directions, respectively.

$$l_{Xsi} = x_{si} - x_{CG} \quad l_{Ysi} = y_{si} - y_{CG} \quad l_{Zsi} = z_{si} - z_{CG}$$

The parameters required to define the propellant sloshing inside a tank are: the slosh mass m_{si}, slosh frequency ω_{si} in (rad/sec), damping ζ_{si}, and the steady-state location of the slosh mass relative to the vehicle. They can be obtained from the tank geometry, propellant density, surface level, the vehicle acceleration, etc. The slosh frequency is proportional to the square root of the steady-state acceleration A_T. The slosh frequency is typically defined at $1g$ acceleration, ω_{s0} in (rad/sec), and it is scaled up and down by multiplying it with the square root of the vehicle acceleration in g's, $\omega_s = \omega_{s0}\sqrt{A_T}$.

The derivation of slosh parameters as a function of tank geometry, propellant level, and density, have been obtained experimentally and are documented in NASA reports [Abramson [47]]. Slosh stability is typically analyzed in the frequency domain by calculating the phase and gain margins using Nichols or Nyquist diagrams. The slosh modes are low damped and they are often phase stable with the bubble resonances opening away from the -1 critical point. Sometimes, however, depending on vehicle CG and the slosh mass location relative to the vehicle's center of rotation (CR), the direction of the resonance is pointing toward or is encircling the critical -1 point. This would indicate a reduced slosh margin and the possibility of closed-loop slosh instability (divergent oscillations at the slosh frequency). Most often the slosh modes are phase stable unless the slosh mass happens to be located between the vehicle's CG and the CR. The derivation of slosh stability is presented in the next section, equations 3.32–3.35.

The disturbance force on the vehicle generated by the acceleration of a single slosh mass $m_s(i)$ that is oscillating relative to the tank can be resolved in three components along the vehicle x-, y-, and z-axes, as shown in equation 3.26.

$$
\begin{aligned}
F_{Zs}(i) &= m_s(i) \left(\omega_{si}^2 z_{si} + 2\varsigma_i \omega_{si} \dot{z}_{si} \right) \cos \chi \\
F_{Xs}(i) &= -m_s(i) \left(\omega_{si}^2 z_{si} + 2\varsigma_i \omega_{si} \dot{z}_{si} \right) \sin \chi \\
F_{Ys}(i) &= m_s(i) \left(\omega_{si}^2 y_{si} + 2\varsigma_i \omega_{si} \dot{y}_{si} \right)
\end{aligned}
\tag{3.26}
$$

The combined slosh forces from all tanks on the vehicle F_{Xsl}, F_{Zsl}, and F_{Ysl} along the x-, z- and y-axes are:

$$
F_{Xsl} = \sum_{i=1}^{Nsl} F_{Xs}(i) \qquad F_{Ysl} = \sum_{i=1}^{Nsl} F_{Ys}(i) \qquad F_{Zsl} = \sum_{i=1}^{Nsl} F_{Zs}(i)
\tag{3.27}
$$

Equation 3.28 calculates the moments on the vehicle in roll, pitch, and yaw due to multiple propellants sloshing by combining the forces from all tanks. It also includes the moments generated by the displacements of the slosh mass coupling with the vehicle acceleration.

$$
\begin{aligned}
L_{XSL} &= \sum_{i=1}^{Nsl} \left\{ F_{Zsi} l_{syi} - F_{Ysi} l_{szi} - m_s(i) A_Z y_{si} \right\} \\
M_{YSL} &= \sum_{i=1}^{Nsl} \left\{ F_{Xsi} l_{szi} - F_{Zsi} l_{sxi} - m_s(i) \left[d_{si}^2 \dot{q} + A_T z_{si} \right] \right\} \\
N_{ZSL} &= \sum_{i=1}^{Nsl} \left\{ F_{Ysi} l_{sxi} - F_{Xsi} l_{syi} - m_s(i) \left[l_{syi}^2 \dot{r} - A_X y_{si} \right] \right\} \\
A_T &= \sqrt{A_X^2 + A_Z^2}
\end{aligned}
\tag{3.28}
$$

where

d_{si} distance between the vehicle CG and the acceleration vector passing through the slosh mass,

A_T is the vehicle measured acceleration in the x–z plane.

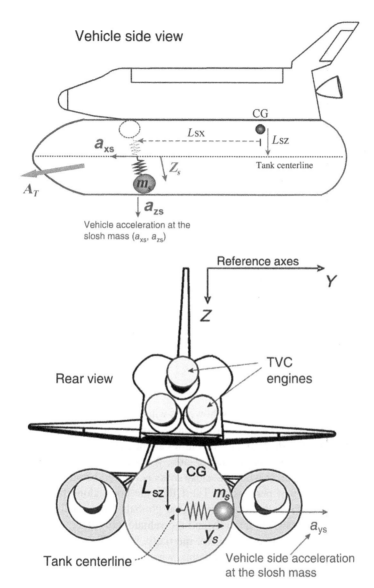

FIGURE 3.12   Spring–Mass Slosh Model

Slosh Mass Stability Relative to its Position

The stability of a slosh mode inside a launch vehicle strongly depends on the location of the slosh mass along the vehicle x-axis. When you plot the open-loop frequency response of the control loop in a Nyquist plot the resonance bubble is phase stable when it is opening away from the -1 point. The control system in this case removes energy from the slosh mode and it will actively dampen it when the

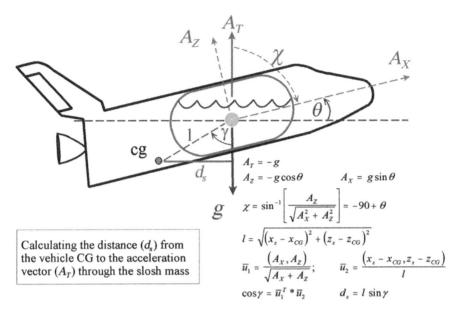

$$A_T = -g$$
$$A_z = -g\cos\theta \qquad A_x = g\sin\theta$$

$$\chi = \sin^{-1}\left[\frac{A_z}{\sqrt{A_x^2 + A_z^2}}\right] = -90 + \theta$$

$$l = \sqrt{(x_s - x_{CG})^2 + (z_s - z_{CG})^2}$$

Calculating the distance (d_s) from the vehicle CG to the acceleration vector (A_T) through the slosh mass

$$\bar{u}_1 = \frac{(A_x, A_z)}{\sqrt{A_x + A_z}}; \qquad \bar{u}_2 = \frac{(x_s - x_{CG}, z_s - z_{CG})}{l}$$

$$\cos\gamma = \bar{u}_1^T * \bar{u}_2 \qquad d_s = l\sin\gamma$$

FIGURE 3.13 Equations for Calculating the Distance (d_s) Between the Acceleration Vector at the Slosh Mass and the Vehicle CG. In This Example the Acceleration Is Up Toward the Vertical Direction due to the Lifting Aerodynamic Forces Against Gravity and the Propellant Level Is Horizontal

loop is closed. There are some cases, however, where slosh is phase unstable because the resonance bubble is opening toward the −1 point and in this case the control system destabilizes the slosh mode, and baffles are needed in these cases to passively dampen the propellant motion inside the tank. We will use a simple model to prove that slosh is phase unstable when the slosh mass is located between the vehicle CG and the CR, or center of percussion. The CR is the point along the vehicle x-axis that does not move laterally when the engine gimbals sinusoidally. As the engine wiggles the CG will translate laterally and the vehicle will rotate, and the combined motions define the point that stands still inertially. If the flight control system uses only attitude and angular rate measurements, when the slosh mass is between the CG and CR, the Nyquist plot will show the slosh resonance bubble opening toward the −1 point, indicating phase instability.

Equation 3.29 shows the translational motion z of the CG perpendicular to the centerline, and the vehicle rotation θ.

$$M\ddot{z} = T\delta + k\xi_s$$
$$J\ddot{\theta} = Tl_g\delta + rk\xi_s \tag{3.29}$$

where

M is the vehicle mass,
$\ddot{z}$ is the acceleration at the CG,

ξ_s is the slosh mass displacement relative to the tank centerline, and
k is the slosh stiffness.

When the stiffness $k = 0$, the lateral acceleration at the CR (which is zero by definition) is given by equation 3.30, and we can solve for the distance l_c of the CR relative to the CG.

$$\ddot{z}_C = \ddot{z} - l_c\ddot{\theta} = \left(\frac{T}{M} - l_c\frac{T l_g}{J}\right)\delta = 0 \Rightarrow l_c = \frac{J}{M l_g} \qquad (3.30)$$

The inertial acceleration of the slosh mass $\ddot{z}_{sl}$ is excited by the stiffness force due to the mass relative displacement (ignoring damping being small).

$$m_{sl}\ddot{z}_{sl} + k\xi_s = 0 \qquad (3.31)$$

The lateral displacement of the tank centerline at the slosh mass attachment is ($z + r\theta$), which is also equal to ($z_{sl} - \xi_s$), and the relative displacement ξ_s from the tank centerline is $\xi_s = z_{sl} - (z + r\theta)$. The simultaneous equations of motion are shown in the transfer function equation 3.32.

$$\begin{bmatrix} J s^2 & 0 & -rk \\ 0 & M s^2 & -k \\ m_{sl}r s^2 & m_{sl}s^2 & m_{sl}s^2 + k \end{bmatrix}\begin{bmatrix} \theta \\ z \\ \xi \end{bmatrix} = \begin{bmatrix} T l_g \\ T \\ 0 \end{bmatrix}\delta \qquad (3.32)$$

We can solve for the θ/δ transfer function as follows in equation 3.33.

$$\frac{\theta(s)}{\delta(s)} = \frac{TM l_g m_{sl}s^2 + \left(Ml_g - m_{sl}r + m_{sl}l_g\right)Tk}{\left[J Mm_{sl}s^2 + \left(JM + Mm_{sl}r^2 + J m_{sl}\right)k\right]s^2} \qquad (3.33)$$

where the zero frequency is:

$$\sqrt{\left(1 - \frac{m_{sl}r}{M l_g} + \frac{m_{sl}}{M}\right)\frac{k}{m_{sl}}} \qquad (3.34)$$

and the pole frequency is:

$$\sqrt{\left(1 + \frac{m_{sl}r^2}{J} + \frac{m_{sl}}{M}\right)\frac{k}{m_{sl}}}$$

If the slosh mass is located at the CR, that is, $r = -J/M l_g$ then the poles are equal to the zeros. There is a pole/zero cancellation and the slosh mode does not get excited because there is no lateral vehicle motion there. If on the other hand the slosh mass is located at the CG, that is, $r = 0$, there is a pole/zero cancellation again because the poles are also equal to the zeros. However, in this case the slosh mode gets excited,

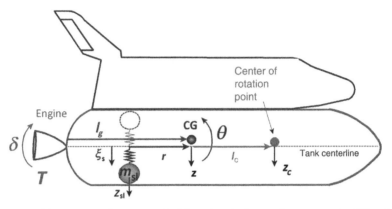

FIGURE 3.14   Slosh Is Phase Unstable When the Slosh Mass Is Between the CG and CR

but it would produce no moment on the vehicle but only lateral forces and therefore, the slosh motions would not be sensed in attitude or angular rate. The slosh mode is phase unstable when the frequency of the poles is lower than the frequency of the zeros and this happens when:

$$0 > r > -\frac{J}{M \, l_g} \qquad (3.35)$$

Otherwise, when $r > 0$ or if: $r < -J/M \, l_g$, the frequency of the zeros is less than the frequency of the poles and the slosh resonance is phase stable, which means that the tank does not require baffles. However, since the slosh mass location and the CG move with time in order to avoid baffles the slosh mass must stay out of the unstable region during the entire flight. Notice also, that this derivation is valid if we ignore the TWD dynamics. Otherwise, it is more complicated to predict the exact conditions of slosh instability. Figure 3.15 shows the variations of the pole and zero frequencies in a launch vehicle as a function of the slosh mass location relative to the CG and the CR. Notice that when the slosh mass is between the CG and the CR the zero frequency is greater than the pole frequency which implies that the slosh mode is phase unstable.

3.9   STRUCTURAL FLEXIBILITY

Structural flexibility is a very important issue in flight control system stability and performance and it must properly be accounted for in the mathematical model. Flight vehicles are designed with minimum weight objectives, hence their structures exhibit some form of flexibility. Some parts of the vehicle can develop considerable amounts of displacement and acceleration as a result of structural flexibility in addition to the displacement and acceleration that arise owing to the rigid-body motion. The primary function of the flight control system is to stabilize and to guide the flexible

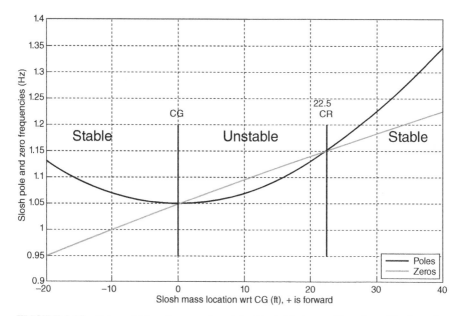

FIGURE 3.15 Pole and Zero Frequencies of the Slosh Mode as a Function of Its Location Along the Tank Centerline

vehicle without violating its operational requirements. It processes data from sensors to provide command signals to engines, control surfaces, and RCS jets. The sensors measure angular or translational motion, which in addition to the rigid-body motion includes motion caused by structural deformation at the locations of the sensors. These deformations affect the command signals to the control effectors. Since the effectors apply forces to the structure, energy can be fed to the structure at various frequencies, including those where flex modes may be excited. Because structural damping is small, it is possible for the effectors to add energy faster than it is dissipated causing excessive structural deflections and possible structural failures. If the deformation characteristics are ignored the flight vehicle may not be properly controlled and in many instances it may exhibit self-excited divergent oscillations that can be destructive. The structural dynamics should, therefore, be considered as an integral part of the control loop and the control system designer must be aware that divergent structural feedback can occur and must ensure that the flex phenomena are properly modeled, analyzed, and compensated.

Problems arising from the dynamic interaction of the control system interaction with the flexible structure are influenced by the sensor locations, the local structural flexibility, mode shapes, frequencies, and damping characteristics. The structure where the sensor is mounted on when excited by engine noise and vibration it may exhibit undesired responses resulting from local flexibility that can affect sensor performance and produce erroneous sensor signals or saturation which may seriously affect the control system operation. Local oscillations may also be reinforced by the control system to instability and limit cycles. In adverse situations, an attempt is made to counteract the interaction by modifying the control system, including filters, stiffening the sensor mounting structure, or relocating the sensors. If these modifications fail to resolve the interaction problem, structural redesign may be necessary. Improper determination and selection of dominant vibration modes in the design have often led to catastrophic structural instability problems in flight vehicles. The design is therefore supplemented with frequency domain analysis, simulations, component tests, system vibration tests, and flight tests.

Flexibility in general limits the control system bandwidth. It also affects the vehicle performance and usually requires filtering. As a rule of thumb there should be at least a factor of ten separation between the control system bandwidth and the first dominant resonance. The control system must be designed to process the sensor signals so that there is a net flow of energy out of the structure. One approach is for the control system to filter sensor signals at resonant structural frequencies, thereby, preventing the effectors from supplying energy at those frequencies. This is called "gain stabilization" and it avoids excessive response in high frequency flex resonances. In general, however, one or more of the lower frequency modes are not sufficiently separated from the control system bandwidth to permit gain stabilization and the alternative approach is "phase stabilization," where the controller is designed so that control forces are phased in such a way as to remove energy from the modes. Most designs employ both methods, with phase stabilization of low frequency modes and gain stabilization of high frequency modes.

In launch vehicles the main source of flex mode excitation is the TVC. In aircraft, the acceleration of the vehicle in combination with the aerodynamic forces can excite the structure into bending motion especially in the wings and the tails. This causes significant aeroelastic phenomena to occur which may have a serious impact on vehicle stability and performance. The deflection of an elevator, for example, causes a torsional flexibility on the wing due to the aerodynamic loading in the opposite direction, which reduces its effectiveness and may even cause elevator reversal if the wing torsional stiffness is too low. Low wing stiffness may also cause wing divergence which is another aeroelastic phenomenon to be avoided. That is, when the aerodynamic center of the wing is in front of its centerline, the lift generated at high dynamic pressures has a tendency to rotate the wing (nose up), which increases the aero moment even further. This torsional moment is resisted by the wing stiffness and if the torsional stiffness is not enough, the wing will twist off. Dynamic instabilities involving coupling between wing bending modes with short-period resonance have also occurred and caused major design revisions. There are other factors to consider in the design of flex structures. Surfaces and other external panels should also be examined for the possibility of panel flutter. In digital control systems we must also consider the effect of high frequency aliasing, where some local vibration modes fold back to excite and destabilize lower frequency modes. Analog filters in the rate gyro and accelerometer signals are often included before sampling to eliminate noise from folding back and exciting structural resonances. Furthermore, the frequency of updating pitch and guidance commands should not coincide with either flexible or rigid-body mode frequencies. The input commands are sometimes filtered to remove signals that excite vibrations.

The oscillatory motion of a vehicle structure can be represented with a combination of elastic modes which are excited by forcing functions generated by the actuators, propellant sloshing, aeroelastic coupling, wind gusts, and other external disturbances. The term "elastic modes" refers to the normalized mode shapes of the flight vehicle in "free–free" vibration. The mode shapes and frequencies of the "free–free" vehicle, also known as "modal data," are obtained from a finite element modeling program, such as NASTRAN$^{®}$. Each flex resonance in the structural model is characterized by a second-order transfer function that has a low damping coefficient and a resonance frequency, as shown in equation 3.36. Each mode j is excited by forces and moments applied on the vehicle structure in different locations. The coefficients on the right-hand side of the equation are obtained from the finite elements model. The modal data for each mode consist of the mode frequency ω_j in (rad/sec), the damping coefficient ζ_j, the generalized mass $m_{g(j)}$, the modal shapes $\varphi(j,k)$, and slopes $\sigma(j,k)$, (eigenvectors) in different locations k on the vehicle structure which are referred to as "nodes." The response to an excitation of the modal displacement $\eta_j(t)$ of a jth mode in terms of frequency amplitude and phase, depend on the forcing functions at the application points k, the mode frequency ω_j, and by the modal shapes $\varphi_x(j,k)$, $\varphi_y(j,k)$, $\varphi_z(j,k)$ along the x-, y-, and z-axes at the force application points, and also by the modal slopes $\sigma_x(j,k)$, $\sigma_y(j,k)$, $\sigma_z(j,k)$ about x, y, and z at the torque excitation points.

The terms on the right-hand side of each bending equation 3.36 are the external forces and torques that excite the flex resonance. They are forces and torques generated by the gimbaling of the engines, propellant sloshing, aerodynamic forces, aerosurface pivoting, etc. The output of each flex mode transfer function is the generalized displacement $\eta_j(t)$ that expresses the response of each natural frequency under the influence of the combined force and torque excitations. This is not a physical variable but an intermediate state in the state-space system. A sensor mounted on the flexible structure is measuring, in addition to the rigid-body motion, a linear combination of the generalized displacements $\eta_j(t)$ from all the flex modes which are observable at that sensor. The elastic modes create high frequency oscillations which are superimposed on the rigid-body motion. The amount and direction by which each mode is affecting the sensor depends on the modal shapes and slopes of the structure at the location of the sensor. Flexibility at the hinges of aerosurfaces or engines also affects the local responses of the structure as a result of actuator activity. It generates additional reaction loads on the actuators as a result of the local oscillation.

The excitation of flex modes are not only defined in terms of direct application of forces and torques on structural nodes, as described in equation 3.36, but the flex modes are also excited by distributed aerodynamic forces generated due to the vehicle motion relative to the atmosphere. This introduces aeroelasticity which defines modal excitation expressed in terms of variations in the angles of attack, sideslip, body rates, and surface deflections. Flex excitation in this case is defined by different sets of aeroelastic coefficients that couple the rigid-body motion to the generalized modal displacements $\eta_j(t)$ as shown in equation 3.38. The aeroelastic coupling may cause the vehicle resonant frequencies undergo substantial and irregular variations along its flight, tending sometimes for resonances to approach one another rather than to increase uniformly with time as it would result from the consumption of a propellant. Aeroelastic coupling can sometimes be destabilizing because it may lower the frequency of the first flex mode closer to the rigid-body bandwidth while reducing the vehicle static stability, and phase stabilization of the lowest frequency vibration modes may be required. To protect against adverse aeroelastic effects fins are sometimes mounted to increase static stability and also stiffen the frequency of structural modes.

Aeroelasticity is defined by a set of aeroelastic coefficients called the generalized aerodynamic force derivatives (GAFD) that couple the structural modes with aerodynamics, and specifically with variations in (α, β, body rates, and surface deflections). It is more dominant in large and flexible aircraft because flexibility is not only excited by the TVC, but also by distributed aerodynamic forces produced due to the vehicle and surfaces' interaction with the airflow. Aeroelasticity also includes the opposite effect. The vehicle flexibility, such as, the surfaces, wings and tails significantly affecting the aerodynamic forces and moments characteristics and the overall vehicle performance, especially at high dynamic pressures. Aeroelastic modeling is not simple and it involves complex CFD models which are developed to study flutter effects and to generate aeroelastic coefficients for control analysis models that define coupling between structure and aerodynamics. In some cases, such as in flexible aircraft and on the Space Shuttle during reentry, aeroelasticity is significant enough to be included in the control analysis. In some launch vehicles, however, the flexure excitation due to the TVC is much more dominant and the aeroelastic effects are often ignored.

Another factor that significantly excites flexibility is the inertia forces produced by the pivoting of the heavy nozzles and the control surfaces. These forces can yield undesirable deformations of the supporting structure and they in turn produce disturbances that must be accounted for. This excitation is defined by the "inertial coupling coefficients" which couple the structural modes with the rotational accelerations of the control surfaces or nozzles. The inertial coupling coefficients, otherwise known as "h-parameters," are generated from a finite elements program together with the modal data. The modes are "free–free" and the control surfaces are included in the FEM and the joints at the hinges are locked (infinitely stiff). The control surface hinges are released in the simulation model by the h-parameters in the equations of motion, see equation 3.38. In the early phases of a design the h-parameters and the aeroelastic coefficients are usually not available because it involves extensive CFD modeling and in the modal equations that follow we present two approaches for modeling structural flexibility. The first approach is easier to implement because it does not include aeroelasticity and it does not require GAFD and h-parameters data, and it is described in equations 3.36 and 3.37 The surfaces are assumed to be rigid and they are attached to the flexible structure by the hinges. The structure does not include the aerosurfaces and it is excited by aerodynamic and inertia forces produced at the hinges due to rigid surface deflections, rates, and accelerations. The second method is shown in equation 3.38, and it is obviously more efficient because it includes both aeroelasticity and inertial coupling coefficients.

3.9.1   The Bending Equation

The generalized displacement $\eta_{j(t)}$ of a flex mode j is obtained from a second-order, low-damped transfer function that is excited by forces and torques which are applied at different locations on the vehicle structure. The forcing functions are mainly due to forces generated by the engines or RCS jets, the control surface rotations, and also due to propellant sloshing. We are presenting two methods of modeling the flex mode excitation: a simple method that does not require GAFD and h-parameters, and a more complex and efficient method that uses GAFD and the inertial coupling coefficients $h_s(k,j)$ to excite flexibility.

3.9.1.1   *Flex Mode Excitation without Aeroelasticity and Inertial Coupling Coefficients*   In the first case the gimbaling engines, the slosh masses, and the aerosurfaces are attached to the flexible structure as separate bodies. Equation 3.36 denotes the excitation of the generalized displacement η_j of the jth structural mode, without GAFD and inertial coupling coefficients. It is excited by the engine and aerosurface forces, the slosh forces, and the reaction forces and torques produced due to the engine and surface accelerations.

$$m_g(j)\left(s^2 + 2\varsigma_j\omega_j s + \omega_j^2\right)\eta_j(s) =$$

$$+ \sum_{i=1}^{Nslosh}\left[F_{Xsi}\,\varphi_{xs}(i,j) + F_{Ysi}\,\varphi_{ys}(i,j) + F_{Zsi}\,\varphi_{zs}(i,j)\right] \qquad \text{Slosh forces}$$

$$+ \sum_{k=1}^{Neng}\left[F_{Xek}\,\varphi_{xe}(k,j) + F_{Yek}\,\varphi_{ye}(k,j) + F_{Zek}\,\varphi_{ze}(k,j)\right] \qquad \text{Engine forces}$$

$$- \sum_{k=1}^{Neng} m_{ek}\left[\varphi_{xe}(k,j)\,a_{xe(k)} + \varphi_{ye}(k,j)\,a_{ye(k)} \right.$$
$$\left. + \varphi_{ze}(k,j)\,a_{ze(k)}\right] \qquad \text{Engine inertia forces}$$
$$\text{(3.36)}$$
$$- \sum_{k=1}^{Neng} I_{ek}\left[\sigma_{ye}(k,j)\,\ddot{\delta}_{ye(k)} + \sigma_{ze}(k,j)\,\ddot{\delta}_{ze(k)}\right] \qquad \text{Engine inertia torques}$$

$$+ \sum_{l=1}^{Nsurf}\left[F_{Xs(l)}\,\varphi_{xcs}(l,j) + F_{Ys(l)}\,\varphi_{ycs}(l,j) + F_{Zs(l)}\,\varphi_{zcs}(l,j)\right] \qquad \text{Surface forces}$$

$$- \sum_{l=1}^{Nsurf} m_{cs(l)}\left[\varphi_{xcs}(l,j)\,a_{xcs(l)} + \varphi_{ycs}(l,j)\,a_{ycs(l)} \right.$$
$$\left. + \varphi_{zcs}(l,j)\,a_{zcs(l)}\right] \qquad \text{Surface inertia Forces}$$

$$- \sum_{l=1}^{Nsurf} I_{hs(l)}\sigma_{hs}(l,j)\,\ddot{\delta}_{fs(l)} \qquad \text{Surface inertia torques}$$

The total deflection of the lth control surface $\delta_{fs(l)}$ including flexibility is obtained from equation 3.37, where ϕ_{hsl} and λ_{hsl} are the orientation angles of the hinge vector.

$$\sigma_{hs}(l,j) = \left(\sigma_{ys}(l,j)\cos\phi_{hsl} + \sigma_{zs}(l,j)\sin\phi_{hsl}\right)\cos\lambda_{hsl} + \sigma_{xs}(l,j)\sin\lambda_{hsl}$$

$$\delta_{fs}(l) = \delta_{cs}(l) + \sum_{j=1}^{N\,\text{mod}} \sigma_{hs}(l,j)\,\eta(j) \tag{3.37}$$

where

m_{ek}, I_{ek}	are the mass and moment of inertia of the kth engine about its gimbal.
a_{yek}, a_{zek}	vehicle accelerations along the y- and z-axes including flexibility at kth engine gimbal.
$a_{ys(l)}$, $a_{zs(l)}$	accelerations at the hinge of lth aerosurface along y and z, including flexibility.
F_{Yek}, F_{Zek}	forces at the kth engine gimbal along the body y- and z-axes.
F_{Ysi}, F_{Zsi}	slosh forces at tank (i) along the body y- and z-axes.
$\varphi_{zek}(j)$, $\varphi_{yek}(j)$	are the mode shapes of the jth mode at the kth engine gimbal, along the z- and y–axes, respectively, in (ft/ft).
$\sigma_{zek}(j)$, $\sigma_{yek}(j)$	are the modal slopes of the jth mode at the kth engine gimbal, about z- and y–axes, respectively, in units of (rad/ft).
$\varphi_{zsi}(j)$, $\varphi_{ysi}(j)$	are the mode shapes of the jth mode at the ith slosh mass location, in (ft/ft).
$\varphi_{zcp}(j)$, $\varphi_{ycp}(j)$	are the mode shapes of the jth mode at the disturbance point in units of (ft/ft), (assuming that the disturbance is applied at a point).
$\varphi_{zasi}(j)$, $\varphi_{yasi}(j)$	are the mode shapes of the jth mode at the ith aerosurface hinge, along the z- and y–axes, respectively, in (ft/ft).
$\sigma_{zasi}(j)$, $\sigma_{yasi}(j)$	are the mode slopes of the jth mode at the ith aerosurface hinge, about the z- and y–axes, respectively, in (rad/ft).
η_j, ω_j	are the generalized modal displacement and mode frequency of the jth mode.
$m_g(j)$	is the generalized mass of the jth mode.
δ_{zek}, δ_{yek}	are the deflections of the kth engine relative to trim about z and y in (rad), including flexibility.
$\delta_{fs(l)}$	is the control surface l deflection relative to trim including flexibility from equation 3.37.
$\sigma_{hs(l,j)}$	is the modal slope about the hinge vector at the center of surface l hinge line, and it is calculated in terms of the slopes about x, y, and z.
ϕ_{hsl}, λ_{hsl}	are the bank and the sweep angles of the lth surface hinge vector.
$m_{cs(l)}$, $I_{hs(l)}$	are the mass and the moment of inertia about the hinge of the lth aerosurface.

The bending modes are mainly excited by the TVC, aerosurfaces, and slosh forces. In the absence of aeroelastic data a bending mode is also excited by the aerodynamic forces created due to surface deflections, as in equation 3.18, and they are applied at the nodes that correspond to the center of the hinges. There are also reaction forces at the hinges generated by variations in vehicle acceleration at those points, see equations 3.5 and 3.6. This model is obviously not as efficient as the aeroelastic model presented in equation 3.38 which also includes the inertial coupling between flexibility and aerosurface accelerations.

3.9.1.2   *Modal Excitation Using GAFD and Inertial Flex Coupling Coefficients*

GAFD and inertia coupling coefficients provide a more accurate modeling of the aeroelastic coupling between aerodynamics and structural flexibility and also the dynamic coupling between control surface accelerations and structural excitation. The GAFD data consists of three types of coefficients.

1. A set of coefficients that define how the vehicle basic aerodynamic forces and moments, such as: C_Z, C_m, C_n, etc. vary as a function of flexibility, that is, generalized modal displacements and modal rates $(\eta_j, \dot{\eta}_j)$.

2. A set of coefficients that define how the generalized modal displacement η_j of the jth flex mode, defined in equation 3.38, is excited by variations in rigid vehicle motion. Specifically by changes in the angles of attack, sideslip, body rates, control surface deflections, and by interactions with the other flex modes. The coefficients also define how the generalized modal displacements vary as a function of the rate of change of the above variables.

3. The third set of coefficients are hinge moment coefficients. They define how the moment at the hinge of the ith control surface is affected by changes in the vehicle angles of attack, sideslip, body rates, accelerations, modal displacements, modal rates, and also by the interactions with other control surface deflections and rates $(\delta_s, \dot{\delta}_s)$. These coefficients are also used in the hinge moment equations which are described in Section 3.10.2.

The GAFD data are calculated using unsteady aerodynamic theory by postprocessing the generalized aerodynamic forces Q_{ij} which is an aerodynamic matrix obtained from the "doublet lattice" process. It requires a CFD model and a finite elements model. The Q_{ij} terms are also used for flutter and loads analysis. The generalized aerodynamic forces are complex matrices and they are functions of the vehicle Mach number, dynamic pressure, at reduced frequencies. The GAFD matrices are calculated at specific frequencies that correspond to the modal frequencies plus one or two additional frequencies below the first flex mode. At each Mach number and reduced frequency a complex generalized force matrix is generated, a matrix for the real part, and a matrix for the imaginary part. In flutter analysis, a Mach number and a reduced frequency are assumed and the flutter solution is calculated. For the development of a control analysis model, a single complex generalized aerodynamic force matrix, independent of frequency, is constructed by extracting the corresponding rows from the reduced frequency-dependent matrices. The real part of this complex matrix consists of displacement coefficients and the imaginary part consists of the velocity coefficients. The inputs to the doublet lattice process are the modal data (mode shapes and mode frequencies) obtained from the finite elements model. The aerodynamic shape of the vehicle including the fuselage is initially modeled by means of flat plates and the doublet lattice process calculates the generalized aerodynamic forces at different Mach numbers. When the fuselage is modeled as a slender body and an interference cylinder is included with the flat plates representing the lifting surfaces, adjustments can be made to the aeromodel to closely match measured data. For example, the rigid-body aerodynamic derivatives, such as Cma, Cnb, etc. derived from the doublet lattice output are compared with the aerodynamic derivatives obtained from wind tunnel data and appropriate model corrections are made. The program finally combines the FEM with the aerodynamic model and creates the generalized aerodynamic forces, also known as the Q_{ij} matrix. The GAFD data used in the equations are extracted from the Q_{ij} matrix.

The aeroelastic model provides a more accurate representation of flexibility because it is based on a detailed CFD aeroelastic model. Equation 3.38 is used instead of 3.36 for the calculation of the modal displacement η_j. The forcing function excitations from the control surfaces are calculated not by direct force application at the hinges, as in 3.36, but via the GAFD and h-parameters. The aeroelastic coefficients describe the modal excitation as a result of the aerodynamic forces generated from the vehicle motion relative to the airflow defined in terms of α_w, β_w, body rates, and surface deflections δ_{sk}, and also as a function of their rates.

The coefficients also include dynamic coupling with other modes η_j. In equation 3.38 the excitation of a jth mode due to the acceleration of the kth control surface is defined by the inertial coupling coefficients or h-parameters $h_s(k,j)$. The inertial coupling coefficients are the off-diagonal blocks in the mass matrix created in the finite elements model which couples the engine or surface accelerations with flexibility. They also define angular acceleration at the kth surface due to jth mode excitation. This formulation is more accurate than equation 3.36 where the surfaces and nozzles are assumed to be rigid and the structure excitation is a result of inertial forces generated from rigid effector accelerations. The flex mode is also excited by pitch and yaw gimbal accelerations coupled by means of similar h-parameters for the engines $h_{ye}(k,j)$ and $h_{ze}(k,j)$. This method is obviously more accurate because unlike equation 3.36 where the effectors are rigid and the control surfaces are coupled as separate rigid bodies. Equation 3.38 denotes flex mode (j) excitation using GAFD and inertial coupling coefficients. It includes flexibility of the effectors structure, with the hinges locked, and their interaction with the vehicle structure are all included in the FEM. The hinges in equation 3.38 are released by the introduction of the h-parameters $h_s(l, j)$ and the coupling with the actuator which generates $\ddot{\delta}_s$. The deflection, rate, and acceleration inputs from the actuator are relative to the hinge and they do not

including any additional flex terms.

$$m_g(j)\left(s^2 + 2\varsigma_j\omega_j s + \omega_j^2\right)\eta_j(s) =$$

$$+ \sum_{i=1}^{N\text{slosh}}\left[F_{Xsi}\,\varphi_{xs}(i,j) + F_{Ysi}\,\varphi_{ys}(i,j) + F_{Zsi}\,\varphi_{zs}(i,j)\right]\qquad\text{Slosh forces}$$

$$+ \sum_{k=1}^{N\text{eng}}\left[F_{Xek}\,\varphi_{xe}(k,j) + F_{Yek}\,\varphi_{ye}(k,j) + F_{Zek}\,\varphi_{ze}(k,j)\right]\qquad\text{Engine forces}$$

$$- \sum_{k=1}^{N\text{eng}}\left[h_{ye}(k,j)\,\ddot{\delta}_{yek} + h_{ze}(k,j)\,\ddot{\delta}_{zek}\right]\qquad\text{Engine inertial coupling}$$

$$- \sum_{l=1}^{N\text{surf}} h_s(l,j)\,\ddot{\delta}_{s(l)}\qquad\text{Control surface coupling}$$

Aeroelastic terms

$$+\bar{Q}\,S_{\text{ref}}^*\begin{bmatrix}+ C_{\eta_j\alpha}\alpha_w + C_{\eta_j\beta}\beta_w + \left(\dfrac{l_{\text{ch}}^*}{2V_0}\right)\left(C_{\eta_j\dot\alpha}\dot\alpha_w + C_{\eta_j q}q\right) + \left(\dfrac{l_{\text{ch}}^*}{2V_0}\right)^2 C_{\eta_j\dot q}\dot q \\[2ex] + \left(\dfrac{l_{\text{sp}}^*}{2V_0}\right)\left(C_{\eta_j\dot\beta}\dot\beta_w + C_{\eta_j p}p + C_{\eta_j r}r\right) + \left(\dfrac{l_{\text{sp}}^*}{2V_0}\right)^2\left(C_{\eta_j\dot p}\dot p + C_{\eta_j\dot r}\dot r\right) \\[2ex] + \sum_{k=1}^{N\text{surf}}\left[C_{\eta(j)\,\delta(k)}\delta_{s(k)} + \left(\dfrac{\bar c_{sk}}{2V_0}\right)C_{\eta(j)\,\dot\delta(k)}\,\dot\delta_{s(k)}\right] \\[2ex] + \sum_{i=1}^{N\text{mode}}\left[C_{\eta(j)\,\eta(i)_i}\eta_i + \left(\dfrac{C_{\eta(j)\,\dot\eta(i)}}{V_0}\right)\dot\eta_i\right]\end{bmatrix}$$

$$(3.38)$$

S^*_{ref} is the surface reference area of the vehicle in (ft^2) used to scale the GAFD data.

$l^*_{\text{ch}}, l^*_{\text{sp}}$ are the chord and span reference lengths in (ft) used to normalize the GAFD data.

$\bar c_{sk}$ is the chord in (ft) of the control surface k used to normalize the GAFD data.

$\delta_{cs(k)}$ is the clockwise rotation in (rad) of a control surface (k) about its hinge vector. The rotation is only due to actuator displacement and it does not include structural flexibility.

$h_{s(k,j)}$ are the inertial coupling coefficients (h-parameters) in (lb-sec^2) which couple the excitation of the jth mode to the kth surface acceleration.

η_j is the generalized modal displacement of the jth flex mode in (ft).

Note that, in equation 3.38 the aeroelastic coefficients apply only for the control surfaces. Usually, there are no GAFD generated for the TVC engines. The inertia coupling coefficients, however, apply to both surfaces and nozzles. Since the modal excitation due to engine and surface slewing is captured by the h-parameters, unlike equation 3.13 which includes the inertial terms in the calculation of gimbal forces, the engine forces in this case should include only the TVC and throttling terms and not the inertial terms. Equation 3.39 denotes engine forces at the gimbal due to gimbaling and throttling, excluding the inertial terms.

$$F_{xe}(k) = T_{ek}\left(-\cos\Delta_{Z_k}\sin\Delta_{E_k}\delta_{yek} - \cos\Delta_{E_k}\sin\Delta_{Z_k}\delta_{zek} + \cos\Delta_{Z_k}\cos\Delta_{E_k}\delta_{Thr_k}\right)$$

$$F_{ye}(k) = T_{ek}\left(-\sin\Delta_{Z_k}\sin\Delta_{E_k}\delta_{yek} + \cos\Delta_{E_k}\cos\Delta_{Z_k}\delta_{zek} + \sin\Delta_{Z_k}\cos\Delta_{E_k}\delta_{Thr_k}\right)$$

$$(3.39)$$

$$F_{ze}(k) = -T_{ek}\left(\cos\Delta_{E_k}\delta_{yek} + \sin\Delta_{E_k}\delta_{Thr_k}\right)$$

where the throttle control $\quad \delta_{Thr_k} = \left(\dfrac{\delta T_{ek}}{T_{ek}}\right)$

However, when the TVC nozzles are sufficiently rigid to be modeled as separate rigid bodies coupled to the structure without h-parameters then the engines/flex coupling should be implemented as shown in equation 3.36, and the engine forces at the gimbals should be calculated from equation 3.13 including the inertial forces.

3.10 LOAD TORQUES

The translational and angular accelerations of the vehicle create disturbance torques at the TVC gimbals or at the hinges of the control surface. These are disturbance torques reacting against the control torques from the actuators. The actuator device, which is either hydraulic or electromechanical position control servo, must have the control force capability to overcome friction, load inertia, aerodynamics, and other load torques in order to position the effector as commanded by the control system. This reaction torque is an external loading that feeds back to the actuator system and in launch vehicles it is referred to as "actuator load torque" or sometimes "Dog-Wags-Tail." In aircraft it is usually called "hinge moments." The vehicle accelerations at the hinges consist of both rigid body and flex components due to local vibrations. For early analysis we usually avoid including the TWD and load-torque dynamics in the dynamic model, often because the information is not yet available.

As already mentioned, the proper implementation of the TWD and load-torque feedback dynamics requires a vehicle model that includes the necessary inputs and outputs in order to be properly coupled with the actuator dynamic models. Otherwise, those effects will not be properly captured and in this case a simple transfer function would be sufficient for actuator modeling. In order to implement the TWD effect the vehicle inputs must include gimbal accelerations in addition to gimbal deflections for each control surface and TVC direction. The vehicle must be coupled with the actuator models that provide the necessary gimbal deflections and gimbal acceleration outputs, as described in Chapter 4. The vehicle model must also provide load-torque outputs at each effector which are fed back to the corresponding actuator load-torque inputs.

The same principle applies for the control surface hinge moments loop, see Figure 4.7. It is a mechanical loop from the vehicle model hinge moment outputs to the actuator load-torque inputs. The inputs to the vehicle model are generated by the actuator subsystems, one actuator per aerosurface or two actuators per TVC engine (pitch and yaw). Each actuator has two inputs: a δ_{comd} input from the mixing logic, and the load-torque input that connects with hinge moment outputs from the vehicle model. Each actuator has three outputs: engine deflections, rates, and accelerations that connect to the vehicle inputs. The positions provide the control forces, and the rates provide aerodamping for the control surfaces. The engine or surface accelerations are important for the implementation of TWD. Further description of the dynamic coupling between vehicle and actuators, and detail actuator models are described in Chapter 4. In the next sections we will examine the load torques at the nozzle gimbals and at the control surfaces separately, beginning with the load torques at the TVC gimbals.

3.10.1   Load Torques at the Nozzle Gimbal

In Section 3.9 we considered two approaches for implementing the dynamic coupling between flexibility and gimbal accelerations: a rigid nozzle approach and a more refined h-parameters approach. The load-torque calculations are also affected by the two approaches. In the first case we assume that the engines are rigid and they couple with a flexible vehicle that vibrates and flexes at the gimbal, but it does not include the engines in the structural model. In the second case the engine nozzles are included in the finite elements model rigidly attached at the gimbals, and the reaction load torques are calculated by means of the pitch and yaw inertial coupling coefficients $h_{ye}(k,j)$ and $h_{ze}(k,j)$ which are provided by the finite elements model.

3.10.1.1   Rigid Engine Coupling   The first approach is shown in equation 3.40 that denotes variations in pitch and yaw load torques $T_{LYe(k)}$, $T_{LZe(k)}$ at engine (k) gimbal. They are generated at the gimbal of the kth engine due to vehicle motion.

$$
T_{LYe}(k) =
$$
$$
m_{ek}\, l_{ek} \left[\begin{array}{l} V_0 \cos \alpha_0 (q - \dot\alpha) - a_{ze(k)} \cos \Delta_{Ek} - a_{xe(k)} \sin \Delta_{Ek} \cos \Delta_{Zk} \\ + A_Z \sin \Delta_{Ek} \delta_{ye(k)} + A_X (\sin \Delta_{Ek} \sin \Delta_{Zk} \delta_{ze(k)} - \cos \Delta_{Ek} \cos \Delta_{Zk} \delta_{ye(k)}) \end{array} \right]
$$
$$
- I_{ek} \left[\dot q + \sum_{j=1}^{N\,\mathrm{mod}} \sigma_{ye}(k,j)\, \ddot\eta(j) \right] \tag{3.40}
$$

$$T_{LZe}(k) = m_{ek} \, l_{ek}[a_{ye(k)} \cos \Delta_{Zk} - a_{xe(k)} \sin \Delta_{Zk} - (A_X \cos \Delta_{Zk} + A_Y \sin \Delta_{Zk})\delta_{ze(k)}]$$

$$- I_{ek} \left[\dot{r} + \sum_{j=1}^{N \, mod} \sigma_{ze}(k,j) \, \ddot{\eta}(j) \right]$$

where

$a_{xe}(k)$, $a_{ye}(k)$, $a_{ze}(k)$ are the vehicle accelerations along x, y, z at gimbal (k), including flexibility. They are defined in equation 3.5, repeated below.

Δ_{Ek} and Δ_{Zk} are the pitch and yaw trim angles of engine (k) as defined in Figure 3.3.

I_{ek} is the nozzle moment of inertia about the gimbal.

$\sigma_{ye(k,j)}$, $\sigma_{ze(k,j)}$ are the pitch and yaw modal slopes of mode (j) at the location of engine (k).

$$a_{xe}(k) = \ddot{x}_{CG} + l_{Zek}\dot{q} - l_{Yek}\dot{r} + \sum_{j=1}^{Nm} \varphi_{xe}(k,j) \, \ddot{\eta}(j)$$

$$a_{ye}(k) = \ddot{y}_{CG} + l_{Xek}\dot{r} - l_{Zek}\dot{p} + \sum_{j=1}^{Nm} \varphi_{ye}(k,j) \, \ddot{\eta}(j) \tag{3.5}$$

$$a_{ze}(k) = \ddot{z}_{CG} + l_{Yek}\dot{p} - l_{Xek}\dot{q} + \sum_{j=1}^{Nm} \varphi_{ze}(k,j) \, \ddot{\eta}(j)$$

- The first group of terms in equation 3.40 represents the load torque at the engine gimbal caused by variations in the translational acceleration (a_{xe}, a_{ye}, a_{ze}) of the vehicle at the kth engine gimbal, along x, y, and z. It also includes a very small term due to change in flight path $\dot{\gamma}$.
- The second group of terms represents variations in load torque due to coupling of the vehicle steady-state nominal accelerations A_X and A_Z with the engine pitch and yaw deflections (δ_{yek}, δ_{zek}) relative to the corresponding trim positions (Δ_{Ek}, Δ_{Zk}). The deflection consists of both rigid rotation plus elastic deformation at the gimbal, as shown in equation 3.12 and Figure 3.4.
- The third group of terms in the load-torque equations (that include the nozzle inertia I_e) represent the variation in load torque due to the inertial rotational acceleration of the vehicle at the gimbal. It is a combination of rigid body plus elastic acceleration at the gimbal.

3.10.1.2 Engine Load Torque Using Inertial Coupling Coefficients

When the pitch and yaw inertial coupling coefficients $h_{ye}(k,j)$ and $h_{ze}(k,j)$ for the engine nozzles are available from the FEM, the variations in pitch and yaw load torques are modified as shown in equation 3.41. In this case, the torque due to flexing at the gimbal is captured by the inertial coupling coefficients and this is a more accurate representation than equation 3.40, where the rigid nozzle is coupled via the modal slope at the support structure.

$$T_{LYe}(k) =$$

$$m_{ek}\, l_{ek} \left[\begin{array}{l} V_0 \cos \alpha_0\, (q - \dot{\alpha}) - a_{ze(k)} \cos \Delta_{Ek} - a_{xe(k)} \sin \Delta_{Ek} \cos \Delta_{Zk} \\ + A_Z \sin \Delta_{Ek} \delta_{ye(k)} + A_X \left(\sin \Delta_{Ek} \sin \Delta_{Zk} \delta_{ze(k)} - \cos \Delta_{Ek} \cos \Delta_{Zk} \delta_{ye(k)} \right) \end{array} \right]$$

$$- I_{ek}\, \dot{q} - \sum_{j=1}^{N\,\mathrm{mod}} h_{ye}(k,j)\, \ddot{\eta}(j) \tag{3.41}$$

$$T_{LZe}(k) = m_{ek}\, l_{ek} \left[a_{ye(k)} \cos \Delta_{Zk} - a_{xe(k)} \sin \Delta_{Zk} - \left(A_X \cos \Delta_{Zk} + A_Y \sin \Delta_{Zk} \right) \delta_{ze(k)} \right]$$

$$- I_{ek}\, \dot{r} - \sum_{j=1}^{N\,\mathrm{mod}} h_{ze}(k,j)\, \ddot{\eta}(j)$$

However, since in this case the dynamic coupling between structure and load torque is handled by the h-parameters, the linear accelerations at the gimbals should not include flexibility, as shown below:

$$a_{xe}(k) = \ddot{x}_{CG} + l_{Zek}\dot{q} - l_{Yek}\dot{r}$$
$$a_{ye}(k) = \ddot{y}_{CG} + l_{Xek}\dot{r} - l_{Zek}\dot{p} \tag{3.5b}$$
$$a_{ze}(k) = \ddot{z}_{CG} + l_{Yek}\dot{p} - l_{Xek}\dot{q}$$

3.10.2   Hinge Moments at the Control Surfaces

The control surface hinge moments are similar to the engine gimbal load torques. They are also implemented using two separate methods: a rigid surface attached to a flexible hinge and a more detailed method that requires h-parameters and GAFD data. In the first approach we assume that the surfaces are rigid panels interacting with a flexible vehicle structure by aerodynamic and inertial forces generated at the hinges. The surfaces are treated as separate bodies and they must not be included in the vehicle mass properties and in the structural models. The second method is a lot more refined because it is based on a finite elements model that includes the control surfaces with the hinges locked. It uses inertial coupling coefficients to calculate the hinge moments due to surface accelerations and the hinge moment coefficients obtained from the GAFD model that calculate aero moments due to flexing.

3.10.2.1   *Rigid Surfaces Coupled with a Flexible Vehicle*   The first method assumes that each aerosurface is rigid and that it is attached to the structure by a hinge that is flexing at the hinge line. Flexibility at the attachments is captured by

the modal slopes at the hinges. Equation 3.42 denotes hinge moment variations at a control surface hinges without h-parameters. They are generated by the vehicle translational and rotational accelerations at the hinges (including flexure accelerations). The hinge moments include contributions due to variations in the angles of attack and sideslip relative to wind, and also their rates. They also include contributions due to aerosurface deflections and the rates of:

$$M_{hs}(k) = m_{sk} l_{hsk} \cos\left(\Delta_{sk}\right) \left[a_{Ns}(k) - \dot{V}_o \delta_{fs}(k) \right] - m_{sk} l_{hsk} a_{Xs}(k) \sin\left(\Delta_{sk}\right)$$

$$+ m_{sk} l_{hsk} V_0 (q - \dot{\alpha}) - I_{hsk} \dot{W}_{sk} - \sum_{j=1}^{N \, \text{mod}} \left[I_{hcsk} \sigma_{hs}(k,j) \ddot{\eta}(j) \right]$$

$$+ \bar{Q} S_{sk} \bar{c}_{sk} \left\{ C_{h\alpha k} \alpha_w + C_{h\beta k} \beta_w + C_{h\delta k} \delta_{fs(k)} + \left(\frac{l_{ch}}{2V_0} \right) C_{h\dot{\alpha}k} \dot{\alpha}_w \right.$$

$$\left. + \left(\frac{l_{sp}}{2V_0} \right) C_{h\dot{\beta}k} \dot{\beta}_w + \left(\frac{\bar{c}_{sk}}{2V_0} \right) C_{h\dot{\delta}k} \dot{\delta}_{fs(k)} \right\} \qquad (3.42a)$$

$$\dot{W}_{sk} = \left[\cos\phi_{hsk} \dot{q} + \sin\phi_{hsk} \dot{r} \right] \cos\lambda_{hsk} + \dot{p} \sin\lambda_{hsk}$$

$$a_{Ns}(k) = a_{Ys}(k) \sin\phi_{hsk} - a_{Zs}(k) \cos\phi_{hsk} \qquad (3.42b)$$

W_{sk}	is the vehicle rate resolved parallel to the hinge vector at the kth control surface.
I_{sk}	is the kth aerosurface moment of inertia about the hinge line.
$a_{Ns(k)}$	is the vehicle acceleration at the hinge line perpendicular to the kth control surface.
$\Delta_{s(k)}$	is the kth aerosurface trim deflection.
$\delta_{fs(k)}$	is the kth aerosurface deflection from trim including flexibility, as in equation 3.19.
S_{sk}, c_{sk}	are the aerosurface reference area in (ft^2) and its reference length in (ft). They are defined in the aerodata and are used to normalize the hinge moment coefficients.
$\phi_{hsk}, \lambda_{hsk}$	are the bank and sweep angles of the kth aerosurface hinge vector.

The surface deflections $\delta_{fs}(k)$ also include the structural deformation terms as defined in equation 3.19. The modal slope σ_{hsk} about the hinge vector of a kth control surface is calculated from equation 3.37 as a function of the hinge orientation angles and the modal slopes about x, y, z.

$$\sigma_{hsk} = \left(\cos\phi_{hsk} \sigma_{ysk} + \sin\phi_{hsk} \sigma_{zsk} \right) \cos\lambda_{hsk} + \sin\lambda_{hsk} \sigma_{xsk} \qquad (3.37)$$

- The first three terms on the right-hand side of Equation 3.42a are reaction moments at the hinge of the kth control surface due to vehicle linear acceleration at the hinge.

- The fourth term is the moment created by the centripetal force due to variations in the flight path angle γ.
- The fifth term is a reaction moment due to the vehicle inertial acceleration $\dot{W}_{sk}$ resolved about the kth hinge vector.
- The sixth term is the moment due to the vehicle rotational flexing acceleration about the hinge of the control surface.
- The last group of terms in the third line represents variations in aerodynamic loading at the hinge due to variations in the angles of attack and sideslip (relative to the wind), surface deflections, and also the rates of the above variables. They are calculated from hinge moment coefficients.

The vehicle accelerations along x, y, and z, at the midpoint of the kth hinge line are defined in equation 3.43. The accelerations include structural accelerations.

$$a_{xcs}(k) = \ddot{x}_{CG} + l_{Zsk}\dot{q} - l_{Ysk}\dot{r} + \sum_{j=1}^{Nm} \varphi_{xcs}(k,j)\ddot{\eta}(j)$$

$$a_{ycs}(k) = \ddot{y}_{CG} + l_{Xsk}\dot{r} - l_{Zsk}\dot{p} + \sum_{j=1}^{Nm} \varphi_{ycs}(k,j)\ddot{\eta}(j) \qquad (3.43)$$

$$a_{zcs}(k) = \ddot{z}_{CG} + l_{Ysk}\dot{p} - l_{Xsk}\dot{q} + \sum_{j=1}^{Nm} \varphi_{zcs}(k,j)\ddot{\eta}(j)$$

The moment arm distances between the center of the kth aerosurface hinge line and the vehicle CG are defined in the following equations:

$$l_{XSk} = X_{Sk} - X_{CG} \quad l_{YSk} = Y_{Sk} - Y_{CG} \quad l_{ZSk} = Z_{Sk} - Z_{CG} \qquad (3.44)$$

3.10.2.2 Using GAFD and Inertial Coupling Coefficients

A more accurate implementation of the hinge moments is by including the inertial coupling coefficients and the GAFD-derived hinge moments data, when available. Equation 3.45 denotes hinge moments at the control surfaces, using hinge moment coefficients from GAFD and inertial coupling coefficients. The finite elements model in this case includes the control surfaces and the hinge moments due to flexing at the surfaces are introduced by the inertial coupling coefficients $h_{s(k,j)}$ which were also used in the bending equations. The hinge moment coefficients are derived from the GAFD model and they generate moments at the aerosurface hinges due to variations in the angles of attack and sideslip relative to the wind, including their rates. There are also contributions due to the vehicle angular rates and accelerations, surface deflections and rates, and terms due to modal displacements (η_j) and rates of displacement.

$$
M_{hs}(k) = m_{sk} l_{hsk} \cos\left(\Delta_{sk}\right) \left(a_{Ns}(k) - \dot{V}_o \delta_s(k)\right) - m_{sk} l_{hsk} a_{Xs}(k) \sin\left(\Delta_{sk}\right)
$$

$$
+ m_{sk} l_{hsk} V_0 \left(q - \dot{\alpha}\right) - I_{hsk} \dot{W}_{sk} - \sum_{j=1}^{N \bmod} \left[h_s\left(k,j\right) \ddot{\eta}(j) \right]
$$

$$
+ \bar{Q} S_{sk} \bar{c}_{sk} \left[C_{h\alpha k} \alpha_w + C_{h\beta k} \beta_w + \left(\frac{l_{ch}^*}{2V_0} C_{h\dot{\alpha} k} \dot{\alpha}_w + \frac{l_{sp}^*}{2V_0} C_{h\dot{\beta} k} \dot{\beta}_w \right) \right]
$$

$$
+ \bar{Q} S_{sk} \bar{c}_{sk} \left[\left(\frac{l_{ch}^*}{2V_0} C_{hkq} q + \frac{l_{sp}^*}{2V_0} \left(C_{hkp} p + C_{hkr} r \right) \right) \right.
$$

$$
\left. + \left(\left(\frac{l_{ch}^*}{2V_0} \right)^2 C_{hk\dot{q}} \dot{q} + \left(\frac{l_{sp}^*}{2V_0} \right)^2 \left(C_{hk\dot{p}} \dot{p} + C_{hk\dot{r}} \dot{r} \right) \right) \right] \quad (3.45)
$$

$$
+ \bar{Q} S_{sk} \bar{c}_{sk} \sum_{l=1}^{N \mathrm{Surf}} \left[C_{h\delta(k,l)} \delta_s(l) + \left(\frac{\bar{c}_{s(l)}}{2V_0} \right) C_{h\dot{\delta}(k,l)} \dot{\delta}_s(l) \right]
$$

$$
+ \bar{Q} S_{sk} \bar{c}_{sk} \sum_{j=1}^{N \mathrm{Mode}} \left[C_{h\eta(k,j)} \eta(j) + \left(\frac{C_{h\dot{\eta}(k,j)}}{V_0} \right) \dot{\eta}(j) \right]
$$

S_{sk} is the reference area in (ft^2) of the kth control surface used to normalize the GAFD data.

c_{sk} is the reference length in (ft) of the kth control surface used to normalize the GAFD.

δ_{sk} is the clockwise rotation of the kth control surface about the hinge vector. It does not include structural flexibility at the hinge.

l_{sp}^*, l_{ch}^* are the reference lengths in (ft) used to normalize the GAFD data. They are normally equal to the vehicle span (l_{sp}) and the mean aerochord (l_{ch}) used in the aero equations.

$h_{s(k,j)}$ is the inertial coupling coefficient that couples the jth flex mode generalized acceleration to the kth surface hinge moment. In matrix form it is the transpose of the h-parameters matrix used in equation 3.38.

W_{sk} is the vehicle rate resolved along the hinge vector of the kth control surface.

$a_{Ns(k)}$ is the vehicle acceleration at the hinge line perpendicular to the kth control surface, as defined in equation 3.46.

The angles ϕ_{hsk} and λ_{hsk} are the bank and the sweep angles that define the orientation of the control surface hinge vector with respect to the vehicle axes, as shown in Figure 3.5.

$$\dot{W}_{sk} = \left[\cos \phi_{hsk} \dot{q} + \sin \phi_{hsk} \dot{r}\right] \cos \lambda_{hsk} + \dot{p} \sin \lambda_{hsk}$$
$$a_{Ns}(k) = a_{Ys}(k) \sin \phi_{hsk} - a_{Zs}(k) \cos \phi_{hsk} \tag{3.46}$$

The vehicle accelerations along x, y, and z, at the midpoint of the kth hinge line are defined by equation 3.47. Notice that the linear accelerations do not include the flex contributions because the flexibility coupling in the hinge moments is introduced by the inertial coupling coefficients $h_s(k,j)$. Also the control surface deflections and rates $\delta_s(k)$, $\dot{\delta}_s(k)$ are relative to the kth hinge and they do not include deformation at the hinge.

$$a_{xcs}(k) = \ddot{x}_{CG} + l_{Zsk} \dot{q} - l_{Ysk} \dot{r}$$
$$a_{ycs}(k) = \ddot{y}_{CG} + l_{Xsk} \dot{r} - l_{Zsk} \dot{p} \tag{3.47}$$
$$a_{zcs}(k) = \ddot{z}_{CG} + l_{Ysk} \dot{p} - l_{Xsk} \dot{q}$$

3.11   OUTPUT SENSORS

The output variables in the linear models represent variations from steady-state trim measurements as seen from typical flight vehicle sensors. They are included in dynamic models and in linear simulations for flight control feedback stabilization and for evaluating the system performance in response to excitations. The sensor measurements are Euler angles, rate gyros, accelerometers, vehicle altitude, velocity, α_w and β_w measurements relative to wind. They consist of combinations from several state variables including structural flexibility. Estimators of α and β are also included in the sensor group.

3.11.1   Vehicle Attitude, Euler Angles

The Euler angles in equation 2.14 are used to calculate variations in vehicle attitude as a function of the body rates. After linearization, equation 3.48 denotes Euler angles' propagation as a function of body rates. It calculates the variations in roll, pitch, and yaw Euler angles (φ, θ, ψ) measured with respect to the trim Euler angles (Φ_0, Θ_0, Ψ_0).

1-2-3 rotation
$$\dot{\phi} = p + \tan \Theta_o \left[q \sin \Phi_o + r \cos \Phi_o + \left(Q_o \cos \Phi_o - R_o \sin \Phi_o \right) \phi \right]$$
$$+ \left(\frac{Q_o \sin \Phi_o + R_o \cos \Phi_o}{\cos^2 \Theta_o} \right) \theta$$
$$\dot{\theta} = q \cos \Phi_o - r \sin \Phi_o - \left(R_o \cos \Phi_o + Q_o \sin \Phi_o \right) \phi$$
$$\dot{\psi} = \frac{\left(R_o \cos \Phi_o + Q_o \sin \Phi_o \right) \tan \Theta_o \theta + r \cos \Phi_o + q \sin \Phi_o + \left(Q_o \cos \Phi_o - R_o \sin \Phi_o \right) \phi}{\cos \Theta_o}$$

3-2-1 rotation

$$\dot{\phi} = p \cos \Theta_o + r \sin \Theta_o + \left(R_o \cos \Theta_o - P_o \sin \Theta_o \right) \theta$$

$$\dot{\theta} = q + \tan \Phi_o \left[p \sin \Theta_o - r \cos \Theta_o + \left(P_o \cos \Theta_o + R_o \sin \Theta_o \right) \theta \right]$$
$$+ \left(\frac{P_o \sin \Theta_o - R_o \cos \Theta_o}{\cos^2 \Phi_o} \right) \phi$$

$$\dot{\psi} = \frac{\left(R_o \cos \Theta_o - P_o \sin \Theta_o \right) \tan \Phi_o \, \phi + \left(P_o \cos \Theta_o + R_o \sin \Theta_o \right) \theta}{\cos \Phi_o}$$

$$(3.48)$$

1-3-2 rotation

$$\dot{\phi} = p + \left(Q_o \sin \Phi_o + R_o \cos \Phi_o \right) \phi + \tan \Psi_o \left[r \sin \Phi_o - q \cos \Phi_o \right]$$
$$+ \left(\frac{R_o \sin \Phi_o - Q_o \cos \Phi_o}{\cos^2 \Psi_o} \right) \psi$$

$$\dot{\theta} = \frac{\left(Q_o \cos \Phi_o - R_o \sin \Phi_o \right) \tan \Psi_o \, \psi + q \cos \Phi_o - r \sin \Phi_o - \left(R_o \cos \Phi_o + Q_o \sin \Phi_o \right) \phi}{\cos \Psi_o}$$

$$\dot{\psi} = r \cos \Phi_o + q \sin \Phi_o + \left(Q_o \cos \Phi_o - R_o \sin \Phi_o \right) \phi$$

3.11.2 Altitude and Cross-Range Velocity Variations

The rate of change in vehicle altitude $\dot{h}$ is calculated from equation 2.17. After linearization, equation 3.49 is integrated to calculate small changes in vehicle altitude δh relative to a nominal altitude h_0, as a function of variations in speed, attitude, alpha, and beta angles.

$$\delta \dot{h} = \left(\cos \alpha_0 \sin \Theta_0 - \beta_0 \cos \Theta_0 \sin \Phi_0 - \sin \alpha_0 \cos \Theta_0 \cos \Phi_0 \right) \delta V$$
$$-V_0 \left(\cos \alpha_0 \cos \Theta_0 \cos \Phi_0 + \sin \alpha_0 \sin \Theta_0 \right) \alpha$$
$$-V_0 \left(\beta_0 \cos \alpha_0 \sin \Theta_0 + \cos \Theta_0 \sin \Phi_0 - \beta_0 \sin \alpha_0 \cos \Theta_0 \cos \Phi_0 \right) \beta \quad (3.49)$$
$$+V_0 \left(\cos \alpha_0 \cos \Theta_0 + \beta_0 \sin \Theta_0 \sin \Phi_0 + \sin \alpha_0 \sin \Theta_0 \cos \Phi_0 \right) \theta$$
$$-V_0 \left(\beta_0 \cos \Theta_0 \cos \Phi_0 - \sin \alpha_0 \cos \Theta_0 \sin \Phi_0 \right) \phi$$

The cross-range velocity in equation 2.18, after linearization becomes as shown in equation 3.50, which calculates variations in cross-range velocity as a function of sideslip and variations in yaw angle.

$$\delta V_{cr} = V_0 \, \cos(\gamma_0) \, \cos(\beta_0 + \Psi_0) \, [\beta + \psi] \qquad (3.50)$$

3.11.3 Gyros or Rate Gyros

The signal measured at a gyro or a rate gyro located on a vehicle consists of a rigid-body motion plus an elastic component that represents local flexibility of the structure at the gyro. The flexibility consists of a superposition from all structural modes that are detectable at the gyro. The measured rates in roll, pitch, and yaw are shown in equation 3.51.

$$p_g = p + \sum_{j=1}^{N \text{ mod}} \sigma_{xg}(k,j)\,\eta(j)$$

$$q_g = q + \sum_{j=1}^{N \text{ mod}} \sigma_{yg}(k,j)\,\eta(j) \qquad (3.51)$$

$$r_g = r + \sum_{j=1}^{N \text{ mod}} \sigma_{zg}(k,j)\,\eta(j)$$

where

p_g, q_g, r_g	are the rates sensed by the gyro.
p, q, r	are the vehicle rigid-body rates.
N_{mod}	is the number of bending modes.
$\sigma_{xg}, \sigma_{yg}, \sigma_{zg}$	are the roll, pitch, and yaw modal slopes of the jth mode at the kth gyro location in (rad/ft).
$\eta(j)$	is the generalized modal displacement of the jth mode.

3.11.3.1   Vehicle Rates in Stability Axes   Equation 2.19 calculates the roll and yaw rates in the stability axes. The small angle variations in stability rates p_s and r_s are obtained by linearizing equation 2.19, Equation 3.52 denotes body to stability axes' transformation for small variations, where the stability rates become functions of body rates and also variations in the angle of attack α.

$$p_s = p_b \cos \alpha_0 + r_b \sin \alpha_0 + \left(R_0 \cos \alpha_0 - P_0 \sin \alpha_0\right) \alpha$$
$$r_s = r_b \cos \alpha_0 - p_b \sin \alpha_0 - \left(R_0 \sin \alpha_0 + P_0 \cos \alpha_0\right) \alpha \qquad (3.52)$$

where

p_b and r_b	are the variations in roll and yaw body rates and
P_0 and R_0	are the steady nominal roll and yaw body rates at trim.

This transformation is useful in developing dynamic models at high angles of attack where roll rotations are commanded about the velocity vector. Stability axis models are often used for flight control design at high angles of attack. The control gains "assume" stability rate feedback and the rate commands from guidance are in stability axes rather than in the body axes. The stability rate demands are converted to simultaneous roll and yaw rate demands that minimize sideslips. For more details read the reentry example.

3.11.3.2   Turn Coordination   In a coordinated turn the vehicle banks at an angle Φ to counteract the centripetal force due to turning and experiences zero lateral acceleration at the CG. Turn coordination is desirable for passenger comfort and also for maximizing aerodynamic efficiency. It is achieved by feeding a signal from the bank angle Φ and the velocity V_0 to the yaw rate R as presented in equation 2.20. If we know ahead of time that the vehicle will be using a turn coordination feedforward loop it may be convenient, for modeling and control design purposes, to include this feedforward loop in the linear dynamic model instead of including it in the flight control system. After linearization equation 3.53 denotes body to stability transformation that includes turn-coordination. The perturbations in the stability rates are defined as a function of the body rates plus other variables.

$$
\begin{aligned}
p_s &= p_b \cos \alpha_o + r' \sin \alpha_0 + \left(R_0 \cos \alpha - P_0 \sin \alpha_0 \right) \alpha \\
r_s &= r' \cos \alpha_o - p_b \sin \alpha_o - \left(R_0 \sin \alpha + P_0 \cos \alpha_0 \right) \alpha \\
\text{where } r' &= r_b + \left[\frac{g \sin \Phi_0 \sin \Theta_0}{V_0} \right] \theta - \left[\frac{g \cos \Phi_0 \cos \Theta_0}{V_0} \right] \phi \\
&\quad + \left[\frac{g \sin \Phi_0 \cos \Theta_0}{V_0^2} \right] \delta V
\end{aligned}
\tag{3.53}
$$

By including the turn coordination terms in the control design model and then using this model to design feedback gains, the gains "assume" that the vehicle control system has a turn-coordination loop already implemented. It separates the two control issues: the turn coordination and the state-feedback design. This feature is useful in designing control gains when the vehicle is at a steady bank angle Φ_0. When the turn coordination is not included, the term r' in equation 3.53 becomes equal to r_b of equation 3.52. The analyst must be careful to avoid including the turn-coordination logic both in the vehicle model and in the flight control system. In simulations and analysis the dynamic models used are in body axes. A body to stability transformation that includes turn-coordination terms is used as a separate block to convert the rate measurements to stability rates, and this is as part of the control software.

3.11.4   Acceleration Sensed by an Accelerometer

Normal, lateral, and axial accelerometers are used to measure the vehicle accelerations along the body z-, y-, and x–axes, respectively. Accelerometer feedback is used by launch vehicle flight control systems to provide load relief and to protect the vehicle structure from excessive normal and lateral aerodynamic loading. The load relief feedback loop alleviates normal and lateral loads by operating in parallel with the attitude control loop. The control gain in the load relief loop is typically phased in-and-out proportionally with the dynamic pressure. Accelerometer feedback is also used in a descending vehicle to control the angles of attack and sideslip and the rate of descent. Equation 3.54 denotes accelerometer measurement along x, y, and z, for

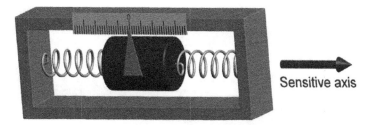

FIGURE 3.16   Accelerometer

small variations. They consist of three components:

- a term due to variations in rigid-body acceleration at the CG,
- two terms due to angular acceleration times the moment arms from the CG,
- two terms due to structural flexibility at the accelerometer.

$$a_{Xaccel} = a_{XCG} + l_{Zacc}\dot{q} - l_{Yacc}\dot{r} + \sum_{j=1}^{N\,mod} \varphi_{Xacc}(j)\,\ddot{\eta}(j) - A_Z \sum_{j=1}^{N\,mod} \sigma_{Yacc}(j)\,\eta(j)$$

$$a_{Yaccel} = a_{YCG} + l_{Xacc}\dot{r} - l_{Zacc}\dot{p} + \sum_{j=1}^{N\,mod} \varphi_{Yacc}(j)\,\ddot{\eta}(j) - A_X \sum_{j=1}^{N\,mod} \sigma_{Zacc}(j)\,\eta(j) \qquad (3.54)$$

$$a_{Zaccel} = a_{ZCG} - l_{Xacc}\dot{q} + l_{Yacc}\dot{p} + \sum_{j=1}^{N\,mod} \varphi_{Zacc}(j)\,\ddot{\eta}(j) + A_X \sum_{j=1}^{N\,mod} \sigma_{Yacc}(j)\,\eta(j)$$

where $l_{Xacc} = X_{accel} - X_{CG}$; $l_{Yacc} = Y_{accel} - Y_{CG}$; $l_{Zacc} = Z_{accel} - Z_{CG}$

where

$\varphi_{xacc}, \varphi_{zacc}, \varphi_{yacc}$	are the mode shapes of the jth elastic mode at the accelerometer along x-, z-, and y–axes, respectively, in (ft/ft).
$\sigma_{Yacc}, \sigma_{Zacc}$	are the mode slopes of the jth elastic mode at the accelerometer in the pitch and yaw directions, in (rad/ft).
$a_{xcg}, a_{ycg}, a_{zcg}$	are the accelerations at the vehicle CG from equation 3.2.

3.11.5   Angle of Attack and Sideslip Sensors

The vane sensors measure the angles of attack and sideslip relative to the wind and the relative wind velocity V_0. This device is generally a probe sticking out in front of the vehicle, as shown in Figure 3.17. Sometimes it consists of tiny holes measuring pressure and from the pressure difference a processor estimates the α_w and β_w angles. These measurements are often used in aircraft as flight control inputs to control the α and β angles. In launch vehicles they can be used instead of accelerometers to provide aerodynamic load relief at high dynamic pressures. However, the measured angles of attack and sideslip α_s and β_s are different from the actual α_w and β_w rigid-body angles relative to the wind because the measurements are corrupted by vehicle rotations and

FIGURE 3.17   Air Data Probe Used for Measuring Airspeed and the Angles of Attack and Sideslip Located in the Front Section of the X-40 Vehicle

structural flexibility at the sensor location. Equation 3.55 denotes angle of attack and sideslip variations measurement at a vane sensor.

$$
\alpha_s = \alpha_w - \left(\frac{l_{xv}q - l_{yv}p}{V_0}\right) + \sum_{j=1}^{N\,\text{mod}} \sigma_{YV}(j)\,\eta(j) + \left(\frac{1}{V_0}\right) \sum_{j=1}^{N\,\text{mod}} \phi_{ZV}(j)\,\dot{\eta}(j)
$$

$$
\beta_s = \beta_w + \left(\frac{l_{xv}r - l_{zv}p}{V_0}\right) - \sum_{j=1}^{N\,\text{mod}} \sigma_{ZV}(j)\,\eta(j) + \left(\frac{1}{V_0}\right) \sum_{j=1}^{N\,\text{mod}} \phi_{YV}(j)\,\dot{\eta}(j)
$$

(3.55)

where

$\varphi_{zv}, \varphi_{yv}$	are the mode shapes of the vehicle at the vane sensor in (ft/ft),
$\sigma_{ya}, \sigma_{z\beta}$	are the pitch and yaw modal slopes at the vane sensor in (rad/ft),
η_j	is the modal displacement of mode (j), in (ft),
l_{xv}, l_{yv}, l_{zv}	are the distances along x, y, and z, between the vane sensor located in front of the vehicle and the vehicle CG, $(l_{xv} = X_v - X_{cg})$.

The first term on the right-hand side of equation 3.55 are the angles of attack and sideslip α_w and β_w relative to the airflow as was defined in Section 3.7. The second term represents a component in the measurement that is caused by the rotational body rates. The last two terms capture the structural flexibility at the sensor, that is, flex rotation and translational displacement rate of the vehicle structure.

3.12   ANGLE OF ATTACK AND SIDESLIP ESTIMATORS

Modern and robust flight control design algorithms use state feedback in order to stabilize the vehicle and to achieve the desired performance characteristics. The vehicle states in the pitch axis are: attitude θ, rate q, and angle of attack α. In the lateral directions the feedback states are: roll and yaw attitude (ϕ, ψ), roll and yaw rates (p, r), and angle of sideslip β. The attitude and the body rates are directly measurable from the IMU and the rate gyros. The angles of attack and sideslip, however, are not always available for feedback and they must be estimated from the accelerometers and the actuator measurements. For a vehicle that uses TVC and elevons for pitch control, alpha can be estimated by solving the normal acceleration equation 3.56.

$$m_v N_Z = \bar{Q} S_{ref} \left(C_{Z\alpha}\, \alpha + C_{Z\delta e}\, \delta_e \right) - T_e \delta_{y_{TVC}} \tag{3.56}$$

where

m_v is the vehicle mass,
N_Z is the normal acceleration at the CG,
$C_{Z\alpha}, C_{Z\delta el}$ are the aerodynamic derivatives with respect to α and δ_{elev},
$\delta_{elev}, \delta_{y\,tvc}$ are the elevon and TVC deflections,
T_e is the engine thrust.

We can solve equation 3.56 for alpha using the following estimator diagram (Figure 3.18). Its inputs are normal acceleration at the CG and actuator deflections angles from the elevon and pitch gimbal.

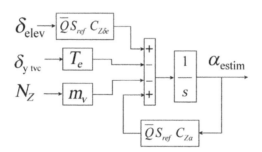

FIGURE 3.18   Angle of Attack Estimator Diagram

Similarly, in the lateral direction the angle of sideslip can be estimated using a beta estimator by solving the lateral acceleration equation 3.57 for beta.

$$m_v N_Y = \bar{Q} S_{ref} \left(C_{Y\beta}\, \beta + C_{Y\delta rud}\, \delta_{rud} + C_{Y\delta ail}\, \delta_{ailer} \right) + T_e\, \delta_{z_{TVC}} \tag{3.57}$$

The β estimator in Figure 3.19 requires knowledge of the lateral acceleration N_Y at the CG, actuator deflections, vehicle weight, and aerodynamic coefficients. In the control system diagram of Figure 3.20 the β-estimate replaces the β-feedback signal in the state-feedback control system.

However, when the accelerometer is not located at the CG, it is also measuring unwanted components due to the vehicle angular acceleration. Filtered signals from the rate gyros are used in this case to compensate the corrupted accelerometer measurements. See examples in Flixan.com.

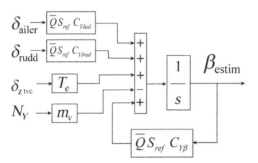

FIGURE 3.19   Angle of Sideslip Estimator Diagram

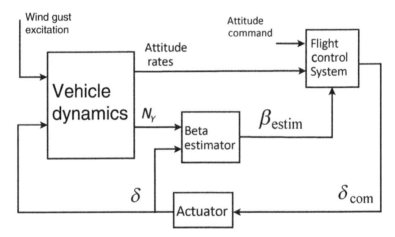

FIGURE 3.20   Closed-Loop State-Feedback System Using a β-Estimator

3.13   LINEARIZED EQUATIONS OF A SPACECRAFT WITH CMGS IN LVLH ORBIT

In Section 2.12 we presented the nonlinear equations of a spacecraft that is controlled with momentum exchange devices (MED), such as a cluster of control moment gyros (CMG) or a reaction wheels assembly (RWA). The model calculates the combined CMG momentum and torque in the body axes, bypassing the detailed geometry, and modeling of the individual devices. If we assume that the spacecraft has a steady pitch rate that is equal to the negative orbital rate $-\omega_0$, we can linearize this spacecraft model relative to the local-vertical-local-horizontal (LVLH) frame. Equation 3.58 calculates the rate of change of angular momentum as a function of CMG and external torques. The attitudes and rates represent small variations relative to the LVLH. The CMG array generates control torques in the body axes and the CMG momentum is also calculated in the body axes.

$$
\begin{bmatrix} I_{XX} & I_{XY} & I_{XZ} \\ I_{XY} & I_{YY} & I_{YZ} \\ I_{XZ} & I_{YZ} & I_{ZZ} \end{bmatrix} \begin{pmatrix} \dot{\omega}_X \\ \dot{\omega}_Y \\ \dot{\omega}_Z \end{pmatrix} = \omega_o \begin{bmatrix} I_{XZ} & 2I_{YZ} & I_{ZZ} - I_{YY} \\ -I_{YZ} & 0 & I_{XY} \\ I_{YY} - I_{XX} & -2I_{XY} & -I_{XZ} \end{bmatrix} \begin{pmatrix} \omega_X \\ \omega_Y \\ \omega_Z \end{pmatrix}
$$

$$
+3\omega_0^2 \begin{bmatrix} I_{ZZ} - I_{YY} & I_{XY} & 0 \\ I_{XY} & I_{ZZ} - I_{XX} & 0 \\ -I_{XZ} & -I_{YZ} & 0 \end{bmatrix} \begin{pmatrix} \varphi \\ \theta \\ \psi \end{pmatrix} + \omega_0^2 \begin{bmatrix} -2I_{YZ} \\ 3I_{XZ} \\ -I_{XY} \end{bmatrix} + T_{\text{CMG}} + \sum T_{\text{ext}}
\tag{3.58}
$$

1. The first term on the right side is the linearized $\omega \times J\omega$ term.
2. The second term on the right side is the gravity gradient torque which is a function of attitude.
3. The third term is a bias torque due to the linearization caused by the products of inertia.
4. T_{CMG} is the CMG control torque, and
5. T_{ext} is other external torques.

The attitude kinematics becomes,

$$
\begin{pmatrix} \dot{\varphi} \\ \dot{\theta} \\ \dot{\psi} \end{pmatrix} = \begin{bmatrix} \omega_0 \psi + \omega_X \\ \omega_0 + \omega_Y \\ -\omega_0 \varphi + \omega_Z \end{bmatrix}
\tag{3.59}
$$

The rate of change in CMG momentum in body axes is obtained by linearizing equation 2.23 as shown in equation 3.60.

$$
\dot{\underline{h}}_{\text{cmg}} = -\left(\underline{\Omega}_0 \times \underline{h}_{\text{cmg}} \right) - \left(\underline{\omega} \times \underline{H}_0 \right) - \underline{T}_{\text{CMG}}
\tag{3.60}
$$

where

$\underline{\Omega}_0$ is the spacecraft average steady-state rate about (x, y, z), and
$\underline{H}_0$ is the nominal CMG momentum in (x, y, z).

If we assume that the rate is entirely in pitch and equal to $-\omega_0$, the variation in CMG momentum equations become:

$$
\begin{aligned}
\dot{h}_X &= \omega_0 h_Z + \omega_Z H_{Y0} - \omega_Y H_{Z0} - T_{CX} \\
\dot{h}_Y &= -\omega_Z H_{X0} + \omega_X H_{Z0} - T_{CY} \\
\dot{h}_Z &= -\omega_0 h_X + \omega_Y H_{X0} - \omega_X H_{Y0} - T_{CZ}
\end{aligned}
\tag{3.61}
$$

This model is used in the examples section 3.10 to design a CMG momentum management control system on a space-station model, and to analyze system stability.

3.14 LINEARIZED EQUATIONS OF AN ORBITING SPACECRAFT WITH RWA AND MOMENTUM BIAS

In Section 2.13 we demonstrated a nonlinear model of a spacecraft in circular orbit in LVLH attitude, controlled with an array of reaction wheels. We can linearize this spacecraft model relative to the LVLH frame, assuming that it has a steady pitch rate that is equal to the negative orbital rate $-\omega_0$. Equation 3.62 calculates the rate of change of angular momentum as a function of RW and external torques.

$$
\begin{bmatrix} I_{XX} & I_{XY} & I_{XZ} \\ I_{XY} & I_{YY} & I_{YZ} \\ I_{XZ} & I_{YZ} & I_{ZZ} \end{bmatrix} \begin{pmatrix} \dot{\omega}_X \\ \dot{\omega}_Y \\ \dot{\omega}_Z \end{pmatrix} = \omega_0 \begin{bmatrix} I_{XZ} & 2I_{YZ} & I_{ZZ} - I_{YY} \\ -I_{YZ} & 0 & I_{XY} \\ I_{YY} - I_{XX} & -2I_{XY} & -I_{XZ} \end{bmatrix} \begin{pmatrix} \omega_X \\ \omega_Y \\ \omega_Z \end{pmatrix}
$$
$$
+3\omega_0^2 \begin{bmatrix} I_{ZZ} - I_{YY} & I_{XY} & 0 \\ I_{XY} & I_{ZZ} - I_{XX} & 0 \\ -I_{XZ} & -I_{YZ} & 0 \end{bmatrix} \begin{pmatrix} \varphi \\ \theta \\ \psi \end{pmatrix} + \omega_0^2 \begin{bmatrix} -2I_{YZ} \\ 3I_{XZ} \\ -I_{XY} \end{bmatrix} + T_{RW} + \sum T_{ext} \quad (3.62)
$$

1. The first term on the right side is the linearized $\omega \times J\omega$ term.
2. The second term on the right side is the gravity gradient torque which is a function of attitude.
3. The third term is a bias torque due to the linearization caused by the products of inertia.
4. T_{RW} is the RW array control torques in body axes, and
5. T_{ext} is other external torques.

Equation 3.63 is the rate of change of the RW momentum resolved in the body axes.

$$
\begin{pmatrix} \dot{h}_x \\ \dot{h}_y \\ \dot{h}_z \end{pmatrix} = - \begin{pmatrix} \omega_{X0} \\ \omega_{Y0} \\ \omega_{Z0} \end{pmatrix} \times \begin{pmatrix} h_x \\ h_y \\ h_z \end{pmatrix} - \begin{pmatrix} \omega_x \\ \omega_y \\ \omega_z \end{pmatrix} \times \begin{pmatrix} H_{X0} \\ H_{Y0} \\ H_{Z0} \end{pmatrix} - \underline{T}_{RW} \quad (3.63)
$$

The torque applied to the spacecraft by the reaction wheels array in body axes is:

$$
T_{RW} = C_w^b \, \underline{T}_{wi} - \underline{\omega}_0 \times C_w^b \, \underline{h}_{wi} - \underline{\omega} \times C_w^b \, \underline{h}_{W0} \quad (3.64)
$$

The torque applied to the spacecraft by each individual wheel $\underline{T}_{rwi}$ is a combination of the motor torque T_{wi} applied about the wheel spin vector $\underline{a}_i$ plus the gyroscopic torque due to the coupling between the wheel momentum and the spacecraft rate. The torque at each individual wheel can be used to excite structural flexibility.

$$
\underline{T}_{rwi} = \underline{a}_{wi} \, T_{wi} - \underline{\omega}_0 \times \underline{a}_{wi} h_{wi} - \underline{\omega} \times \underline{a}_{wi} h_{wi0} \quad (3.65)
$$

The rate of change of momentum of the ith reaction wheel about its spin axis is:

$$\dot{h}_{wi} = -T_{wi} \tag{3.66}$$

The measured ith wheel angular rate relative to the spacecraft is equal to its inertial rate minus the spacecraft rate resolved in the ith wheel spin axis, that is,

$$w_i = \frac{h_{wi}}{J_{wi}} - \underline{a}_{wi}^T \underline{\omega} \tag{3.67}$$

where

$\underline{\omega}_0$	is the nominal spacecraft body rate about x-, y-, z-axes: ω_{X0}, ω_{Y0}, ω_{Z0}
$\omega_x, \omega_y, \omega_z$	is the variation in vehicle rate from nominal.
H_{X0}, H_{Y0}, H_{Z0}	is the nominal reaction wheel momentum of the RW assembly.
h_x, h_y, h_z	is the variation in RW momentum from nominal in body axes.
$\underline{T}_{wi}$	is the motor torque vector in reaction wheel axes.
$\underline{h}_{w0}$	is the nominal wheels momentum vector: $h_{w01}, h_{w02}, h_{w03}, \ldots h_{w0n}$.
$\underline{a}_{wi}$	is the unit vector of the ith RW spinning direction.
C_w^b	is the wheel axes to body transformation matrix.

3.15   LINEARIZED EQUATIONS OF SPACECRAFT WITH SGCMG

In Section 2.14 we developed a nonlinear model of a spacecraft controlled with an array of SGCMGs. Assuming that the spacecraft is in circular orbit, the angular acceleration in the body frame is the same as equation 3.58. Equation 3.68 is the linearized rate of change in CMG momentum which is equal to the CMG torque in body axes, and it is obtained by linearizing the second equation in 2.46. $\underline{T}_{cmg}$ is the torque generated by the CMG array and applied to the vehicle. It is opposite to the torque applied to the CMG cluster.

$$\begin{pmatrix} \dot{h}_x \\ \dot{h}_y \\ \dot{h}_z \end{pmatrix} = - \begin{pmatrix} \omega_{X0} \\ \omega_{Y0} \\ \omega_{Z0} \end{pmatrix} \times \begin{pmatrix} h_x \\ h_y \\ h_z \end{pmatrix} - \begin{pmatrix} \omega_x \\ \omega_y \\ \omega_z \end{pmatrix} \times \begin{pmatrix} H_{X0} \\ H_{Y0} \\ H_{Z0} \end{pmatrix} - \underline{T}_{cmg} \tag{3.68}$$

The linearized SGCMG model is applicable for small gimbal variations relative to some nominal gimbal angle positions: $(\delta_{01}, \delta_{02}, \delta_{03}, \ldots \delta_{0n})$. The CMG torque is mainly due to the precession of momentum h_0 coupling with the gimbal rate $\dot{\delta}_i$. It is also a function of the spacecraft rates (from equation 2.40) and the gimbal angle δ_0. It is converted to spacecraft axes by the transformation in equation 2.41.

$$
\underline{T}_{\text{cmg}} = - \sum_{k=1}^{N_{\text{cmg}}} P_i \left[\begin{array}{c} \overset{T_g}{h_0 \dot{\delta}_i + (J_s - J_g) \left(\dot{\theta}_0 \cos \delta_0 + \dot{\varphi}_0 \sin \delta_0 \right) \dot{\delta}_i} \\ (J_g - J_o) \left(\dot{\varphi}_0 \cos \delta_0 - \dot{\theta}_0 \sin \delta_0 \right) \dot{\delta}_i \end{array} \right]_i \tag{3.69}
$$

The gimbal acceleration for each CMG in the linearized model is obtained by linearizing equation 2.57.

$$
J_g \ddot{\delta}_i = T_g - h_0 \left(\dot{\theta}_0 \cos \delta_0 \, \delta + \dot{\varphi}_0 \sin \delta_0 \, \delta + \sin \delta_0 \dot{\theta} - \cos \delta_0 \dot{\varphi} \right) \tag{3.70}
$$

The relative gimbal angle δ for each CMG is obtained by integrating

$$
\begin{aligned}
\dot{\delta} &= \dot{\delta}_i - \dot{\xi} \\
\dot{\xi} &= \omega_x \sin \beta \sin \gamma - \omega_y \sin \beta \cos \gamma + \omega_z \cos \beta \\
\dot{\theta} &= \omega_x \cos \gamma + \omega_y \sin \gamma \\
\dot{\varphi} &= -\omega_x \cos \beta \sin \gamma + \omega_y \cos \beta \cos \gamma + \omega_z \sin \beta
\end{aligned} \tag{3.71}
$$

where

$\ddot{\delta}_i$	is the inertial acceleration of a gimbal,
δ_0	is the nominal gimbal angle of a CMG (assumed constant in linear model),
δ	is the small variation of a CMG gimbal angle from δ_0,
$\dot{\xi}$	is the variation in spacecraft rate about the gimbal axis $\underline{m}$,
$\dot{\theta}, \dot{\varphi}$	are the variations in spacecraft rates about the $\underline{r}$ and $\underline{q}$ axes,
$\dot{\theta}_0, \dot{\varphi}_0$	are the nominal spacecraft rates about the $\underline{r}$ and $\underline{q}$ axes,
$\omega_{X0}, \omega_{Y0}, \omega_{Z0}$	are the nominal spacecraft body rates about x, y, z,
$\omega_x, \omega_y, \omega_z$	are the variation in vehicle rate about the body axis,
H_{X0}, H_{Y0}, H_{Z0}	are the nominal CMG array momentum about x, y, z,
h_x, h_y, h_z	are the variations in CMG momentum about x, y, z,
h_0	is the CMG constant momentum about its spin axis.

The servo system for each gimbal must provide the necessary torque T_g that will regulate the gimbal rate $\dot{\delta}$ according to the gimbal rate command $\dot{\delta}_{\text{comd}}$ received from the steering logic. The steering logic converts the torque commands from the attitude control system to gimbal rate commands. By accurately controlling the gimbal rates the SGCMG array achieves the demanded control torques.

4

ACTUATORS FOR ENGINE NOZZLES AND AEROSURFACES CONTROL

Flight vehicles are controlled by TVC engines that rotate about a gimbal or by control surfaces rotating about a hinge. The purpose of the servo-actuator system is to provide the force that is needed to rotate the engine or the control surface in the direction needed to maneuver and stabilize the vehicle. Since the actuator dynamics greatly affect and couple with the vehicle dynamics we will dedicate this chapter to model them and to provide some detailed examples. In launch vehicles, a small rotation of the thrust vector angle about the gimbal (typically between $\pm5°$ and $\pm10°$ in pitch and in yaw) is sufficient to create normal and lateral control forces to stabilize and steer the vehicle, overcome the wind-gust disturbances, and to balance the aerodynamic moments. In an aircraft the control surface deflections are larger, of the order of $\pm40°$ relative to the trim position. The control surface is typically the trailing edge of a wing, tail, or vertical stabilizer that rotates about a hinge line parallel to it and the actuator provides the force to rotate it against the aerodynamic loads. In some cases it is the leading edge that rotates, and in some aircraft it is the entire surface that rotates and not just the edge. The rotation of the surface changes the airflow around the aircraft and creates the aerodynamic forces and moments needed to trim and control the flight vehicle.

Figure 4.1a shows a TVC nozzle for a Saturn-5 rocket connected to the vehicle by a gimbal. The thrust force is transmitted to the vehicle through the gimbal which allows the engine to rotate in pitch and yaw. The actuator is a hydromechanical servo device that provides the force to control the deflection (δ) of the engine in one direction. A TVC engine is controlled by two orthogonal actuators that rotate the nozzle in pitch

Performance Evaluation and Design of Flight Vehicle Control Systems, First Edition. Eric T. Falangas.
© 2016 by Eric T. Falangas. Published 2016 by John Wiley & Sons, Inc.

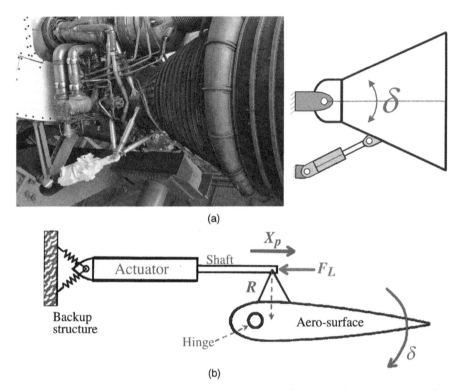

FIGURE 4.1   (a) TVC Engine for a Saturn-5 Rocket; (b) Aerosurface Actuator Pushing Against a Flexible Backup Structure

and yaw. One end of the actuator mechanism is attached to the vehicle and the other end is a piston rod that extends or retracts and is attached on the pivoting nozzle, as shown in the Saturn-5 engine. The position of the actuator shaft can be varied and is controlled by a closed-loop servo system that provides hydraulic or electromechanical forces on the shaft to rotate the nozzle. Each actuator has a sensor that measures the shaft extension and controls its position. The angular position of the nozzle is almost proportional to the piston extension, although not exactly because the measurement becomes corrupted by the deformation of the support structure, as we shall see. The nozzle deflection is controlled in two directions by adjusting the extension of the actuator pistons. During flight the control system calculates the required changes in the engine's pitch and yaw rotations and the actuators are commanded to either extend or retract in order to achieve the demanded nozzle deflections and control forces to guide and stabilize the vehicle. Figure 4.1b shows an actuator system for an aerosurface. By extending, it supplies the force to rotate the surface about the hinge line. The actuator has to "fight" against the load-force F_L of the surface resisting the motion mainly due to aerodynamic forces, vibrations, or normal acceleration of the vehicle. The actuator extension x_p is typically used to estimate the surface rotation.

The measurement, however, is slightly corrupted by the deformation of the support structure, as shown.

The selection of an actuator device is determined mainly by the power requirement of the load. The power is determined by the aerodynamic forces and the speed of response. A flight vehicle can be better controlled when its static stability is marginal. If the vehicle is passively unstable and it diverges too fast, then the actuator has to respond sufficiently fast in order to catch up with the instability and to prevent it from diverging. If the vehicle is too stable, on the other hand, it requires a lot of actuator power in order to maneuver it. Other factors to be considered for the selection of an actuator include the dynamic characteristics, the power sources available, the reliability of the equipment, and other physical and economic limitations.

In this section we will present three actuator models which are typically used in flight control system analysis to pivot control surfaces or TVC engines. We will also demonstrate the dynamic coupling between the flight vehicle and the actuator systems and describe the different types of stiffnesses. The actuator models in addition to the actuator dynamics and the local control system that controls the shaft extension, they also include the rotational dynamics of the effector load relative to the hinge, and the stiffness of the support structure, shaft, and load linkage. The actuators are eventually combined with the vehicle dynamics, and the vehicle/actuator interconnected model must include one actuator for each effector input. The coupling between the actuator and vehicle dynamics is important to be implemented correctly for the proper function of "tail-wags-dog" (TWD) and "load-torque" dynamics that determine the dynamic interaction between local vehicle flexibility, actuator dynamics, and the flight control system, that can affect local structural stability and actuator performance. The dynamic models are based on actuator parameters consisting of: piston area, load moment of inertia, amplifier gains, friction coefficients, gear ratios, piston and backup stiffness, geometry, etc. The actuator systems have two inputs: (1) engine deflection command δ_c in (rad) coming from the flight control system, and (2) load torque T_L in (ft-lb), which is an external loading torque on the actuator generated by the vehicle motion. The actuator models have three outputs that become inputs to the vehicle dynamic model: (1) effector rotation angle (δ) in (rad), (2) effector rate in (rad/sec), and (3) acceleration in (rad/sec^2).

4.1 ACTUATOR MODELS

The actuators most commonly used in flight vehicles for positioning control surfaces or engine nozzles fall into three categories: (a) electrohydraulic, (b) electromechanical, and (c) pneumatic. Pneumatic have limited usage due to low bandwidth and lack of fine positioning accuracy, and they are not considered here. In this section we will present the equations of motion describing the operation of three commonly used actuator models. We begin with a simple actuator model where most of the servo system details are reduced to gains and integrators. It is simple because the servo closed-loop dynamics are simplified to a first-order transfer function driving the load dynamics. It includes, however, more structure than a simple transfer function block because it captures the main functions of an actuator which are: the actuator shaft positioning servo, the load dynamics, external loading, stiffness of the actuator and of the supporting structure, and the position measurement. This model can be used in the early to mid phases of a flight vehicle analysis when the actuator modeling details are not fully defined. Then we will describe a hydraulic actuator and an electromechanical actuator (EMA) system in more detail including actuator parameters with an example.

4.1.1 Simple Actuator Model

A simple actuator model is shown in Figure 4.2a. In the forward loop we have a closed-loop servo system that controls piston velocity and it has been simplified to a first-order transfer function. The input to the servo is amps and the output is piston velocity which is integrated to piston position X_P. The total system stiffness K_T consists of three types of stiffnesses which are combined in series.

1. The stiffness of the backup structure. That is the point where the actuator is attached to the vehicle support structure,
2. The actuator stiffness consisting of piston plus oil (or electrical) stiffness, and
3. The load stiffness due to torsional flexibility at the linkage where the shaft is attached to the load.

The piston force F_L pushes against the total system stiffness K_T and through the moment arm R to create a control torque that drives the effector load inertia I_L. X_P is the position of the shaft's end point relative to the actuator and X_L represents the linear displacement at the nozzle attachment as a result of a load rotation angle (δ). This is not necessarily equal to X_P because of the three compliances. In fact it is the difference between the piston extension and linear nozzle displacement ($X_P - X_L$) that creates the force F_L pushing against K_T.

The control torque due to the shaft extension is not the only torque driving the load. There is also an external load torque T_L that is produced by the vehicle inertial accelerations (both rigid and flex) and by aerodynamic forces. This external to actuator load torque comes from the vehicle system outputs and creates a mechanical feedback loop. The rotations of the load (δ, $\dot{\delta}$, $\ddot{\delta}$) are relative to the gimbal, consisting of a component due to the piston extension X_P, plus a component due to external loading pushing against the combined structural stiffness K_T. There is also viscous damping at the gimbal B_G, static and stiction friction, and a rotational stiffness at the gimbal K_G caused by pipes and other mechanical hardware connected to the nozzle.

The actuator position measurement X_{fb} is located on the actuator device (not the gimbal) and it is intended to measure the actuator shaft position. If you ignore the structural stiffness the effector deflection is nearly proportional to the actuator position. The sensor measures the extension of the rod relative to the actuator device, but this does not necessarily measure the total rotation of the nozzle or surface, because the vehicle support structure where the actuator is attached on deforms under force,

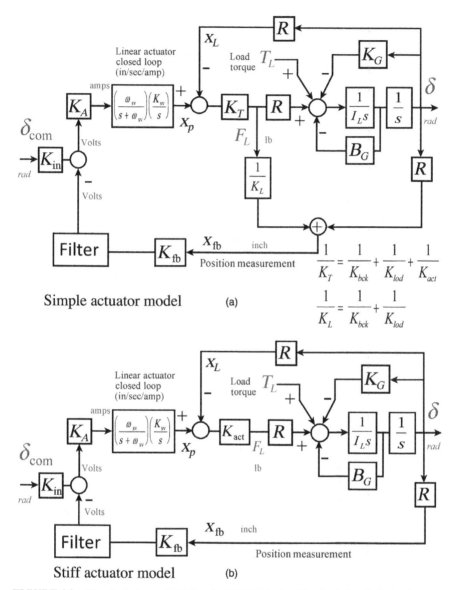

FIGURE 4.2   Simple Actuator Model and a "Stiff" Version That Includes Only the Actuator Shaft Stiffness (K_{act}); Backup and Load Stiffness Are Infinite

the linkage mechanism that attaches to the load also deforms, and the actuator rod itself compresses under load. The actual rotation of the effector, therefore, is not only due to the rotation at the hinge (which is measurable) but it includes a component due to structural deformations that makes it different from measurement, and this error must be captured in the sensor model. The position sensor measurement cannot obviously "see" the deformation due to load linkage and backup stiffness which are combined in K_L. The measurement, however, is affected by the actuator rod stiffness (K_{act}). When it compresses, the end of the rod is not exactly where the measurement thinks that it is. In Figure 4.2a the position measurement signal X_{fb} consists of two components:

1. A component ($R\delta$) due to the effector rotation, and
2. A component $(K_T/K_L)(x_p - R\delta)$, where K_L is the combined backup and load stiffness.

To illustrate this, let us suppose, for example, that the actuator piston position is locked at zero, and an external load T_L is applied to the system. The effector/load will deflect at an angle δ as a result from all three stiffnesses (actuator, load, and backup stiffness combined in K_T). The position sensor in the actuator, instead of measuring zero, because the piston does not move, it is measuring a small displacement caused by the deformation of the piston, plus some additional compression due to oil or electrical stiffness which are all lumped in the actuator stiffness K_{act}. The sensor does not see the backup structure deformation and the deformation at the load. We must, therefore, subtract from the total flex rotation (δ) the components due to backup and load stiffness (F_L/K_L) and the remaining measurement is due to the actuator stiffness K_{act}. In the ideal situation, when ($K_{act} = \infty$) and ($K_T = K_L$), the feedback measurement becomes equal to the actual shaft position ($X_{fb} = X_p$). The backup and load stiffness are infinite in the actuator and they have been taken out of the actuator, as shown in Figure 4.2(b), because they are actually included in the vehicle dynamics.

4.1.2 Electrohydraulic Actuator

Electrohydraulic actuators are most commonly used for gimbaling the TVC in launch vehicles, and have gained the widest acceptance in the aerospace industry. They have great power capability and can deliver larger torques than electrical equipment of comparable size and weight. For continuous operation, they offer a minimum (equipment/power) ratio. Where intermittent operation is required, a hydraulic system can provide large amount of power from a small volume of accumulator. Their dynamic characteristics are expressed by small time constants, and they develop higher (peak torque/inertia) ratios. The most common form of utilization of hydraulic servos in vehicle control loops consists of a high pressure supply (pump), an electrohydraulic servo valve, a hydraulic actuator (cylinder), a feedback transducer, and a servo amplifier. The hydraulic power supplies currently used are of two main types. The first type employs a variable displacement pump whose output flow is controlled by means of a servo sensing the high pressure side of the hydraulic system. A relief valve is also connected from the high pressure side to the low pressure side to minimize pressure transients above the operating pressures of the system. For normal operations the valve remains closed, opening only when the pressure exceeds a value overcoming the preload on the relief valve. The second type of power supply uses a fixed displacement pump with a relief valve to maintain the supply pressure within set limits,

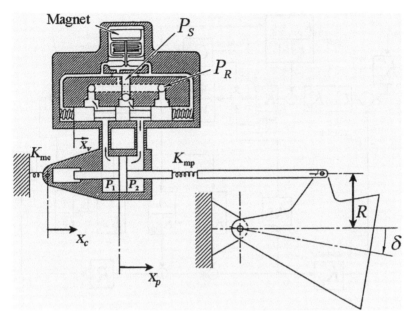

FIGURE 4.3   Schematic of a Hydraulic Servo Actuator

as well as to meet the normal flow requirements. In this system the relief valve is normally open so that supply pressure and valve opening maintain flow through the relief valve equal to the flow output of the fixed displacement pump. When there is a flow demand the relief valve closes and the supply pressure is reduced. The dynamics of both power supply and relief valve exhibit a fairly flat response with small phase shift within the bandwidth of the overall servo loop, therefore, the supply pressure will be assumed to be constant at the value of zero flow demand. Electrohydraulic valves are designed for flow or pressure control. These units are highly complex devices and exhibit high-order nonlinear responses. Still, in the frequency range of interest, they can be represented by a first- or second-order transfer functions.

An electrohydraulic actuator system is shown in Figure 4.3 and the block diagram in Figure 4.4. It comprises a high pressure hydraulic supply (pump), a servo amplifier, a hydraulic actuator, a servo valve, and a position transducer. The electrohydraulic servo valve is a flow control type. There is a leakage orifice across the load piston and a differential pressure feedback loop from a load pressure transducer to the servo valve. The hydraulic cylinder is attached to the vehicle at a support point that has a backup stiffness K_{mc}. On the other side of the cylinder the piston is attached to the load (a TVC engine). The piston has a spring constant K_{mp}. This type of electrohydraulic actuator can be described by the following equations.

The error voltage signal (V_e) at the input of the servo amplifier is: $V_e = V_c - V_f$, where V_c and V_f are the command and feedback signals, respectively. The feedback transducer is integrally built within the cylinder and it measures the nozzle position

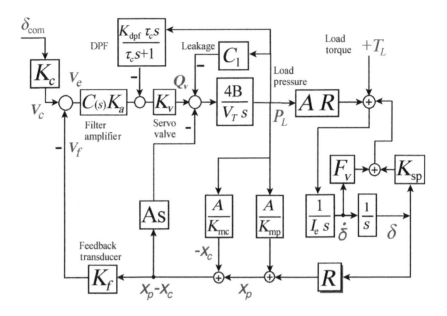

FIGURE 4.4   Electrohydraulic Actuator System

by measuring the piston displacement relative to the cylinder. It feeds it back into the servo amplifier.

$$V_c = K_c\delta_{com}; \quad V_f = K_f(x_p - x_c) \quad\quad (4.1 \ \& \ 4.2)$$

The error signal V_e drives a unit gain compensator $C(s)$, and the output from the compensator is:

$$V_{ef}(s) = C(s)V_e(s) \quad\quad (4.3)$$

The servo valve is a flow control type coupled to an actuator having a leakage orifice across the load piston to provide damping. The output piston rod is hinged to the engine nozzle that moves about its gimbal point, see Figure 4.3. We can assume that the flow from the servo valve into one side of the cylinder is equal to the flow out of the other side. For low load pressures, the flow of hydraulic fluid (Q_v) from the servo valve is:

$$Q_v = K_2 x_v \sqrt{P_s - P_r} - l_v P_r \qu\quad (4.4)$$

where

x_v is the valve spool displacement (ft),
P_s is the supply pressure to the valve (lb/ft^2),

P_r is the reservoir or exhaust pressure (lb/ft²),
K_2 is a constant,
P_L is the differential pressure across the piston,
$l_v P_L$ is the leakage across the spool (ft³/sec).

The total flow of hydraulic fluid, (Q_v) from the servo valve into the actuator cylinder is the sum of the piston displacement flow, leakage, and compressibility flow.

$$Q_v = A(\dot{x}_p - \dot{x}_c) + C_1 P_L + \left(\frac{V_t}{4B}\right) \dot{P}_L \qquad (4.5)$$

Where:

A is the piston cross-sectional area (ft²)
P_L is the load pressure in (lb/ft²)
C_1 is the orifice leakage coefficient in (ft³/sec)/(lb/ft²)
δ_c is the commanded engine rotation in (rad)
x_p is the piston rod displacement, (ft)
x_c is the cylinder mount displacement with respect to the vehicle, in (ft)
R is the moment arm in (ft) from the engine pivot to the piston rod, see figure (4.3).
K_f is the feedback transducer gain (volt/ft)
V_t is the volume of fluid under compression (ft³)
B is the bulk modulus of the hydraulic fluid (lb/ft²).

The load pressure (P_L) displaces the piston rod and the engine. The displacement (x_p) with respect to the cylinder, and the displacement of the cylinder mount relative to the vehicle structure (x_c) is obtained from the following equations:

$$AP_L = K_{mp}\left(x_p - R\delta\right); \quad \text{and} \quad x_c = -\frac{AP_L}{K_{mc}} \qquad (4.6), (4.7)$$

Where:

K_{mp} is the spring constant of the piston rod (lb/ft),
K_{mc} is the spring constant of the cylinder mount (lb/ft),
R is the moment arm of the actuator from engine gimbal (ft),
δ is the engine rotation in rad.

From equations 4.5–4.7 we obtain the following equation, where Q_v represents the valve flow rate in (ft^3/sec) which is proportional to the spool displacement.

$$\dot{P}_L = \frac{(-C_1 P_L + Q_v - AR\dot{\delta})}{\left[\frac{V_1}{4B} + A^2 \left[\frac{1}{K_{mp}} + \frac{1}{K_{mc}}\right]\right]} \tag{4.8}, (4.9)$$

$$\text{where} \quad Q_v = K_v \left\{ K_a V_{ef} - K_{dpf}\left(\frac{\tau_c s}{\tau_c s + 1}\right) P_L \right\}$$

The second term in equation 4.9 is a dynamic pressure feedback term from the load pressure to the servo valve and it is used to control system damping, where

K_v is the servo-valve flow gain in (ft^3/sec/ma),
K_a is the servo-amplifier gain in (ma/volt),
K_{dpf} is the dynamic pressure feedback gain in (ma/lb/ft^2),
τ_c is the dynamic pressure feedback time constant in (sec),
A is the piston cross-sectional area (ft^2).

The moment equation of the engine nozzle rotation about its pivot is as follows:

$$I_e \ddot{\delta} = (RA)P_L + T_L - F_v \dot{\delta} - K_{sp}\delta \tag{4.10}$$

where

I_e is the engine inertia in (slug-ft^2),
F_v is the viscous friction coefficient (ft-lb)/(rad/sec),
K_{sp} is the engine bearing spring constant (ft-lb/rad),
δ is the engine rotation in rad,
T_L is the load torque due to the vehicle motion (ft-lb).

The first term on the right-hand side of equation 4.10 represents the torque on the nozzle due to the actuator piston pressure. The second term is the external load torque created by the vehicle motion (rigid body and flexibility). The third term is a damping torque due to friction at the gimbal, and the fourth term is the torque due to spring stiffness at the pivot.

4.1.3 Electromechanical Actuator

The EMA are becoming more popular in the last 25 years mainly because of simplicity, reliability, and reduced need for maintenance in comparison with the hydraulic systems. Figure 4.5 shows an EMA system that uses a spinning motor controlled by dc voltage and control electronics. A current amplifier supplies the dc current required to drive the motor. The motor generates the power to rotate the load. At the end of the motor's rotor there is a small gear driving a bigger gear producing a

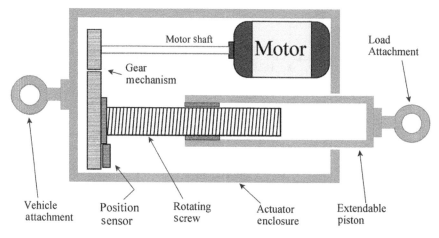

FIGURE 4.5   Electromechanical Actuator System

higher torque. The bigger gear is connected to a screw gear that has a spiral screw type of mechanism that rotates. When the screw mechanism rotates it converts the rotational motion from the motor to translation that extends or retracts the shaft and pushes against the load. The end of the shaft is connected via a linkage mechanism to the load that can pivot about a hinge with respect to the vehicle. The load is either an aerosurface that rotates about a hinge or a TVC engine that gimbals. One side of the actuator is attached to a stiff point on the vehicle structure and the other end of the shaft is attached via a linkage mechanism to the load and causes it to pivot about the hinge.

As the shaft is pushing against the load it generates a force that rotates the nozzle or the control surface similar to the hydraulic actuator. The EMA system described here has two gear ratios. The first gear ratio (N_{gear}) counts the number of motor spins for one rotation of the screw gear and the second gear ratio (N_{screw}) defines the number of shaft rotations per unit length extension of the shaft. The servo loop is closed by means of a position measurement located in the actuator. A rate feedback from the motor velocity is also used in order to provide the required damping characteristics in the position control loop.

The EMA system block diagram is shown in Figure 4.6. A current amplifier supplies the dc current required to drive the motor. There is also a position sensor that counts the number of rotations of the screw to determine the shaft position. This measurement is used in the position control servo system. A rate feedback from the motor velocity is also used in the control loop to achieve the desired closed-loop transfer function characteristics. The surface deflection command (δ_{com}) is in (rad) and is converted to actuator position command in (ft). The Proportional Integral Derivative (PID) controller generates the supply voltage (v_i) which drives the dc motor as a function of position error. It is a function of: shaft position error and integral, and also rate feedback from the motor speed. It includes a low-pass compensator $C(s)$.

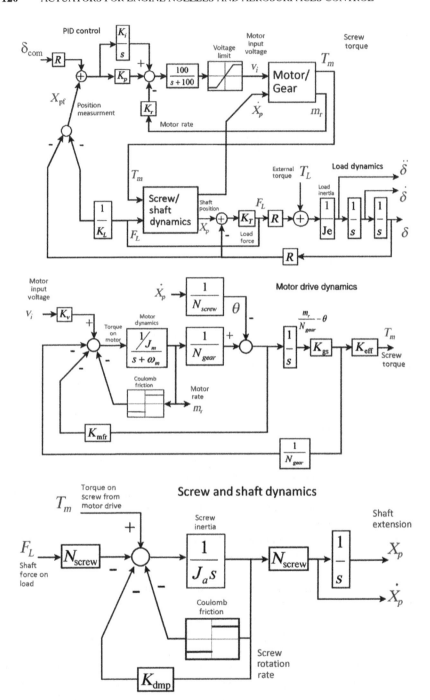

FIGURE 4.6   Electromechanical Actuator Block Diagram with Subsystems

In equation 4.11, the motor dynamics is approximated with a high bandwidth (ω_m) first-order lag that includes the control electronics and the back-emf. The inputs are torques and the output is the rotor rate ($d\mu/dt$) in (rad/sec). There are four torques that affect the motor: an electrical control torque that drives the motor, and three reaction torques. The control torque driving the motor is due to the input voltage (v_i) and the motor torque gain (K_v) in (ft-lb/volt). The opposing torques affecting the motor are: (a) a torque applied to the load scaled down by the gear reduction (N_{gear}), (b) a coulomb friction torque (K_C) always in the opposite direction to the rotor speed, and (c) a damping torque (K_{mfr}) due to relative speed (rotor load) rates. The motor rate equation is:

$$\dot{\mu} = \frac{1/I_m}{(s+\omega_m)}\left[K_v v_i - \frac{T_m}{N_{gear}} - K_{mfr}\dot{\varepsilon} - K_c sign(\dot{\mu})\right] \tag{4.11}$$

The motor drives a gear mechanism that rotates the screw at an angle (θ). The torque on the screw (T_m) generated by the motor is:

$$T_m = K_{gs}\left(\frac{\mu}{N_{gear}} - \theta\right); \qquad \varepsilon = \left(\frac{\mu}{N_{gear}} - \theta\right) \tag{4.12}$$

where

ω_m	is the motor bandwidth in (rad/sec),
I_m	is the motor inertia (slug-ft^2),
ε	is the relative position (rotor load),
μ	is the motor rotation angle in (rad),
θ	is the screw rotation angle in (rad),
N_{gear}	is the motor to screw gear ratio,
N_{screw}	is the shaft screw ratio (ft/rad),
T_m	is the torque on the screw,
K_{gs}	is the gear stiffness in (ft-lb/rad),
K_{mfr}	is the motor gear damping friction, (ft-lb/rad/sec) due to relative speed,
K_C	is the motor Coulomb Friction (ft-lb).

The direction of the screw rotation causes the shaft to either extend or to retract. Equation 4.13 calculates the screw rotation angle (θ) as a function of the supplied torque (T_m) and the load force (F_L). The shaft position x_p is linearly related to the screw angle by the screw gear ratio.

$$J_a\ddot{\theta} = K_{eff}T_m - N_{screw}F_L - K_{dmp}\dot{\theta} - K_c sign(\theta)$$
$$F_L = K_T\left[x_p - x_L\right]; \quad x_L = R\delta \tag{4.13}$$
$$x_p = N_{screw}\theta$$

where

N_{screw} is the screw gear ratio in (ft/rad). It represents the length of the piston extension when the screw mechanism rotates 1 rad.

F_L is the force across the actuator shaft in (lb) due to the load.

J_a is the combined inertia of the gears plus the screw.

R is the moment arm, shaft to hinge in (ft).

K_{dmp} is the shaft friction (ft-lb/rad/sec).

x_p is the piston extension and is related to the spiral shaft rotation (θ) by the screw gear ratio.

Equation 4.14 calculates the load angular acceleration as a function of the applied torques.

$$J_e \ddot{\delta} = F_L R - B_e \dot{\delta} - K_g \delta + T_{ext\_load} \tag{4.14}$$

where

J_e is the load moment of inertia of the engine or the control surface about the hinge,

δ is the load deflection in (rad),

K_g is the stiffness at the hinge (ft-lb/rad),

T_{ext_load} is the external torque due to aerodynamics or the vehicle acceleration forces.

The variable ($x_L = R\delta$) represents the distance traveled at the end of the shaft (the point where it is attached to the load) when the load rotates at an angle (δ). It is not just the piston extension x_p that creates the piston force F_L, but it is the difference ($x_P - x_L$) pushing against the total system stiffness K_T. The rotation angle (δ) represents the total rotation of the load at the gimbal which consists of two components: a component due to the piston extension (x_P), plus a component due to structural deformation (K_T). Equation 4.15 calculates the position feedback measurement x_{fb}.

$$x_{fb} = x_p \frac{K_T}{K_L} + x_L \left(1 - \frac{K_T}{K_L} \right) = x_p \frac{K_T}{K_L} + \frac{F_L}{K_{act}} \tag{4.15}$$

It represents what the position sensor measures inside the actuator. It consists of two components, a component due to actual shaft position x_p, and a component due to compression inside the actuator caused by the actuator internal stiffness K_{act}. The system has a total stiffness K_T. Part of the total stiffness is due to backup structure and load stiffness K_L, and the other part is due to the shaft and electrical compression K_{act} inside the actuator. The sensor measures shaft rotations due to compression K_{act} but it does not see the deformation due to K_L. The stiffness is related by the following equation:

$$\frac{1}{K_T} = \frac{1}{K_L} + \frac{1}{K_{act}} \tag{4.16}$$

Consider the two extreme situations. First, when the load is very light, there is no friction, and the load does not resist the motion of the shaft x_P. The shaft force F_L is reduced to zero. The extension due to (δ), $(x_L = x_P)$, and the feedback position measurement at the output of the summing junction becomes $x_{fb} = x_P$. This is exactly what you would expect for the position sensor to measure. Let us now assume that there is infinite stiffness in the actuator, and K_{act} is infinite. The total stiffness now reduces to backup plus load flexure $(K_T = K_L)$. Assume also that the shaft is locked at zero position $(x_p = 0)$. If we apply an external load torque to the load T_L, the piston force is: $(F_L = -T_L/R)$. The feedback sensor measurement x_{fb} will be zero as you would expect because the sensor cannot see the deformation due to K_L outside the actuator box.

$$x_{fb} = x_L - \frac{K_T}{K_L}x_L = 0$$

In actuality, however, when you are holding x_p to zero and are applying an external load T_L the sensor at the shaft will measure the small compression of the motor and shaft. The sensor measurement x_{fb} will be:

$$x_{fb} = x_L \left(1 - \frac{K_T}{K_L}\right) = x_L \frac{K_T}{K_{act}} = \frac{T_L/R}{K_{act}}$$

This is exactly what you would expect from the sensor to see; an extension measurement due to the external load acting against K_{act} alone.

4.2 COMBINING A FLEXIBLE VEHICLE MODEL WITH ACTUATORS

The actuator dynamics in general are not included with the vehicle equations of motion, but for many reasons they are modeled and analyzed as separate subsystem blocks that are combined with the vehicle dynamics. This offers flexibility in selecting different types of actuators, adjusting the complexity of the model, allows the capability to include actuator nonlinearities in simulations, and to perform a separate analysis on the actuator subsystem alone. The actuator models include the rotational dynamics of the effectors and the associated stiffnesses. The proper modeling of the dynamic coupling between the engine or surface rotation, the actuator, and the vehicle backup structure plays a very important role in the operation of the "TWD" and "load-torque" dynamic effects, and in the closed-loop control system stability and performance. Factors such as hydraulic fluid compressibility, gimbal friction, nozzle or control surface flexibility, backup structure stiffness, load inertia, nonlinearities, and other parameters should be included correctly in the actuator model. The dynamic coupling between the actuator nonlinearities and flexibility at the effector mounting structure is very critical and it often causes "TWD" oscillatory limit cycles if not properly designed. It is a good practice to keep the engine or the control surface resonant frequency above the TWD frequency. The stiffness in the actuator model must be properly accounted for according to the vehicle model. If the flexibility of

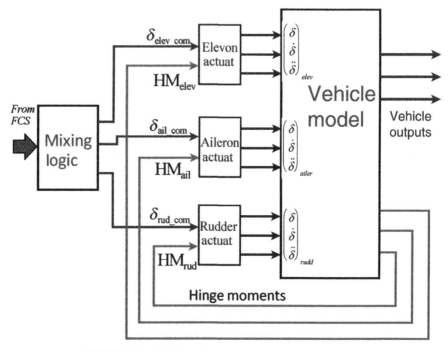

FIGURE 4.7   Vehicle and Actuator Interconnection Diagram

the support structure is already included in the vehicle flex model it should obviously not be included again in the actuator system.

Since the rotational dynamics of the effector load about the hinge is included in the actuator model, it is obviously not included in the vehicle, but the vehicle model receives the rotational positions, velocities, and acceleration signals from the actuators. The signals that drive the vehicle model are not inertial but they represent the relative motion of the effector at the hinge or gimbal, with or without local flexibility. The vehicle model generates the control and reaction forces from the actuator deflection signals, see equations 3.13 and 3.18. Figure 4.7 shows the dynamic model of a flight vehicle that is controlled by three aerosurfaces (elevon, aileron, and rudder) and each surface is controlled by a separate actuator subsystem. A mixing-logic matrix receives the roll, pitch, and yaw acceleration demands from the flight control system and converts them to elevon, aileron, and rudder actuator commands. There is a hinge moments mechanical loop between the vehicle output and the actuators load-torque input. The vehicle, mixing-logic matrix, and actuator subsystems are combined together using MATLAB®, Flixan, or other system interconnection software. The actuator blocks are not just SISO transfer functions. In addition to the shaft control hydraulics they also include the engine or control surface rotational dynamics about the hinge, as already described. The actuator consists of a shaft that is driving the load dynamics through some stiffness, and the load is a TVC engine or an aerosurface pivoting about a hinge. The actuator system outputs consist of deflection, rate, and rotational acceleration at the gimbal that drive the vehicle model. The surface deflections generate the aero forces (or in the case of a gimbaling engine, the TVC forces) that control the vehicle. The load accelerations generate the TWD forces. The rates generate damping forces (not often used). Each actuator model, in addition to the deflection command input, has a second HM input to receive the external load torques from the vehicle hinge moment outputs. The moments at the hinge of each surface represent the external loading generated by the vehicle accelerations. They are described in Section 3.10.2, and are fed back to the actuator HM inputs by closing a mechanical feedback loop, as shown in Figure 4.7. In the case of a TVC engine the vehicle model provides two load-torque outputs for pitch and yaw gimbaling that feedback to two separate actuators. The load torque reacts against the command torque and the actuator has to overcome it in order to control the effector position.

The analyst must be very careful in selecting the stiffness coefficients in the actuator model in order to avoid including them twice, in both the actuator and the structural models. There are three types of stiffness involved in modeling the servo system that controls the position of a control surface or a nozzle relative to the hinge and they combine together in series to define the local structural resonance.

1. The stiffness of the backup structure which is the stiffness of the support structure at the point where the actuator is attached to the vehicle,
2. The actuator shaft stiffness consisting of piston plus oil or electrical stiffness, and
3. The load linkage stiffness due to structural flexibility at the linkage and the aerosurface itself or the engine nozzle.

All three stiffnesses combine together to form a total stiffness K_T that is associated with the local resonance. When we are dealing with rigid vehicle models, the backup and load stiffnesses are obviously not included in the vehicle and, therefore, all three stiffnesses must be included in the actuator model and in this case the actuator model itself captures the local resonance of the load oscillating at the pivot. The deflection δ, rate, and acceleration outputs consist of, not only rigid surface rotation, but also the effects due to the combined spring constant K_T from the three stiffnesses. The position feedback measurement in the actuator control loop is the shaft extension. However, the position sensor measurement includes an error due to the shaft deformation. The measurement does not "see" the deformation of the backup structure and load linkage that contribute to the deflection δ. To summarize, all three stiffnesses should be included in the actuator model when: (a) the vehicle model is rigid, or (b) the vehicle is partially flexible, that is, the back-up structure and the actuator linkage stiffnesses are "rigid" in the finite-elements-model. The combined stiffness coefficient K_T and the load moment of inertia define the resonance of the effector about the hinge. However, when the flight vehicle is flexible and the stiffnesses of the support structure and linkage are included in the structural model, they should not be included again in the actuator, otherwise, the stiffness K_T will be reduced further. In this case, the stiff actuator model should be used that includes only shaft or piston stiffness. This type of modeling is important when analyzing the actuator loop stability and performance because the flight control system interacts with the actuator dynamics, the load dynamics, and the backup structure, and they often excite each other to instability, mostly due to stiction, backlash, and other non-linearities.

There is also the situation where the entire vehicle structure is included in the finite element model; that includes: the aerosurfaces, nozzles, and actuator linkage dynamics. This includes the backup and load stiffnesses and they must not be included in the actuator. In this case the structural model is calculated with the hinges or gimbals locked, assuming that there are rigid links across the two ends of the actuator shaft. In the actuator model the backup and load stiffnesses must be set to infinity and only the actuator piston stiffness is included. The locked hinges are released in the simulation equations by the actuator models that provide the rigid rotations, rates, and accelerations to the vehicle equations. The effector flexing at the hinge is calculated from the inertial coupling coefficients $h_s(k,j)$, also known as "h-parameters" which are generated with the structural model, as shown in equations 3.38. The actuator models must be "stiff" because they provide only the "rigid" component of the deflection resulting from the extension of the shaft. The flex deflection caused by structural bending at the hinge is already included in the finite element model. Only the actuator shaft stiffness in series with the oil stiffness (for hydraulic actuators), or the shaft stiffness in series with the electrical stiffness (for electro-mechanical actuators), that is K_{act} instead of K_T, should be included in the "stiff" actuator model. In the stiff actuator model, Figure 4.2b, the backup and load stiffnesses are infinite. The feedback path via $(1/K_L)$, shown in Figure 4.4a becomes zero. The stiffness K_{act} is significantly bigger than K_T in the "soft" actuator model.

4.3   ELECTROMECHANICAL ACTUATOR EXAMPLE

In this example we will demonstrate an electromechanical actuator system for an elevator aero-surface similar to the model presented in Section 4.1.3. The purpose of this example is to familiarize the readers with some of the control and structure interaction issues as we attempt to push the actuator bandwidth beyond the structural capability of the vehicle. We will first analyze the actuator system in the frequency domain. Identify the issue and attempt to fix it using a filter. Then we will analyze the actuator system in the time domain including non-linearities. The Simulink model of the EMA system is shown in Figure 4.8 and it is already described in Section 4.1.3. Its inputs are surface deflection command in (deg) and external load-torque in (ft-lb). The outputs are: deflection, deflection rate, and acceleration. The position measurement feedback (X_{Pfb}) includes the error due to local deformation. Figure 4.9 shows the actuator PID position control system which generates the voltage for the torque motor that controls the position of the shaft. It includes a low-pass filter, rate and voltage limits, and an anti-winding compensator to prevent the integrator from saturating. Figure 4.10 shows the motor subsystem. The torque produced by the motor is a function of the input voltage minus friction and the shaft position feedback. Figure 4.11 shows the actuator shaft extension which is a function of the

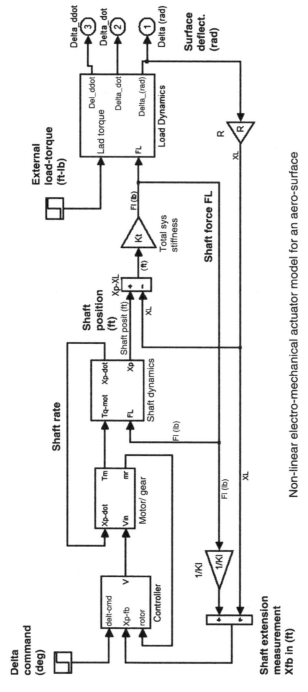

FIGURE 4.8 Non-linear EMA System

Non-linear electro-mechanical actuator model for an aero-surface

127

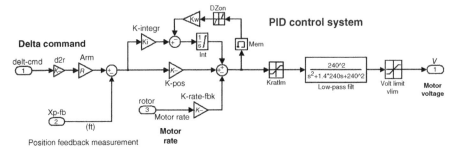

FIGURE 4.9   Position Control System of the Actuator

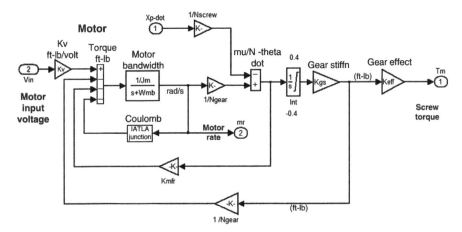

FIGURE 4.10   DC Motor Generates the torque to rotate the Screw Gear

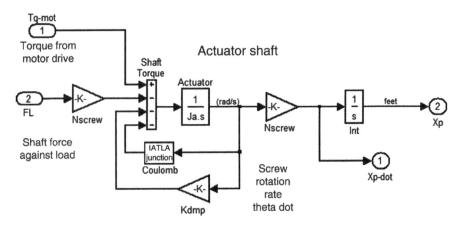

FIGURE 4.11   Shaft Dynamics Produces the Shaft Extension

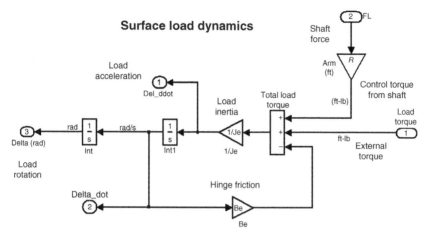

FIGURE 4.12 Surface Rotation is a function of Control Torque plus External Load Torque

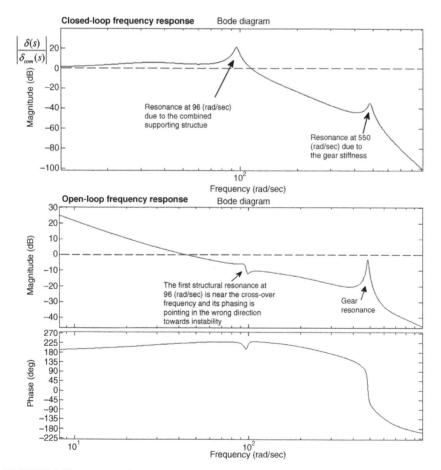

FIGURE 4.13 (a) Closed-Loop Frequency Response showing the first structural resonance amplified by feedback. (b) Open-Loop frequency response shows the 1st structural resonance is too close to the bandwidth and it is phased in the wrong direction

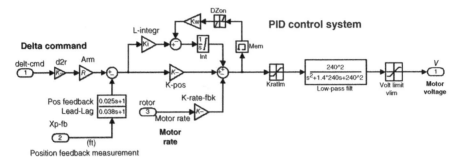

FIGURE 4.14   A Lead-Lag Filter is Introduced in the Position Measurement Feedback loop to Compensate Against the Structural Resonance

motor torque minus the torque from the load. Figure 4.12 shows the load dynamics of the aerosurface which rotates about the hinge. It is driven by the control torque and also by the external load-torque. The external load torque is a feedback from the vehicle model and it is generated by aerodynamics and the vehicle acceleration.

This actuator has a dynamic problem! Its bandwidth is too close to the first resonance of the support structure and its closed-loop frequency response $\delta(s)/\delta_{com}(s)$ has a big spike at the structural resonance, see Figure 4.13a. The open-loop Bode frequency response in Figure 4.13b shows that the first resonance is too close to the cross-over frequency, plus it is phased in the wrong direction towards instability. This system although stable, it excites, however, the structure at the 96 (rad/sec) mode, and we shall attempt to compensate it by introducing a lead-lag filter in the control feedback, as shown in Figure 4.14. This is only to improve closed-loop performance. Figures 4.15 and 4.16 show the Nichols and Bode plots of the open-loop frequency responses. The Nichols shows the gain and phase margins of the compensated versus the uncompensated systems. The phase margin at the first resonance was improved by the filter at the expense of some gain margin. It also increases the gain at the second structural resonance which is phase-stable. Figure 4.17 shows the closed-loop frequency responses of the compensated versus the uncompensated systems. The compensated system provides a little more attenuation on both resonances.

Figures 4.18 through 4.21 show simulations of the closed-loop systems response to a 5° deflection command.

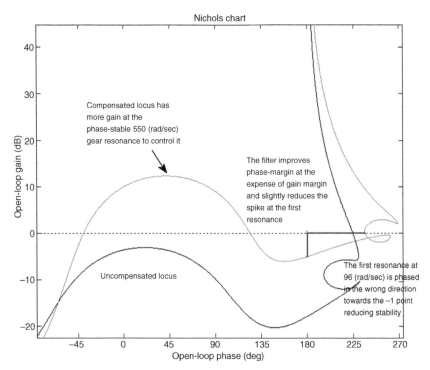

FIGURE 4.15   The Lead-Lag Compensator Increases the Phase-Margin at the Expense of Some Gain Margin

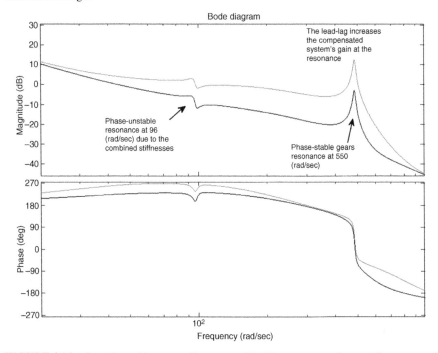

FIGURE 4.16   Open-Loop Frequency Response of the Uncompensated versus Compensated Actuator Systems

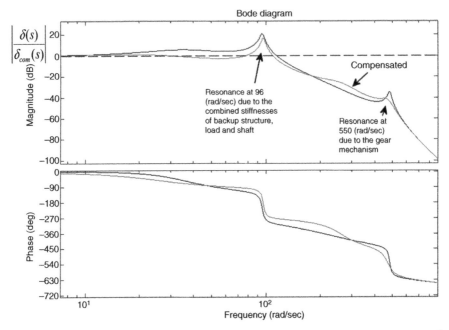

FIGURE 4.17 Closed-Loop Frequency Response: Compensated versus Uncompensated Systems, Shows a small improvement with the filter addition

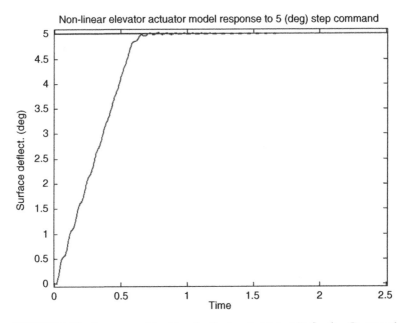

FIGURE 4.18 Response of the Elevator Surface to 5 (deg) Deflection Command

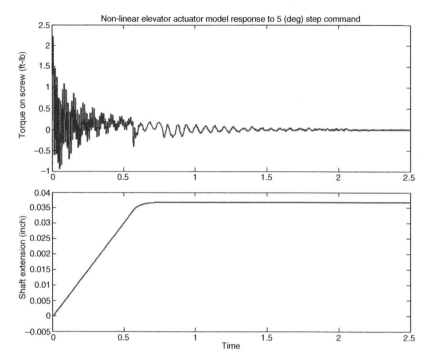

FIGURE 4.19 High Frequency Oscillations due to the Gear Stiffness are Visible in the Torque; The Actuator Shaft Extends to the Commanded Position

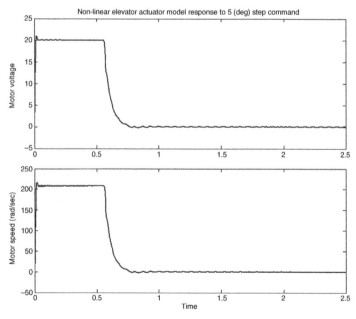

FIGURE 4.20 The Motor Voltage Saturates at 20 V, and so does the Motor Speed at 208 (rad/sec)

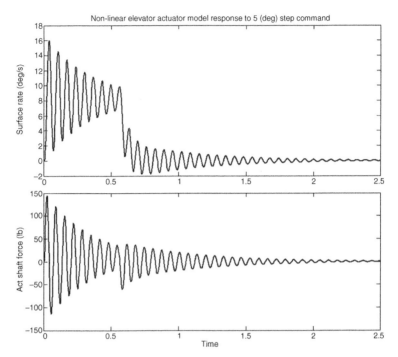

FIGURE 4.21 The Combined Stiffness (Backup Structure, Load, and Shaft) Causes the Surface to Oscillate at 96 (rad/sec). The Shaft Force also Oscillates at the Same Frequency since the Surface Oscillates

The following file loads the actuator parameters in Matlab®. The Simulink files for this example can be obtained from Flixan.com/ Actuators/ Examples/ EMA Examples

```
% EMA Type-B Actuator Model Parameters
r2d= 180/pi; d2r =pi/180;
Kload= 480000;            % Load Stiffness (lb/ft) Kload
Kback= 480000;            % Backup Structure Stiffn Kback
Kact= 2*10^7;             % Actuator Shaft Stiffn + Electric Kact
Kti=(1/Kload)+(1/Kback)+(1/Kact); Kt=1/Kti; % Total Stiffness
Kli=(1/Kload)+(1/Kback); Kl=1/Kli;  % Stiffness of Load + Backup
Kp= 12000;               % PID gain
Ki= 380000;              % PID gain
Krfb= 0.012;             % Rate Feedback Gain
Kratlm= 22.0;            % Controller Rate Limit
Kw=18;                   % Anti-windup gain
Vlim= 136;               % Controller Output Voltage Limit volts
```

```
R= 0.4;                    % Moment Arm (ft) R= 0.4
Je= 4.0;                   % Rudder Inertia Je= 4.0 (sl-ft^2)
Ja= 0.00042;               % Gear plus Screw Inertia Ja= 0.00042
Jm= 0.02;                  % Motor Torque Inertia
Be= 28;                    % Load Damping Be= 28
Kv= 200;                   % Motor Torque Gain (ft-lb/volt)
Kvl= 0.05;                 % Motor Loading (rad/sec per ft-lb)
Wmb= 1000;                 % Motor Bandwidth (rad/sec)
Kmfr= 8;                   % Motor Friction
Keff= 0.9;                 % Motor Gear Effectiveness 0.9
Kgs= 80;                   % Gear Stiffness Kgs=80
Nscrew= 0.0026;            % Screw Ratio Nscrew=0.0026 (inch/rad)
Ngear= 8.2;                % Gear Ratio Ngear= 8.2
Kdmp= 0.001;               % Shaft Damping Kdmp= 0.01
```

5

EFFECTOR COMBINATION LOGIC

The effector combination or mixing logic is a matrix (K_{mix}) that connects between the flight control system outputs and the actuator inputs. Its purpose is to convert the FCS acceleration demands to TVC or aerosurface deflections or thrust variations commands. Flight vehicles are in general controlled by different types of effectors that produce moments and forces in 3 or more directions, which are mainly 3 rotations and optionally some translations. The effectors are thrust vector control (TVC) engines, thrust varying (throttling) engines, aerosurfaces, and reaction control jets (RCS) that provide the "muscle" power to maneuver the vehicle. The mixing logic combines the vehicle effectors together as a system and becomes an integral part of the flight control system (FCS). In the event of an effector failure it is the mixing logic matrix that must be adjusted instead of the FCS gains. Figure 5.1 shows a mixing logic matrix for a typical flight vehicle that is controlled by different types of effectors. The inputs are acceleration demands coming from the flight control system, and they are converted to TVC pitch and yaw deflections, throttle commands, and aerosurface deflections that drive the control actuators. The FCS demands are functions of commands minus measurements that control the vehicle attitude and flight direction. They are mainly 3 rotational acceleration demands and may also include some translational demands, such as, accelerations along X, Y and Z. Translational control is used when translation or velocity control is required separately of rotations, such as, during vehicle separation, hovering at low speeds, or controlling the rate of descent. This is possible, of course, when the vehicle has the effector capability to generate translations, for example, a throttling engine, jets, body-flap, or a speed-brake to provide linear control along those directions.

Performance Evaluation and Design of Flight Vehicle Control Systems, First Edition. Eric T. Falangas.
© 2016 by Eric T. Falangas. Published 2016 by John Wiley & Sons, Inc.

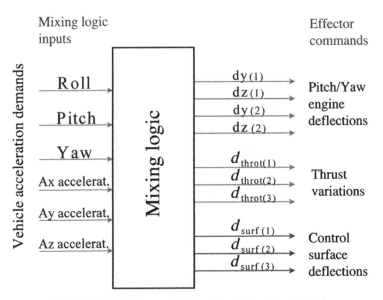

FIGURE 5.1   Effector Combination Mixing-Logic Matrix

The effector sizing is based on requirements defined by the vehicle performance goals. The effectors as a system must be capable of providing the required accelerations for maneuverability and the control authority to react against disturbances in the controlled directions, which are at least 3 rotations, plus some translations, as it will be discussed in Chapter 7. The mixing logic algorithm described in this section optimizes the actuator effectiveness, because it takes into consideration the vehicle geometry, thrusts, angle of attack, mass properties, aero-surface coefficients and the capability of each effector in the required directions. It maximizes the vehicle response in the commanded directions using minimum deflections. It uses pseudo-inversion to determine an optimal combination of the controls that achieve the demanded accelerations while reducing cross-coupling between the control axes. When the matrix is connected open-loop in series with the vehicle model, as shown in Figure 5.2, it attempts to diagonalize the plant which means that the vehicle accelerations approximate the accelerations requested by flight control. This, of course, is true when we ignore the aerodynamics of the base vehicle. The matrix will provide the proper accelerations. However, the vehicle will eventually diverge if it is open-loop unstable. That is why we need feedback stabilization. This pre-multiplication of the vehicle with the effector mixing matrix creates a plant model that is more efficient for control design because it already includes some decoupling between axes. An efficient mixing logic should be time-varying because the control authority of the effectors changes as a function of geometry, dynamic pressure, angle of attack, thrust, and CG location.

5.1   DERIVATION OF AN EFFECTOR COMBINATION MATRIX

We now present the derivation of a mixing-logic matrix for a vehicle that is controlled by gimbaling TVC engines, throttling engines or reaction control jets, and aerosurfaces.

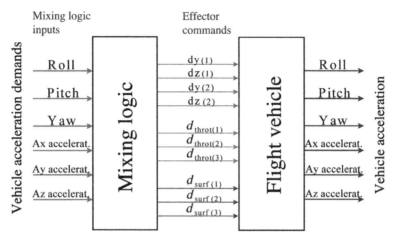

FIGURE 5.2   When the Mixing-Logic Matrix Is Connected in Series with the Vehicle Model (Open-Loop) the Vehicle Accelerations Should be Approximately Equal to the Accelerations Demanded by the FCS

5.1.1   Forces and Moments Generated by a Single Engine

Equation 5.1 calculates the forces generated by a single thruster engine (i) that is mounted on a vehicle at fixed orientation angles (or trimmed at those angles): Δ_E in pitch (elevation angle with respect to the x-y plane), and Δ_Z in yaw (azimuth angle about the body z-axis), see Figure (5.3). The forces along the body x-, y-, and z-axes are:

$$
\begin{aligned}
F_{Xe(i)} &= T_{e(i)} \cos(\Delta_E) \cos(\Delta_Z) \\
F_{Ye(i)} &= T_{e(i)} \cos(\Delta_E) \sin(\Delta_Z) \\
F_{Ze(i)} &= -T_{e(i)} \sin(\Delta_E)
\end{aligned}
\tag{5.1}
$$

Let us define the throttle control $D_{th(i)}$ for engine (i) to be the ratio of thrust variation divided by the nominal engine thrust.

$$
D_{th(i)} = \frac{\delta T_{e(i)}}{T_{e(i)}}, \text{ where } \delta T_{e(i)} \text{ is the thrust variation}
\tag{5.2}
$$

The product $D_{th(i)} \times T_{e(i)} = \delta T_{e(i)}$ is the variation of engine thrust force above or below its nominal thrust value $T_{e(i)}$. Equation 5.3 calculates the force variation at the gimbal of an engine (i), resolved along the vehicle x-, y-, and z-axes, due to the combined effects of gimbaling and throttling.

$$
\begin{aligned}
F_{xe(i)} &= T_{e(i)} \left[-s(\Delta_E)c(\Delta_Z)\delta y_{(i)} - c(\Delta_E)s(\Delta_Z)\delta z_{(i)} + c(\Delta_E)c(\Delta_Z)D_{th(i)} \right] \\
F_{ye(i)} &= T_{e(i)} \left[-s(\Delta_E)s(\Delta_Z)\delta y_{(i)} + c(\Delta_E)c(\Delta_Z)\delta z_{(i)} + c(\Delta_E)s(\Delta_Z)D_{th(i)} \right] \\
F_{ze(i)} &= T_{e(i)} \left[-c(\Delta_E)\delta y_{(i)} - s(\Delta_E)D_{th(i)} \right]
\end{aligned}
\tag{5.3}
$$

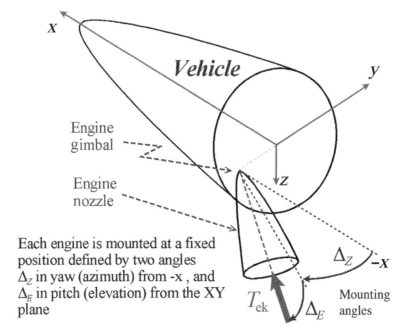

FIGURE 5.3   Engine Orientation Angles (Δ_E and Δ_Z) with Respect to the Vehicle Body Axis

Let us define the distances $l_{xe(i)}$, $l_{ye(i)}$, $l_{ze(i)}$ between the ith engine gimbal and the vehicle CG, as follows:

$$l_{xe(i)} = X_{e(i)} - X_{CG} \quad l_{ye(i)} = Y_{e(i)} - Y_{CG} \quad l_{ze(i)} = Z_{e(i)} - Z_{CG} \tag{5.4}$$

The roll, pitch, and yaw moments on the vehicle resulting from the forces generated by a single engine (i) are obtained from the following matrix equation:

$$\begin{bmatrix} L_{e(i)} \\ M_{e(i)} \\ N_{e(i)} \end{bmatrix} = \begin{bmatrix} 0 & -l_{zei} & l_{yei} \\ l_{zei} & 0 & -l_{xei} \\ -l_{yei} & l_{xei} & 0 \end{bmatrix} \begin{bmatrix} F_{xe(i)} \\ F_{ye(i)} \\ F_{ze(i)} \end{bmatrix} \tag{5.5}$$

We will now calculate the moment and force variations in the vehicle body axes generated by each individual effector and combine them together as a system. This is due to gimbaling, throttling, and also due to the control surface deflections. The contribution of each effector will be included and we will derive an expression for the total vehicle moments and forces as a function of the contributions from all effectors. One more detail that will be considered in the mixing-logic calculations is the maximum capability of each effector. This consideration is important because the various engines or aerosurfaces may have different maximum deflection angles or throttling capabilities. We must derive, therefore, a mixing law that will take into consideration the effector capabilities according to their peak contributions in each direction, by spreading the control authority evenly among the effectors proportionally, according to their capabilities. For example, if two engines have equal thrust but different gimbaling capabilities, the engine with the larger rotational capability should be allowed to deflect at a larger angle than the engine with the smaller rotation range. Ideally, they should all reach their saturation limits together when the control demand is exceeded. This maximizes the control effectiveness.

5.1.2 Moments and Forces Generated by a Single Engine Gimbaling in Pitch and Yaw

Consider an engine (i) which is mounted at fixed elevation and yaw angles $\Delta_E(i)$ and $\Delta_Z(i)$, respectively, see Figure 5.3. The engine is further gimbaling at small angles $\delta y(i)$ and $\delta z(i)$ in pitch and yaw directions with respect to the mounting positions. The moment variations on the vehicle are obtained from equation 5.6.

$$
\begin{pmatrix} L_{g(i)} \\ M_{g(i)} \\ N_{g(i)} \end{pmatrix} = T_{e(i)} \begin{pmatrix} 0 & -l_{zei} & l_{yei} \\ l_{zei} & 0 & -l_{xei} \\ -l_{yei} & l_{xei} & 0 \end{pmatrix} \begin{pmatrix} -c(\Delta_Z)s(\Delta_E) & -c(\Delta_E)s(\Delta_Z) \\ -s(\Delta_Z)s(\Delta_E) & +c(\Delta_E)c(\Delta_Z) \\ -c(\Delta_E) & 0 \end{pmatrix} \begin{pmatrix} \delta_{y(i)} \\ \delta_{z(i)} \end{pmatrix}
$$

$$(5.6)$$

This equation can be normalized by dividing the pitch and yaw engine deflections with the max deflection capabilities in both directions, so that the normalized inputs can vary between $\{0$ and $\pm 1\}$ as follows:

$$
\begin{pmatrix} L_{g(i)} \\ M_{g(i)} \\ N_{g(i)} \end{pmatrix} = T_{e(i)} \begin{pmatrix} 0 & -l_{zei} & l_{yei} \\ l_{zei} & 0 & -l_{xei} \\ -l_{yei} & l_{xei} & 0 \end{pmatrix} \begin{pmatrix} -c(\Delta_Z)s(\Delta_E)\delta_{y\,max} & -c(\Delta_E)s(\Delta_Z)\delta_{z\,max} \\ -s(\Delta_Z)s(\Delta_E)\delta_{y\,max} & +c(\Delta_E)c(\Delta_Z)\delta_{z\,max} \\ -c(\Delta_E)\delta_{y\,max} & 0 \end{pmatrix}
$$
$$
\times \begin{pmatrix} \delta_{y(i)}/\delta_{y\,max} \\ \delta_{z(i)}/\delta_{z\,max} \end{pmatrix}
$$

By multiplying out the matrices in the above equation, it can be expressed in a simplified form as follows:

$$
\begin{pmatrix} L_{g(i)} \\ M_{g(i)} \\ N_{g(i)} \end{pmatrix} = \begin{pmatrix} | & | \\ V_{gyi} & V_{gzi} \\ | & | \end{pmatrix} \begin{pmatrix} \delta_{y(i)}/\delta_{y\,max} \\ \delta_{z(i)}/\delta_{z\,max} \end{pmatrix}
$$
$$(5.7)$$

where $V_{gy(i)}$ and $V_{gz(i)}$ are column vectors that correspond to the pitch and yaw engine deflections, respectively.

Similarly, the forces applied at the gimbal due to gimbaling in pitch and yaw of an engine (i) can be resolved along the body x-, y-, and z-axes and normalized by dividing the pitch and yaw deflections with the max deflections as shown in equation 5.8, written also in column vector form:

$$
\begin{pmatrix} F_{X(i)} \\ F_{Y(i)} \\ F_{Z(i)} \end{pmatrix} = T_{e(i)} \begin{pmatrix} -c(\Delta_Z)s(\Delta_E)\delta_{y\,\mathrm{max}} & -c(\Delta_E)s(\Delta_Z)\delta_{z\,\mathrm{max}} \\ -s(\Delta_Z)s(\Delta_E)\delta_{y\,\mathrm{max}} & +c(\Delta_E)c(\Delta_Z)\delta_{z\,\mathrm{max}} \\ -c(\Delta_E)\delta_{y\,\mathrm{max}} & 0 \end{pmatrix} \begin{pmatrix} \delta_{y(i)}/\delta_{y\,\mathrm{max}} \\ \delta_{z(i)}/\delta_{z\,\mathrm{max}} \end{pmatrix}
$$

$$
\begin{pmatrix} F_{X(i)} \\ F_{Y(i)} \\ F_{Z(i)} \end{pmatrix} = \begin{pmatrix} | & | \\ U_{gyi} & U_{gzi} \\ | & | \end{pmatrix} \begin{pmatrix} \delta_{y(i)}/\delta_{y\,\mathrm{max}} \\ \delta_{z(i)}/\delta_{z\,\mathrm{max}} \end{pmatrix}
$$

(5.8)

where $U_{gy(i)}$ and $U_{gz(i)}$ are column vectors that correspond to the pitch and yaw engine deflections, respectively.

5.1.3   Moments and Forces of an Engine Gimbaling in a Single Skewed Direction

A vehicle with multiple engines may have sufficient degrees of freedom to be maneuvered by gimbaling some of the engines in a single direction instead of two (pitch and yaw). Gimbaling in a single direction instead of two saves on actuators, weight, and cost. Figure 5.4 shows a gimbaling engine that rotates only about a single axis that is skewed at an angle $\gamma(i)$. The engine is mounted at fixed $\Delta_Y(i)$ and $\Delta_E(i)$ orientation (or trim) angles and it can gimbal from its mounting position along the direction defined by angle $\gamma(i)$. The deflection angle δ_γ along $\gamma_{e(i)}$ can be resolved in pitch and yaw components, and the roll, pitch, and yaw moments on the vehicle generated by the ith single gimbaling engine which is gimbaling at a skewed direction $\gamma(i)$ are defined in equation 5.10.

$$
\delta_{y(i)} = \delta_{\gamma(i)} \cos(\gamma_{e(i)}) \qquad \delta_{z(i)} = \delta_{\gamma(i)} \sin(\gamma_{e(i)})
$$

(5.9)

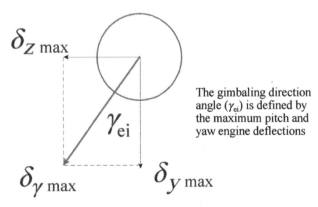

The gimbaling direction angle (γ_{ei}) is defined by the maximum pitch and yaw engine deflections

FIGURE 5.4   Engine Gimbaling in a Single Skewed Direction (γ_{ei})

The roll, pitch, and yaw moments on the vehicle generated by a single gimbaling engine (i) are obtained from the following normalized equation, written also in column vector form:

$$
\begin{pmatrix} L_{g(i)} \\ M_{g(i)} \\ N_{g(i)} \end{pmatrix} = \delta_{\gamma\,max} T_{e(i)} \begin{pmatrix} 0 & -l_{zei} & l_{yei} \\ l_{zei} & 0 & -l_{xei} \\ -l_{yei} & l_{xei} & 0 \end{pmatrix} \begin{pmatrix} -c(\Delta_Z)s(\Delta_E)c(\gamma_{ei}) - c(\Delta_E)s(\Delta_Z)s(\gamma_{ei}) \\ -s(\Delta_Z)s(\Delta_E)c(\gamma_{ei}) + c(\Delta_E)c(\Delta_Z)s(\gamma_{ei}) \\ -c(\Delta_E)\cos(\gamma_{ei}) \end{pmatrix}
$$
$$
\times \left(\delta_{\gamma(i)} / \delta_{\gamma\,max} \right) \tag{5.10}
$$
$$
\begin{pmatrix} L_{g(i)} \\ M_{g(i)} \\ N_{g(i)} \end{pmatrix} = \begin{pmatrix} | \\ V_{g\gamma(i)} \\ | \end{pmatrix} \left(\delta_{\gamma(i)} / \delta_{\gamma\,max} \right)
$$

where $V_{g\gamma(i)}$ is a column vector, and the normalized input $\{\delta_\gamma / \delta_{\gamma\,max}\}$ varies from $\{0$ to $\pm 1\}$. The forces in the x, y, and z directions are also obtained from a similar equation 5.11.

$$
\begin{pmatrix} F_{X(i)} \\ F_{Y(i)} \\ F_{Z(i)} \end{pmatrix} = \delta_{\gamma\,max} T_{e(i)} \begin{pmatrix} -c(\Delta_Z)s(\Delta_E)c(\gamma_{ei}) - c(\Delta_E)s(\Delta_Z)s(\gamma_{ei}) \\ -s(\Delta_Z)s(\Delta_E)c(\gamma_{ei}) + c(\Delta_E)c(\Delta_Z)s(\gamma_{ei}) \\ -c(\Delta_E)\cos(\gamma_{ei}) \end{pmatrix} \left(\delta_{\gamma(i)} / \delta_{\gamma\,max} \right)
$$
$$
\begin{pmatrix} F_{X(i)} \\ F_{Y(i)} \\ F_{Z(i)} \end{pmatrix} = \begin{pmatrix} | \\ U_{g\gamma(i)} \\ | \end{pmatrix} \left(\delta_{\gamma(i)} / \delta_{\gamma\,max} \right) \tag{5.11}
$$

5.1.4   Moments and Forces Generated by a Throttling Engine or an RCS Jet

Similarly, the change in moments on the vehicle generated by the thrust variation in the ith engine or RCS jet that is mounted at fixed orientation angles $\Delta_E(i)$ and $\Delta_Z(i)$ with respect to the vehicle axis, see Figure 5.3, can be obtained from equation 5.12.

$$
\begin{pmatrix} L_{T(i)} \\ M_{T(i)} \\ N_{T(i)} \end{pmatrix} = T_{e(i)} \begin{pmatrix} 0 & -l_{zei} & l_{yei} \\ l_{zei} & 0 & -l_{xei} \\ -l_{yei} & l_{xei} & 0 \end{pmatrix} \begin{pmatrix} c(\Delta_E)c(\Delta_Z) \\ c(\Delta_E)s(\Delta_Z) \\ -s(\Delta_E) \end{pmatrix} \left(D_{Th(i)} \right) \tag{5.12}
$$

where $T_{e(i)}$ is the nominal engine thrust and $D_{Th(i)}$ is the throttle control input. The throttle input can vary between zero and $\pm D_{Thmax}$, where the maximum throttle input $|D_{Thmax}| < 1$. The product $T_{e(i)} \times D_{Th(i)}$ represents the thrust force variation, above or below the nominal thrust value $T_{e(i)}$. The throttle input can be normalized in a similar fashion as in the TVC equation so that the normalized throttle input ratio $(D_{Th(i)}/D_{Th\max})$ varies between $\{0 \text{ and } \pm 1\}$. The roll, pitch, and yaw moment variations due to a throttling engine or an RCS jet are:

$$
\begin{pmatrix} L_{T(i)} \\ M_{T(i)} \\ N_{T(i)} \end{pmatrix} = D_{Th\max} T_{e(i)} \begin{pmatrix} 0 & -l_{zei} & l_{yei} \\ l_{zei} & 0 & -l_{xei} \\ -l_{yei} & l_{xei} & 0 \end{pmatrix} \begin{pmatrix} c(\Delta_E)c(\Delta_Z) \\ c(\Delta_E)s(\Delta_Z) \\ -s(\Delta_E) \end{pmatrix} \left(\frac{D_{Th(i)}}{D_{Th\max}} \right)
$$

$$
\begin{pmatrix} L_{T(i)} \\ M_{T(i)} \\ N_{T(i)} \end{pmatrix} = \begin{pmatrix} | \\ V_{T(i)} \\ | \end{pmatrix} \left(\frac{D_{Th(i)}}{D_{Th\max}} \right)
$$

$$(5.13)$$

Similarly, the forces in the x, y, and z directions are defined in equation 5.14.

$$
\begin{pmatrix} F_{X(i)} \\ F_{Y(i)} \\ F_{Z(i)} \end{pmatrix} = D_{Th\max} T_{e(i)} \begin{pmatrix} c(\Delta_E)c(\Delta_Z) \\ c(\Delta_E)s(\Delta_Z) \\ -s(\Delta_E) \end{pmatrix} \left(\frac{D_{Th(i)}}{D_{Th\max}} \right) = \begin{pmatrix} | \\ U_{T(i)} \\ | \end{pmatrix} \left(\frac{D_{Th(i)}}{D_{Th\max}} \right) \quad (5.14)
$$

where $V_{T(i)}$ and $U_{T(i)}$ are column vectors for a throttling engine (i).

5.1.5   Moment and Force Variations Generated by a Control Surface Deflection from Trim

The change in vehicle moments generated by the deflection δ_{csi} of a control surface (i) from trim is obtained from equation 5.15.

$$
\begin{pmatrix} L_{CS(i)} \\ M_{CS(i)} \\ N_{CS(i)} \end{pmatrix} = \overline{Q} S_{ref} \begin{pmatrix} l_{sp} C_{l\delta cs(i)} \\ l_{ch} C_{m\delta cs(i)} \\ l_{sp} C_{nl\delta cs(i)} \end{pmatrix} \delta_{cs(i)} \quad (5.15)
$$

where l_{ch} is the mean aerodynamic chord, and l_{sp} is the wingspan aeroreference lengths. Equation 5.15 is also normalized using the maximum control surface deflection capability $\delta_{csi\max}$. This makes the normalized control surface input vary between zero and ± 1. The normalized equations for the moments and forces from the ith aerosurface deflection become:

$$
\begin{pmatrix} L_{CS(i)} \\ M_{CS(i)} \\ N_{CS(i)} \end{pmatrix} = \overline{Q} S_{ref} \, \delta_{csi\max} \begin{pmatrix} l_{sp} C_{l\delta cs(i)} \\ l_{ch} C_{m\delta cs(i)} \\ l_{sp} C_{nl\delta cs(i)} \end{pmatrix} \left(\frac{\delta_{csi}}{\delta_{csi\max}} \right) = \begin{pmatrix} | \\ V_{AS(i)} \\ | \end{pmatrix} \left(\frac{\delta_{csi}}{\delta_{csi\max}} \right)
$$

$$
\begin{pmatrix} F_{X(i)} \\ F_{Y(i)} \\ F_{Z(i)} \end{pmatrix} = \overline{Q} S_{ref} \, \delta_{csi\max} \begin{pmatrix} -C_{A\delta as1} \\ C_{Y\delta as1} \\ C_{Z\delta as1} \end{pmatrix} \left(\frac{\delta_{csi}}{\delta_{csi\max}} \right) = \begin{pmatrix} | \\ U_{AS(i)} \\ | \end{pmatrix} \left(\frac{\delta_{csi}}{\delta_{csi\max}} \right)
$$

$$(5.16)$$

where $V_{AS(i)}$ and $U_{AS(i)}$ are column vectors for the ith aerosurface.

5.1.6   Vehicle Accelerations due to the Combined Effect from all Actuators

The total moment and forces on the vehicle are obtained by the superposition of the individual moments and forces generated from each effector. That is, the TVC engines, the throttling engines, the RCS jets, and the control surface deflections. The combined effectors moment matrix is obtained by stacking up the column vectors $V_{X(i)}$ from each individual effector. Equation 5.17 converts the normalized deflections from each individual effector to vehicle rotational accelerations or changes in angular rates (δP, δQ, δR). Equation 5.18 converts the normalized deflections to vehicle translational accelerations. The equations are also written in matrix form.

$$
\begin{pmatrix} \delta P \\ \delta Q \\ \delta R \end{pmatrix} = \begin{pmatrix} I_{XX} & -I_{XY} & -I_{XZ} \\ -I_{XY} & I_{YY} & -I_{YZ} \\ -I_{XZ} & -I_{YZ} & I_{ZZ} \end{pmatrix}^{-1} \times
$$

$$
\{V_{gy1} \ \ V_{gz1} \ \ V_{gy2} \ \ V_{gz2} \ \ V_{gr1} \ \ V_{gr2} \ \ V_{T1} \ \ V_{T2} \ \ V_{T3} \ \ V_{AS1} \ \ V_{AS2}\}
\begin{bmatrix}
\delta_{y1}/\delta_{y1\,max} \\
\delta_{z1}/\delta_{z1\,max} \\
\delta_{y2}/\delta_{y2\,max} \\
\delta_{z2}/\delta_{z2\,max} \\
\delta\gamma_{1}/\delta\gamma_{1\,max} \\
\delta\gamma_{2}/\delta\gamma_{2\,max} \\
D_{Th1}/D_{Th1\,max} \\
D_{Th2}/D_{Th2\,max} \\
D_{Th3}/D_{Th3\,max} \\
\delta_{cs1}/\delta_{cs1\,max} \\
\delta_{cs2}/\delta_{cs2\,max}
\end{bmatrix}
\quad (5.17)
$$

$$
or \quad \left(\underline{\delta R_o}\right) = I_n^{-1}[V]\left(\underline{\delta/\delta_{max}}\right)
$$

Similarly, the translational accelerations due to the normalized effector deflections are obtained from the following F = ma equation:

$$
\begin{pmatrix} A_X \\ A_Y \\ A_Z \end{pmatrix} = \left(\frac{1}{m_v}\right) \times
$$

$$
\{U_{gy1} \ \ U_{gz1} \ \ U_{gy2} \ \ U_{gz2} \ \ U_{gr1} \ \ U_{gr2} \ \ U_{T1} \ \ U_{T2} \ \ U_{T3} \ \ U_{AS1} \ \ U_{AS2}\}
\begin{bmatrix}
\delta_{y1}/\delta_{y1\,max} \\
\delta_{z1}/\delta_{z1\,max} \\
\delta_{y2}/\delta_{y2\,max} \\
\delta_{z2}/\delta_{z2\,max} \\
\delta\gamma_{1}/\delta\gamma_{1\,max} \\
\delta\gamma_{2}/\delta\gamma_{2\,max} \\
D_{Th1}/D_{Th1\,max} \\
D_{Th2}/D_{Th2\,max} \\
D_{Th3}/D_{Th3\,max} \\
\delta_{cs1}/\delta_{cs1\,max} \\
\delta_{cs2}/\delta_{cs2\,max}
\end{bmatrix}
\quad (5.18)
$$

$$
or \quad \left(\underline{\delta A_c}\right) = m_v^{-1}[U]\left(\underline{\delta/\delta_{max}}\right)
$$

By combining equations 5.17 and 5.18 together we obtain the following matrix equation 5.19 which calculates the vehicle accelerations resulting due to the normalized effector deflections input vector δ/δ_{max}, where each input varies from 0 to ±1.

$$\left(\frac{\delta R_o}{\delta A_c} \right) = [A] \left(\underline{\delta/\delta_{max}} \right) \qquad where : [A] = \begin{bmatrix} I_n^{-1} V \\ m_v^{-1} U \end{bmatrix} \qquad (5.19)$$

The elements of the following diagonal matrix $[D_{max}]$ consist of the maximum deflections and throttles of each effector, as follows:

$$D_{max} = diag \left\{ \delta_{y1\,max} \quad \delta_{z1\,max} \quad \delta_{y2\,max} \quad \delta_{z2\,max} \quad \delta_{y3\,max} \quad \delta_{z3\,max} \quad \delta_{\gamma1\,max}, \cdots \right.$$
$$\left. D_{Th1\,max} \quad D_{Th2\,max} \quad \delta_{cs1\,max} \quad \delta_{cs2\,max} \right\}$$

The mixing-logic matrix is obtained by solving the pseudo-inverse of the above matrix equation 5.19. A solution exists in equation 5.20 when the number of effectors is greater than or equal to the number of degrees of freedom to be controlled (i.e., the number of rotations plus the number of translations). That is, when all control directions are spanned by the effectors, and the matrix $[A]$ has full rank. A typical mixing-logic matrix has the following form and it translates the demanded rotational (DR_o) and translational (DA_c) accelerations to effector deflection commands $\underline{\delta}_{com}$.

$$\underline{\delta}_{com} = D_{max} A^T \left[AA^T \right]^{-1} \left(\frac{\delta R_o}{\delta A_c} \right) \quad or \quad \begin{bmatrix} \delta_{y1} \\ \delta_{z1} \\ \delta_{y2} \\ \delta_{z2} \\ \delta_{y1} \\ \delta_{y2} \\ D_{Th1} \\ D_{Th2} \\ D_{Th3} \\ \delta_{cs1} \\ \delta_{cs2} \end{bmatrix} = \begin{bmatrix} X & X & X & X & X & X \\ X & X & X & X & X & X \\ X & X & X & X & X & X \\ X & X & X & X & X & X \\ X & X & X & X & X & X \\ X & X & X & X & X & X \\ X & X & X & X & X & X \\ X & X & X & X & X & X \\ X & X & X & X & X & X \\ X & X & X & X & X & X \\ X & X & X & X & X & X \end{bmatrix} \begin{pmatrix} DP \\ DQ \\ DR \\ DA_X \\ DA_Y \\ DA_Z \end{pmatrix}$$

$$(5.20)$$

If we substitute $(\delta = \delta_{com})$ in equation 5.19 we end up with equation 5.21 which implies that the change in vehicle accelerations achieved are equal to the accelerations demanded by the flight control system. This can be achieved in both rotational (δR_0) and translational (δA_c) directions, assuming, of course, that the vehicle has the necessary effectors that can provide accelerations along the directions demanded. This is equivalent to premultiplying the vehicle dynamics with the mixing-logic matrix, as shown in Figure (5.2), where the resulting acceleration is equal to the acceleration demanded by the FCS.

$$\left(\frac{\delta R_o}{\delta A_c}\right) = \left(\frac{\delta R_o}{\delta A_c}\right)_{Dem}$$

(5.21)

5.2   MIXING-LOGIC EXAMPLE

The mixing-logic function is better illustrated with an example. The flight vehicle used is a rocket plane shown in Figure 5.5 that is controlled by different types of effectors. It has two main engines of 60,000 (lb) of thrust each, that gimbal in pitch and yaw and they also throttle ±30% from nominal thrust. It has five aerosurfaces: two inboard and two outboard elevons and a vertical rudder. It also has two sets of analogue RCS thrusters, one pair thrusting in the ±Z direction and the second pair thrusting in the ±Y direction. The RCS thrusts are proportional to the throttle commands that vary between 0 and ±1 producing a maximum thrust of ±3000 (lb). This vehicle obviously has enough effectors to be controlled in all six directions, three rotational and three translational. Selecting, however, an effector combination logic by inspection is not very easy and we shall, therefore, use the mixing-logic algorithm.

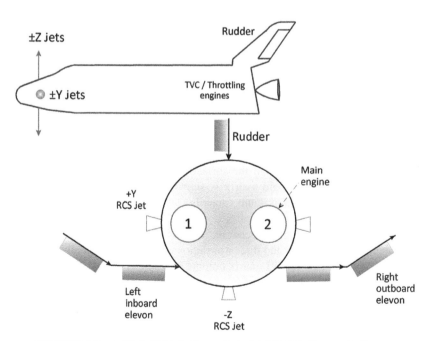

FIGURE 5.5   A Flight Vehicle Controlled with Multiple Types of Effectors

A=

```
1.0e+004 *

     0    0.0001        0        0        0    0.0001        0        0        0        0
     0   -0.0000        0        0        0    0.0000        0   -0.0001        0        0
     0         0        0    0.0001        0         0        0        0        0        0
     0         0        0   -0.0000        0         0   -0.0000        0   -0.0000   0.0000
     0         0        0        0        0    0.0001        0        0        0        0
     0    0.0000        0        0        0   -0.0000        0   -0.0000        0        0
     0         0   0.0000    0.0001        0         0   0.0000        0    0.0000   0.0000
0.0000    0.0001        0        0        0   -0.0001        0    0.0000        0        0
     0         0   1.8638        0        0         0  -1.8638        0        0  -0.0000
     0         0  -0.0015   -0.0000        0         0  -0.0164        0    0.0000  -0.0000
```

B=

```
Columns 1 through 11                                                                                 Columns 12 through 13

      0         0        0        0        0         0        0         0         0        0         0          0         0
 0.4866   -0.4866   0.1393   0.1393   0.0002   -0.0002        0    0.0137    0.2282   0.0304   -0.0304    -0.2282    0.1157
      0         0        0        0        0         0        0         0         0        0         0          0         0
-0.3865   -0.3865        0        0  -0.0154   -0.0154  -0.0221    0.0000   -0.0027  -0.0130   -0.0130    -0.0027   -0.0000
      0         0        0        0        0         0        0         0         0        0         0          0         0
 0.0096   -0.0096  -0.3715  -0.3715   0.0516   -0.0516        0    0.0216    0.0086   0.0013   -0.0013    -0.0086   -0.0255
-0.0003   -0.0003        0        0  -0.0003   -0.0003   0.0000         0   -0.0000  -0.0000   -0.0000    -0.0000    0.0000
      0         0   0.0004   0.0004        0         0        0    0.0000   -0.0000  -0.0000         0     0.0000    0.0000
      0         0        0        0        0         0        0         0         0        0         0          0         0
-5.0813   -5.0813        0        0   6.2502    6.2502   0.2585   -0.0000   -0.4142  -0.6812   -0.6812    -0.4142   -0.0158
```

C =

```
Columns 1 through 9                                            Column 10

     0   -0.0018        0        0        0    0.0006        0   -0.5258        0          0
     0         0        0   -0.0010        0         0  -0.0157         0   -0.0001     0.0000
     0    0.0000        0        0        0   -0.0003        0   -0.0216        0          0
     0         0        0        0        0         0   0.5651         0    0.0013    -0.0000
     0   -0.0021        0        0        0    0.0015        0   -2.3670        0          0
     0         0        0   -0.0029        0         0 -12.7690         0    0.0336    -0.0003
```

D =

```
Columns 1 through 9                                            Columns 10 through 13

 0.4866   -0.4866   0.1393   0.1393   0.0002   -0.0002        0    0.0137    0.2282   0.0304   -0.0304   -0.2282    0.1157
-0.3865   -0.3865        0        0  -0.0154   -0.0154  -0.0221    0.0000   -0.0027  -0.0130   -0.0130   -0.0027   -0.0000
 0.0096   -0.0096  -0.3715  -0.3715   0.0516   -0.0516        0    0.0216    0.0086   0.0013   -0.0013   -0.0086   -0.0255
 0.1404    0.1404        0        0   8.0412    8.0412  -0.0000   -0.0000   -0.0226  -0.0257   -0.0257   -0.0226   -0.0206
      0         0   8.0412   8.0412        0         0        0    0.4021   -0.1852        0         0    0.1852    0.3087
-8.0412   -8.0412        0        0   0.1404    0.1404   0.4021         0   -0.6175  -1.0291   -1.0291   -0.6175         0
```

FIGURE 5.6   Vehicle Dynamic Model in State-Space Form

Figure 5.6 shows a state-space representation of the vehicle dynamic model. It has 13 inputs that correspond to the 13 effectors: 2 pitch and 2 yaw main engine deflections, 2 main engine throttles, 2 RCS throttles, and 5 aerosurfaces. The outputs are six accelerations: three rotational and three translational. The dynamic pressure is low in this flight condition and the surfaces can use some help from the RCS and the TVC. The mixing logic matrix is shown in Figure 5.7, and it is derived from the vehicle data. The maximum capability of each effector is also taken into consideration. The mixing-logic matrix combines the 13 effectors to produce the accelerations demanded in all 6 directions. The inputs to the mixing matrix are the six demands, three rotational and three translational that would normally come from the flight control system. The 13 matrix outputs go to the effectors. They are engine and surface deflection commands in (rad), and also throttle commands which vary between 0 and ±1. Note that the throttle input is not thrust but it represents the percentage of thrust variation above or below nominal and its magnitude should not exceed 1. The value of the actual thrust is already included in the dynamic model.

Figure 5.8 is an open-loop simulation model similar to the diagram shown in Figure 5.2. It contains the vehicle state-space model in series with the mixing matrix

Flight control acceleration demands rotational and translational

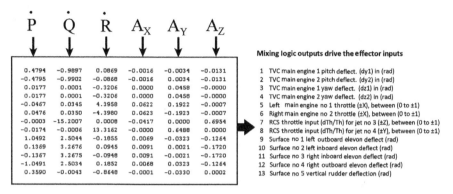

$$\dot{P} \quad \dot{Q} \quad \dot{R} \quad A_X \quad A_Y \quad A_Z$$

Mixing logic outputs drive the effector inputs

0.4794	-0.9897	0.0869	-0.0016	-0.0034	-0.0131
-0.4795	-0.9902	-0.0868	-0.0016	0.0034	-0.0131
0.0177	0.0001	-0.3206	0.0000	0.0458	-0.0000
0.0177	0.0001	-0.3206	0.0000	0.0458	-0.0000
-0.0467	0.0345	4.3958	0.0622	0.1922	-0.0007
0.0476	0.0350	-4.3980	0.0623	-0.1923	-0.0007
-0.0003	-15.2007	0.0008	-0.0417	0.0000	0.6954
-0.0174	-0.0006	13.3162	-0.0000	0.6488	0.0000
1.0492	2.5044	-0.1855	0.0069	-0.0323	-0.1264
0.1369	3.2676	0.0945	0.0091	0.0021	-0.1720
-0.1367	3.2675	-0.0948	0.0091	-0.0021	-0.1720
-1.0491	2.5034	0.1852	0.0068	0.0323	-0.1264
0.3590	-0.0043	-0.8648	-0.0001	-0.0330	0.0002

1 TVC main engine 1 pitch deflect. (dy1) in (rad)
2 TVC main engine 2 pitch deflect. (dy2) in (rad)
3 TVC main engine 1 yaw deflect. (dz1) in (rad)
4 TVC main engine 2 yaw deflect. (dz2) in (rad)
5 Left main engine no 1 throttle (±X), between (0 to ±1)
6 Right main engine no 2 throttle (±X), between (0 to ±1)
7 RCS throttle input (dTh/Th) for jet no 3 (±Z), between (0 to ±1)
8 RCS throttle input (dTh/Th) for jet no 4 (±Y), between (0 to ±1)
9 Surface no 1 left outboard elevon deflect (rad)
10 Surface no 2 left inboard elevon deflect (rad)
11 Surface no 3 right inboard elevon deflect (rad)
12 Surface no 4 right outboard elevon deflect (rad)
13 Surface no 5 vertical rudder deflection (rad)

The mixing logic matrix translates the acceleration demands to effector deflections and throttle commands

FIGURE 5.7   Mixing-Logic Matrix Inputs and Outputs

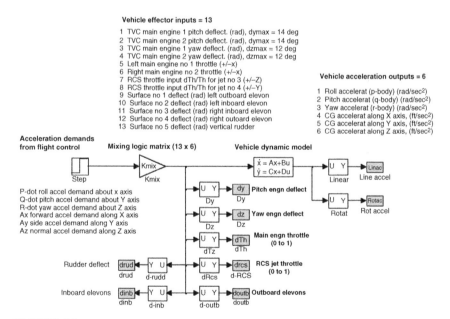

FIGURE 5.8   Open-Loop Simulation Used to Test the Effectiveness of the Mixing-Logic Matrix

and it is used for testing the mixing matrix. The input is a vector of six acceleration demands supposedly coming from the FCS, three rotational and three translational as already described. The mixing-logic matrix transforms the demands to 13 effector commands that become inputs to the state-space dynamic model of Figure 5.6. The dynamic model produces vehicle accelerations, three rotational and three translational.

If the mixing-logic matrix is properly designed, the accelerations generated by the dynamic model will be equal to the step accelerations demanded by the FCS. We run the simulation for a short time because it will eventually diverge due to the aerodynamics, because it is open-loop and is receiving a step command. A short time is sufficient to measure the accelerations produced. The simulation shows that the plant is perfectly diagonalized when it is premultiplied with the mixing matrix. From the accelerations' view point it becomes a perfect 6×6 identity matrix. Any combination of acceleration demands produces identical vehicle accelerations, as shown in the simulation results.

In Figure 5.9 the vehicle model is commanded open-loop to accelerate in +roll and −yaw simultaneously. The translational acceleration demands are zero. The results

System and effector responses to 1 (deg/sec²) roll acceleration, and to −1 (deg/sec²) yaw acceleration simultaneous demands

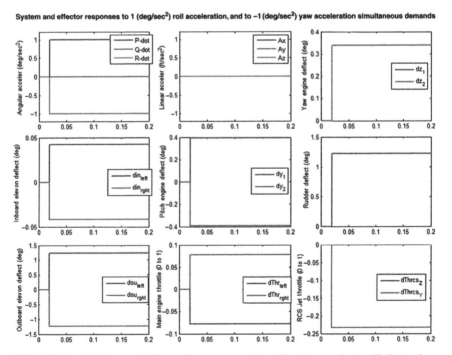

FIGURE 5.9   Vehicle and Effector Responses to Two Simultaneously Applied Acceleration Demands: 1 (deg/sec²) Roll Acceleration Demand and −1 (deg/sec²) Yaw Acceleration Demand

show that the accelerations produced are equal to the demanded accelerations in all directions.

- The two yaw TVC gimbals (δz) rotate in the positive yaw direction, and also the rudder rotates positive to generate the required negative yaw acceleration. The negative yaw acceleration is also assisted by the throttling yaw RCS jets that produce a force in the $-Y$ direction. The $-$yaw acceleration is also slightly assisted by the differential throttling of the two main engines (left engine throttles down, right engine throttles up).

- The $+$roll acceleration is produced by differentially deflecting the 2 main engines in pitch (δy), the inboard elevons, and the outboard elevons (left side down, right side up). The outboard elevons deflect more than the inboard, because they are obviously more effective in roll.

Figure 5.10 shows the open-loop system and effector response to one rotational and two translational step acceleration demands which are applied simultaneously. That is: 1 (deg/sec^2) pitch acceleration demand, 1 (ft/sec^2) axial acceleration demand,

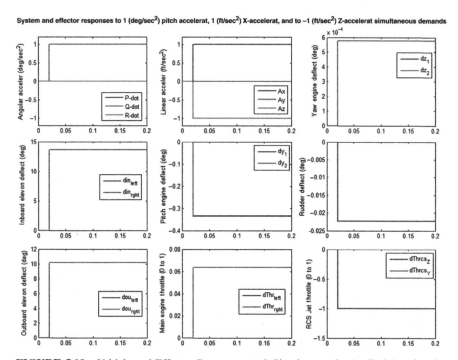

FIGURE 5.10 Vehicle and Effector Responses to 3 Simultaneously Applied Acceleration Demands: 1 (deg/sec^2) Pitch Acceleration Demand, 1 (ft/sec^2) Axial Acceleration Demand, and -1 (ft/sec^2) Normal Acceleration Demand

and -1 (ft/sec^2) normal acceleration demand. All accelerations produced are equal to the demanded accelerations.

- The negative normal acceleration demand causes all four elevons to deflect symmetrically in the positive direction (down). The inboard elevons are more effective and they deflect further than the outboard. The RCS jet is throttling heavily in the $-Z$ direction assisting in the $-A_z$ acceleration.
- The increase in the axial acceleration $+A_x$ is produced by the positive throttling of the two main engines.
- The +pitch acceleration is produced by deflecting the two TVC engines symmetrically in the negative pitch direction ($-\delta y$).
- The forward firing RCS thruster in the $-Z$ direction is also helping in pitch acceleration.

5.3   SPACE SHUTTLE ASCENT ANALYSIS EXAMPLE

The material presented in the previous sections will be illustrated with a Space Shuttle example where we will analyze the flight control system stability and performance during ascent. The flight condition is at 55 sec after lift-off during first stage where the dynamic pressure is at maximum (Max-Q). During stage-1 the shuttle propulsion system consists of three main engines (SSME), and two solid rocket boosters (SRB). The vehicle attitude and its flight direction are controlled by gimbaling the five

FIGURE 5.11   Space Shuttle During Ascent

engines in pitch and yaw. The attitude, rate, and acceleration are measured by an IMU, rate gyros, and accelerometer sensors, respectively, which are located near the top section of the SRBs. Roll attitude and rates are measured by sensors located in the orbiter vehicle. Inside the orange external tank (ET) there are two large propellant tanks containing the liquid oxygen (LOX) located in the front section of the ET and liquid hydrogen (LH2) in the aft section. During Max-Q the tanks are partially filled and propellant sloshing is a potential problem. If slosh is not properly taken care, it may destabilize the vehicle by creating oscillatory disturbance forces that would cause the TVC engines to limit cycle and to throw the vehicle off-course. The control surfaces are not used during ascent but they are scheduled open-loop to minimize loading on the aerosurface actuators. We will present a detailed, step-by-step analysis, of the Shuttle vehicle in this critical Max-Q condition, beginning with rigid-body analysis and then gradually introduce more complexity in the dynamic model, such as tail-wags-dog (TWD) and slosh. Structural flexibility is also significant and is included in the analysis because it is excited by the gimbaling of the TVC nozzles. This includes the dynamic coupling between the nozzles, actuators, and the supporting flex structure. Aeroelasticity, however, is not significant during ascent and it is not included. The pitch and lateral axes will be analyzed separately. The detailed analysis, data, and files for this example are not included in the book but they can be downloaded from Flixan.com/Shuttle Ascent Example.

5.3.1 Pitch Axis Analysis

The pitch axis analysis begins by developing rigid-body dynamic models, the flight control system (which is continuous to begin with), and include simple actuator models for the two SRBs and the three SSMEs. TWD and actuator load torques are not included at this point, because our main purpose is to make sure that the vehicle is stabilizable and achieves a reasonable performance. The performance, however, in terms of command following is not expected to be great at Max-Q because the load-relief feedback from the normal accelerometer degrades the response to guidance commands for the purpose of reducing the normal loads. The performance in this flight condition is measured by the vehicle capability to respond to wind gusts by turning toward the airflow and reducing the aerodynamic angles, rather than how well it performs to commands. The guidance increment commands are typically zero or very small during high-Q and the vehicle maintains a constant attitude at zero alpha and beta. Our analysis at this point will be limited to checking out the vehicle stability margins by performing classical frequency response analysis using Bode and Nichols plots and a transient response to wind-gust excitation. The open-loop block diagram system shown in Figure 5.12 is used for the frequency response calculation having the loop opened at the TVC output which is also the actuators input. A common SSME actuator transfer function block is used for the three SSMEs and a common SRB actuator is used for the two SRBs. The TVC gains, which are normally in the flight control system output, are shown between the actuators and the vehicle dynamics.

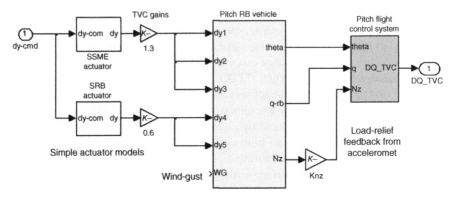

FIGURE 5.12   Simple Block Diagram Used for Open-Loop Frequency Response Analysis

5.3.1.1 Pitch Flight Control System   Figure 5.13 shows a simplified version of the Space Shuttle pitch flight control system during ascent. The inputs are vehicle pitch attitude error Θ in (rad), pitch rate q in (rad/sec) coming from the SRB rate gyro, and normal acceleration N_z in (ft/sec^2). The output DQ_TVC is the command that drives the pitch actuators via the TVC. The gains are designed for this Max-Q flight condition and the load-relief gain is at its max level in order to enhance the load-relief function. Flexibility filters are included which attenuate vibrations from the sensor signals. Notice that there is rate gyro cross-feed into the accelerometer feedback. This is because the accelerometer is not located at the vehicle CG and it is picking up rotational components instead of pure normal acceleration. The properly compensated signal from the SRB rate gyro takes out the rotational component from the accelerometer signal which performs the load-relief function.

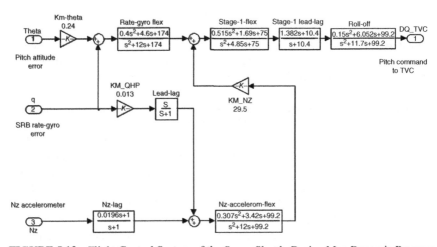

FIGURE 5.13   Flight Control System of the Space Shuttle During Max Dynamic Pressure

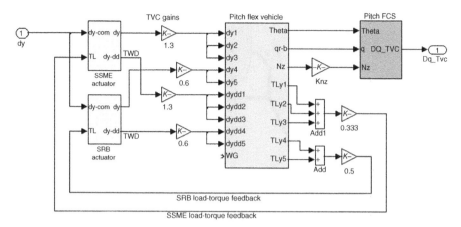

FIGURE 5.14 Analysis Model That Includes Tail-Wags-Dog and Load-Torque Feedback, Used to Calculate the Pitch System Frequency Response

5.3.1.2 Including the TWD Dynamics

Having achieved an acceptable stability using the rigid-body model, our next step is to upgrade the vehicle and actuator models by including gimbal accelerations that excite the TWD dynamics. We will also include the load-torque feedback which is a mechanical feedback loop from the gimbals to the actuator load-torque inputs. This is a gimbal torque generated by the swiveling of the nozzles and it is reacting against the control torques generated by the actuators. The frequency response analysis is repeated using an improved open-loop Simulink model, shown in Figure 5.14. The new model uses upgraded actuators which include load-torque inputs and gimbal acceleration outputs that couple with the vehicle dynamics as it is described in Chapter 4. The vehicle model is also upgraded by turning on the TWD in the TVC engines. This introduces gimbal acceleration inputs, in addition to the gimbal deflection inputs, that will connect to the actuator outputs and create the TWD dynamic effect. The outputs from the vehicle block now include load torques at the five TVC gimbals that close the load-torque feedback loops with the actuators, as shown in Figure 5.14. This model is now used to calculate the open-loop frequency response plot in Figure 5.16 which shows the TWD frequency dip at 26 (rad/sec). The TWD frequency is defined to be the frequency at which the normal component of the thrust force is cancelled at the gimbal by the inertial reaction force, see Fig. 5.15. When the engine oscillates at that frequency the magnitude of the engine inertia reaction force at the hinge is equal and opposite to the magnitude of the lateral component of thrust and the forces cancel. This TWD phenomenon introduces a complex pair of zeros in the transfer function "$\theta(s)/\delta(s)$", which also causes a 180° phase reversal to occur at frequencies greater than the TWD frequency.

Below this so called "TWD" frequency, the resultant lateral force at the gimbal is predominantly due to thrust being in phase with the gimbal angle. That is, an increase in gimbal angle results in an increase in lateral force. Above this frequency, the engine inertia forces produce the dominant lateral force which is in phase with the gimbaling

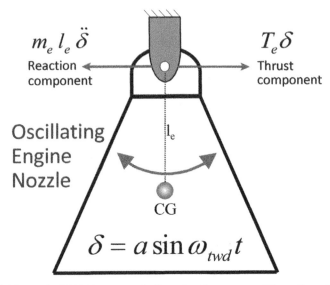

FIGURE 5.15 At the TWD Frequency the Reaction Component of Force Cancels the Thrust Component

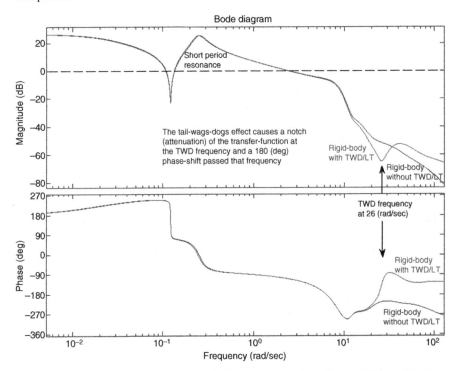

FIGURE 5.16 Frequency Response Analysis of the Open-Loop System With and Without Tail-Wags-Dog and Load-Torque Feedback. The TWD causes additional attenuation and phase-shift at the TWD frequency.

acceleration, or 180° out of phase with the gimbal angle. A system designed without the TWD consideration may perform unsatisfactorily above the TWD frequency.

In particular, some of the higher frequency flex modes may be driven into divergent oscillations by this phase reversal if adequate structural damping or filter attenuation is not present. The TWD frequency should, therefore, be higher than the control system bandwidth. Fortunately, the TWD phenomenon provides a significant amount of attenuation at around the TWD frequency which helps in attenuating the flex modes. For a typical launch vehicle it is calculated from equation 5.22. In this example, using a combined engine thrust ($T_e = 6 \times 10^6$ lb), a total engine mass ($m_e = 2800$ slugs), and an average distance ($l_e = 3.1$ ft) between the engine CG and its pivot point, which is assumed positive when the pivot is ahead of the engine center of mass, we obtain a TWD frequency equal to 26.3 (rad/sec). The Nichols plot in Figure 5.17 shows the vehicle stability margins and the effects of the TWD in gain and phase.

$$\omega_{TWD} = \sqrt{\frac{T_e}{m_e l_e}} = \sqrt{\frac{6 \times 10^6}{2800 \times 3.1}} = 26.3 \quad (\text{rad/sec}) \qquad (5.22)$$

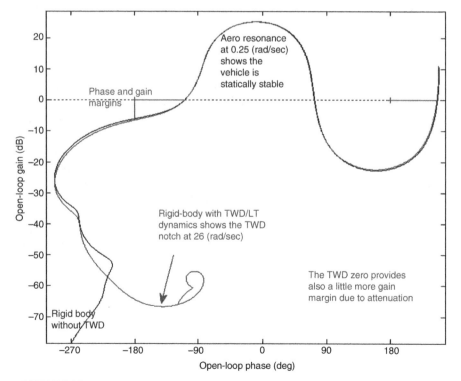

FIGURE 5.17 Nichols Plot Showing the Phase and Gain Margins of the Rigid-Body Space Shuttle System, With and Without TWD and Load-Torque Feedback

5.3.1.3 Load Relief The purpose of the accelerometer feedback in the flight control system, in Figure 5.14, is to reduce the normal acceleration load during a steady wind-shear disturbance. The load-relief activity is phased in and out proportionally with the dynamic pressure by adjusting the gain KM_NZ. Obviously, at high dynamic pressures the command following function of the FCS is compromised for the benefit of reducing alpha and the normal load. Figure 5.18 shows the response of the system to a steady wind-shear disturbance and the effects of the load-relief feedback to reduce alpha and N_Z. It is compared against the response when the load-relief gain is set to zero. The load relief provides nearly 40% more steady-state reduction in alpha and N_Z which are indicators of normal aerodynamic loading. This is achieved at the expense of an increase in attitude error, which is commanded at zero. It means that the vehicle may temporarily deviate from its commanded direction during this high-Q region as a result of the wind disturbance but its flight path will eventually be corrected by the closed-loop guidance when the dynamic pressure is reduced. The load-relief feedback, however, aggravates the transient responses in alpha and N_Z which is not very desirable.

5.3.1.4 Including Propellant Sloshing in the Analysis Model The next step is to include the propellant sloshing dynamics for the liquid oxygen and the liquid hydrogen tanks, as was described in Sections 2.6 and 3.8. Figure 5.19 shows a simplified diagram of the Space Shuttle vehicle with the SSME and SRB thrust vectoring engines and the two sloshing tanks. The oscillating slosh masses represent the sloshing portion of the propellant. Notice that the slosh mass deflections and the liquid surfaces are not exactly perpendicular to the tank centerline. This is because the acceleration vector A_T is not exactly aligned with the vehicle x-axis. Figure 5.20 is the open-loop frequency response of the new system against the previous case that did not include slosh. It shows the TWD zero and the two slosh resonances. Notice that the liquid oxygen resonance is big but the liquid hydrogen resonance is smaller because the slosh mass is also smaller. The overall response is slightly attenuated because of the additional slosh weight included. Figure 5.21 is a Nichols plot of the pitch axis open-loop system. It shows the phase and gain margins and the three pitch resonances. The big loop is caused by the short-period resonance at 0.25 (rad/sec). The LOX resonance is also big at 3.13 (rad/sec). The small resonance is from the LH2 propellant at 2.57 (rad/sec). They are all phase stable because the bubbles are opening in the direction opposite to the −1 point, which on a Nichols plot is the $(\pm 180°, 0\ dB)$ point.

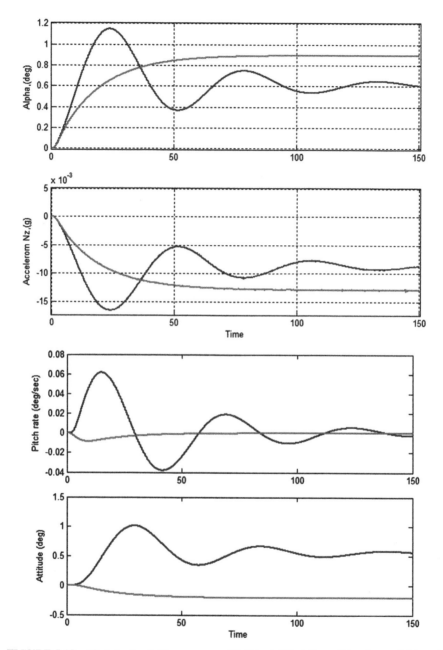

FIGURE 5.18   Pitch System's Response to a 30 (ft/sec) Wind-Shear Disturbance With and Without Load Relief

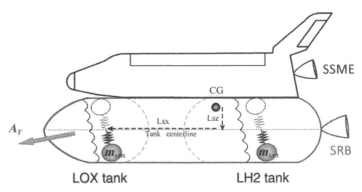

FIGURE 5.19 Simple Diagram of the Space Shuttle During First Stage Showing the TVC Engines and Sloshing Tanks

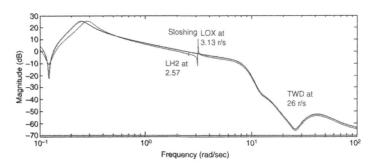

FIGURE 5.20 Frequency Response of the Pitch System Including Propellant Sloshing and TWD Dynamics

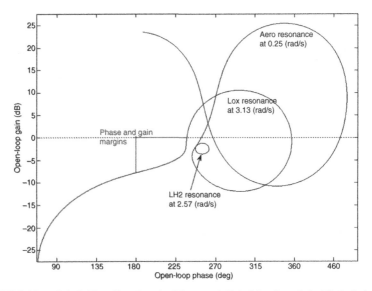

FIGURE 5.21 Nichols Plot Showing the Phase and Gain Margins of the Pitch Axis Space Shuttle System Including Propellant Sloshing, TWD and Load-Torque Feedback

5.3.1.5 *Adding Structural Flexibility* Our next step is to include structural flexibility in the vehicle dynamic model and to repeat the stability analysis by calculating the open-loop frequency response, as before. Several dominant modes were selected from a finite elements model that contains hundreds of flex modes. The strongest modes are graphically selected interactively in the mode selection process using a bar plot that shows the modal strength in the intended directions, see Figure 5.22. The selected pitch modes are shown in light gray. The first six modes in a typical Nastran output are rigid-body modes and they are not included among the selected structural modes. Figure 5.23 shows the open-loop frequency response of the pitch axis system with flexibility against the previous case that did not include flexibility. Note that the structural flexibility filters are not yet included in the flight control system and some of the flex modes are unstable because they peak above the 0 dB line and some encircle the critical + point. A closed-loop simulation of this system would show oscillatory flex mode instability.

5.3.1.6 *Including the Structural Flexibility Filters* The pitch axis stability analysis concludes by including structural flexibility filters in the flex vehicle model. The filters will further attenuate the structural modes and prevent instabilities. The continuous flight control system was also replaced with a discrete FCS sampled at 40 msec. The Bode and Nichols plots in Figure 5.24 show the resonances and the control system phase and gain margins. Notice that all flex modes are now gain stable. That is, their peaks are below the 0 dB horizontal line. However, the LOX slosh mode and the short-period frequencies are only phase stable.

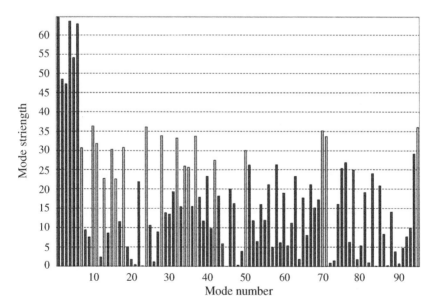

FIGURE 5.22 Flex Modes Are Selected Based on the Modal Strength That Is Calculated Between Actuator and Sensor Locations and Specific Directions; The First Six Rigid-Body Modes Are Not Included

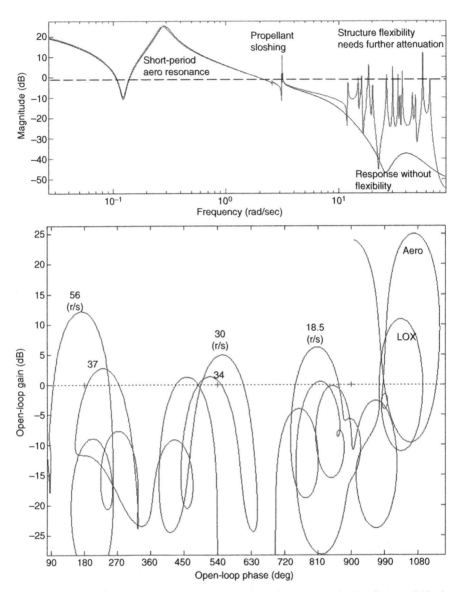

FIGURE 5.23 Bode and Nichols Plots of the Pitch Axis System Including Structural Flexibility and a Flight Control System Without Flex Filters, Showing That Some of the Flex Modes Are Unstable

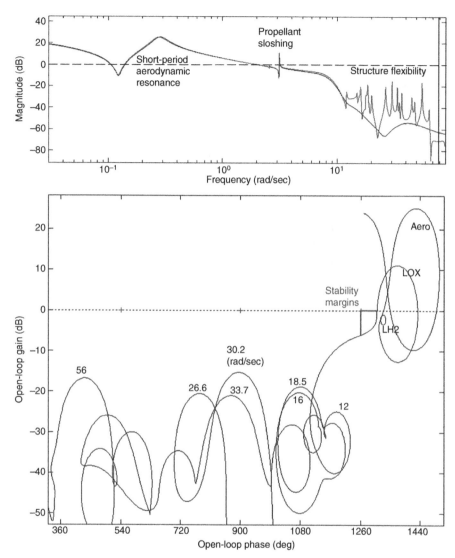

FIGURE 5.24   Bode and Nichols Plots of the Space Shuttle Pitch Axis System Including Structural Flexibility, Flex Filtering, and a Discrete Flight Control System

5.3.2   Lateral Axes Flight Control System

The lateral flight control system controls vehicle attitude in the roll and yaw directions, which are coupled due to the aerodynamics generated by the vertical stabilizer and the cross product of inertia I_{xz}. The control system, therefore, has a cross-feed from the yaw TVC command to the roll rate input command, as shown in Figure 5.25. The inputs to the flight control system are: yaw and roll attitude errors (Ψ_e, Φ_e) in

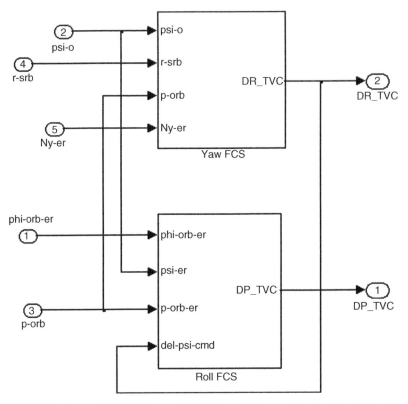

FIGURE 5.25 Lateral Shuttle Ascent Flight Control System Showing the Roll and Yaw Cross-Coupling

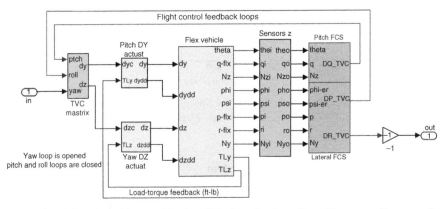

FIGURE 5.26 Block Diagram Used to Calculate the Open-Loop Frequency Response in Yaw, Yaw Loop Is Opened at the TVC Input, Pitch and Roll Loops Are Closed

(rad), yaw, and roll rate (r, p) in (rad/sec), and lateral acceleration (N_y) in (ft/sec^2). The outputs are roll and yaw flight control demands (DP_TVC and DR_TVC) that drive the TVC matrix which converts it to TVC deflection commands. The roll and yaw flight control systems are shown in detail in Figure 5.27. The block diagram system in Figure 5.26 is used for analyzing control loop stability (yaw loop in this case). It calculates the frequency response of the system with the yaw loop opened at DR_TVC. Roll stability is calculated likewise by closing the pitch and yaw loops and opening the roll loop at the DP_TVC output. Figures (5.28) and (5.29) are Bode and Nichols plots showing the stability margins of the roll and yaw loops respectively, including Slosh, Flexibility, and TWD.

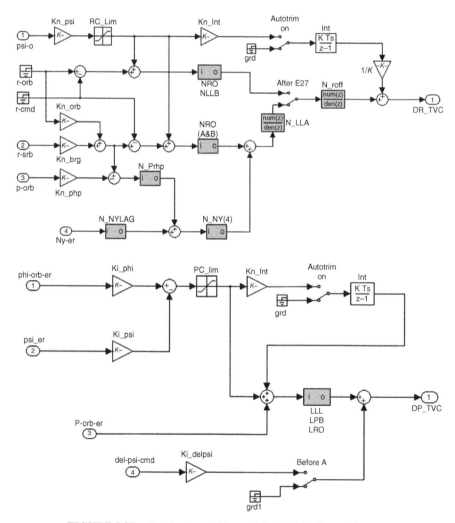

FIGURE 5.27   Shuttle Ascent Yaw and Roll Flight Control Systems

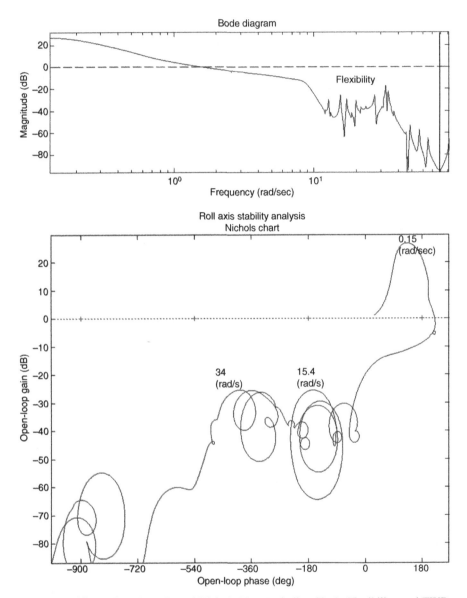

FIGURE 5.28   Roll Axis Body and Nichols Plots Including Slosh, Flexibility, and TWD, Showing Stability Margins

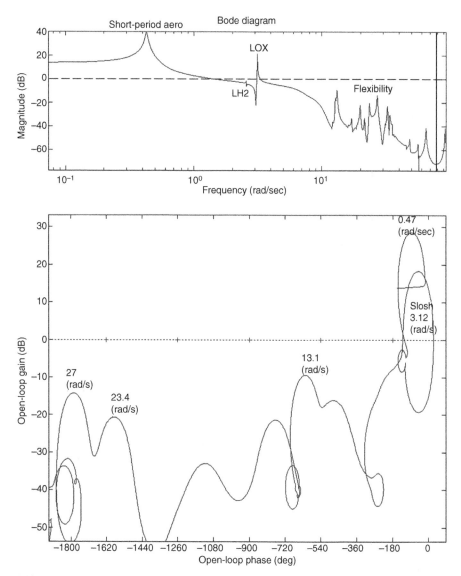

FIGURE 5.29   Yaw Axis Body and Nichols Plots Including Slosh, Flexibility, and TWD, Showing Stability Margins

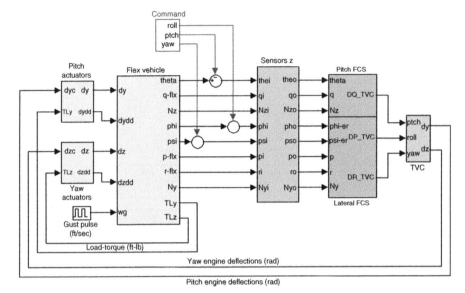

FIGURE 5.30   Block Diagram of the Combined Pitch and Lateral Closed-Loop System

5.3.3   Closed-Loop Simulation Analysis

The closed-loop block diagram system in Figure 5.30 is used for analyzing the Space Shuttle system response to attitude commands and to wind-gust disturbances. The vehicle dynamic model includes both: pitch and lateral dynamics coupled, with TWD, load torques, flexibility, and propellant sloshing. In the simulation case shown below, a wind-gust pulse is applied perpendicular to the vehicle x-axis, and skewed at 45° between the $+Y$ and the $+Z$ axes to excite both pitch and lateral directions.

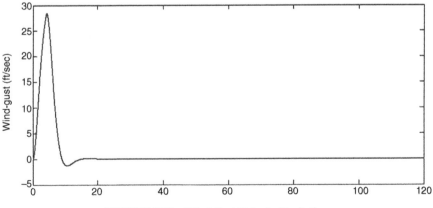

FIGURE 5.31   Wind-Gust Velocity Excitation

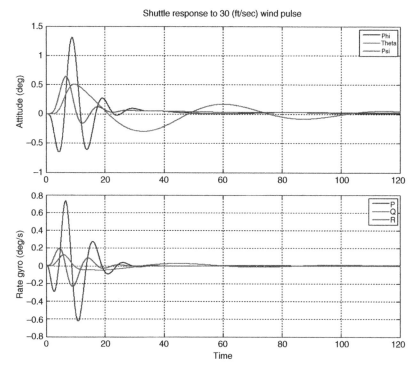

FIGURE 5.32   Roll, Pitch, and Yaw Vehicle Attitude and Body Rates

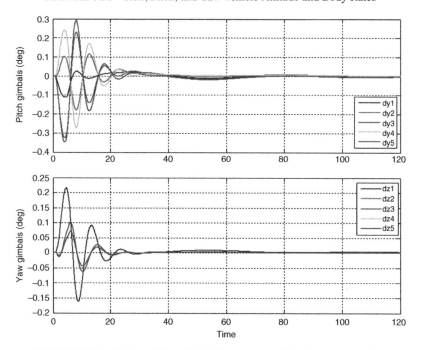

FIGURE 5.33   Pitch and Yaw Gimbal Responses of the Five TVC Engines

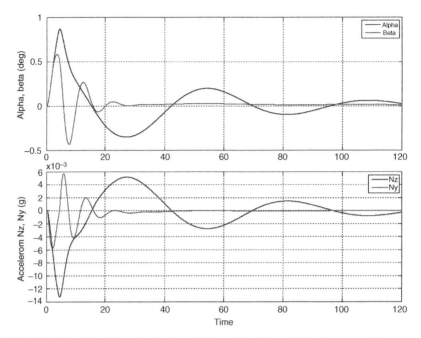

FIGURE 5.34 Aerodynamic Angles and Normal and Lateral Accelerometer Responses

The wind-gust disturbance causes transients in the angles of attack and sideslip which oscillate but the oscillations are eventually reduced by the flight control system. It also causes transients in the vehicle attitude, rates, and accelerometer responses. The engines respond by gimbaling in pitch and yaw. Notice that the wind disturbance causes a significant amount of roll oscillations. This problem will be addressed later in the control design Chapter 9.

6

TRIMMING THE VEHICLE EFFECTORS

Trimming is the process of balancing the moments and forces on a flight vehicle generated by aerodynamics and propulsion, with moments and forces which are generated by the control effectors, such as the engines and aerosurfaces. We begin by trimming the aerosurfaces of a standard aircraft in level flight, and then we present a method for trimming the effectors of a generic flight vehicle with multiple types of effectors along a pre-defined trajectory.

6.1 CLASSICAL AIRCRAFT TRIMMING

Trimming an aircraft requires solving the equations of motion in order to balance the forces and moments in the steady state. For a classical 1-g cruising, for example, to maintain a level flight in the longitudinal direction using an elevator, the lift force must balance the weight and the pitch moment must be zero.

$$C_{\text{Ltrim}} = C_{L0} + C_{L\alpha}\,\alpha_{\text{trim}} + C_{L\delta e}\delta_{\text{e-trim}}$$
$$C_m = 0 = C_{m0} + C_{m\alpha}\,\alpha_{\text{trim}} + C_{m\delta e}\delta_{\text{e-trim}} \tag{6.1}$$

If we recognize that,

$$\frac{C_{m\alpha}}{C_{L\alpha}} = \frac{\partial C_m}{\partial C_L} \quad \text{and} \quad C_{\text{Ltrim}} = \frac{W}{Q\,S_r} \tag{6.2}$$

Performance Evaluation and Design of Flight Vehicle Control Systems, First Edition. Eric T. Falangas.
© 2016 by Eric T. Falangas. Published 2016 by John Wiley & Sons, Inc.

where W is the aircraft weight. We can solve the equations for the elevator trim angle and the trim angle of attack for the classical 1-g cruise condition as shown in equation 6.3.

$$
\delta_{\text{e-trim}} = \frac{C_{m0} + \frac{\partial C_m}{\partial C_L}\left(\frac{W}{\overline{Q}S} - C_{L0}\right)}{-C_{m\delta e} + \frac{\partial C_m}{\partial C_L}C_{L\delta e}} \quad \text{and} \quad \alpha_{\text{trim}} = -\frac{\left(C_{m0} + C_{m\delta e}\delta_{\text{e-trim}}\right)}{C_{m\alpha}} \quad (6.3)
$$

In complex nonlinear vehicle models and in flight conditions that involve more states and controls the solution for obtaining steady-state values of controls and state vectors becomes difficult to obtain analytically. A trim program is used in this case that attempts to solve the simultaneous nonlinear equations of motion and to find a steady-state solution for the controls and states. It involves a numerical optimization algorithm that iteratively adjusts the independent variables in the equations until some solution criterion is satisfied. The solution may not be unique, for example, steady-state level flight at a given engine power can in general correspond to two different airspeeds and angles of attack.

One of the things to be determined is the steady-state conditions, how many of the state and control variables must be varied independently and what constrains must exist on the remaining variables. A computer program is written so the independent variables are adjusted by a numerical algorithm that solves the nonlinear equations, while the remaining variables are determined from the constraint equations. For a steady-state translational flight, for example, we should be able to specify altitude and velocity or flight path angle within the limits imposed by the engine power and aircraft configuration. We should also set the states ϕ, p, q, and r equal to zero, and expect the numeric algorithm to solve for a unique combination of control inputs (throttle, elevator, aileron, rudder) and for the remaining state variables, such as, alpha and theta or gamma to be determined by the algorithm. The sideslip angle β must also be adjusted by the algorithm to take out any side force. Another case to consider is steady-state turn flight where the variables ϕ, p, q, r are no longer zero. The turn can be specified by the rate of change of yaw heading, $\dot{\psi}$. The trim algorithm must solve the equations for the steady attitude θ and ϕ assuming a coordinated turn.

The trim algorithm determines the steady-state flight conditions by solving the nonlinear equations for the state and control vectors that make the state derivatives $(\dot{U}, \dot{V}, \dot{W})$ and $(\dot{P}, \dot{Q}, \dot{R})$ identically zero. A scalar cost function must be defined, like the square of the accelerations, and a function minimization algorithm is used to adjust the control variables and the adjustable state variables in order to minimize this scalar cost. Examples of function minimization algorithms are the IMSL routine, ZXMWD, and the SIMPLEX algorithm. In typical level flight longitudinal trim situation, we select the speed and altitude, set $q = 0$, and adjust the throttle, elevator controls, and the angle of attack state until it trims.

6.2   TRIMMING ALONG A TRAJECTORY

In situations where we are dealing with generic flight vehicles that use multiple types of effectors, it is not always easy to define ahead of time which state variables are to be set and which ones are to be adjusted. In many cases the vehicle state variables are defined before trimming by point mass trajectories. This simplifies the trimming approach by solving only for the controls along each trajectory. A trajectory is a table of flight variables calculated as a function of time, consisting of altitude, vehicle mass, angles of attack and sideslip, dynamic pressure, Mach number, velocity, acceleration, thrust. Trajectories are created by specialists, initially using point mass 3-degrees of freedom (3-DOF) trajectory optimization programs such as "POST®" or "OTIS®" ignoring rotational dynamics. Some of the criteria for shaping a trajectory are: heating, fuel efficiency, payload weight maximization, and structural loading. It captures the vehicle mission, the flight environment, and the maneuverability requirements and it also includes the vehicle moments and forces versus time. So we shall allow the trajectory experts to define the flight environment, which also defines the performance requirements, and the control experts must analyze the vehicle capability to trim and perform along the target trajectory. In addition to trimming, one of the concerns of the flight control analyst is static stability (or instability) which must be within an acceptable range, both statically and dynamically. The vehicle must also have sufficient control authority to maneuver along the required mission and to produce the necessary accelerations in order to counteract the estimated aerodynamic disturbances, as we shall see in Chapter 7.

As already mentioned, trimming is a static analysis, mainly concerned with using the vehicle effectors (surfaces, engines, and RCS) to balance the three basic moments (roll, pitch, and yaw) along a predefined trajectory, and in some cases to also include translational DOF, such as axial, normal, and side accelerations (A_x, A_z, and A_y). The base moments and forces, which are calculated by the trajectory program and are included in the trajectory, are generated by aerodynamics due to the angles of attack and sideslip and also by the engine thrusts. They must be balanced with the control moments and forces generated by the effectors, which are gimbaling or throttling engines, reaction control jets, and aerosurfaces. Additional data needed for trimming are basic aerodynamic coefficients, aerosurface increment coefficients, thrust vector control (TVC), and throttling engine data.

As a system the effectors must have the control authority to trim the vehicle and also to retain some additional control authority for other functions. If the effectors do not have the required authority to trim, then either the trajectory has to be modified, or the effectors, or the vehicle aerodynamics, or all of the above must be modified until an acceptable trimming condition is achieved. This is usually an iterative process that requires several attempts and perhaps modifications in the vehicle shape and resizing the aerosurfaces or the TVC. As a guideline, the control authority required for trimming should not exceed half of the maximum control capability of each effector, because there should be some extra control capability reserved for maneuvering the vehicle and for reacting against wind gusts and other disturbances. The uncertainties in the vehicle parameters and in the location of the CG may also cause further uncertainty in the trim angles. Balancing the three moments is usually the main objective when trimming the vehicle along a predefined trajectory. Sometimes it is also necessary for the effectors' system to have the control authority to trim along some of the linear accelerations, mainly along the x and z directions. For example, it may be necessary to balance the normal acceleration independently of the pitch moment during landing or when separating from another vehicle by using elevon, body flap, and vertical thrusters simultaneously. This would prevent overpitching the vehicle during separation. It is also possible to trim the axial acceleration along a specific trajectory independently of pitch, by means of a speed brake or by varying the engine thrust. Trimming is also important for sizing the effectiveness of the control surfaces, the TVC gimbaling capability, the engine thrusts, throttling capability, and the RCS jet thrusts. It is also used to determine optimal installation angles for the TVC engines, zero positions for the aerosurfaces, and the maximum weight and CG variations that can be tolerated when carrying payloads.

In this section we are presenting an iterative and interactive method for calculating optimal effectors trim positions along a trajectory, for a vehicle that includes multiple types of effectors. This method is based on pseudo-inversion of a matrix and it attempts to allocate control authority to effectors according to their control capability in specific directions. It calculates the trim position for each effector at each point along the trajectory as a function of time. In an ideal situation, when the trajectory is calculated correctly and the trimming along it is perfect, the moments and forces applied on the vehicle from the propulsion and base aerodynamics should perfectly match the accelerations which are defined in the trajectory, and the control deflections should be zero because no correction is needed. Otherwise, if the rotational and translational accelerations (calculated due to α, β, and thrust) do not match the trajectory's accelerations, the control effectors must be used to provide the additional moments and forces required to match the accelerations defined in the trajectory. The purpose of trimming is to properly adjust the aerosurfaces, the TVC deflection angles, and thrusts, as necessary according to their capabilities, in order to provide the additional aerodynamic and propulsion forces and moments on the vehicle and to match the angular and linear accelerations defined in the trajectory along the trim directions.

Equation 6.4 is a "Force = Mass × Acceleration" type of equation. On the right-hand side (RHS) there is a six-dimensional vector consisting of three rotational and three translational accelerations, which are defined in the trajectory. They are multiplied with the vehicle mass and moments of inertia to convert them to three moments and three forces. The moments and forces on the left side of this equation are due to the aerodynamics, propulsion, and known disturbance forces, and in order to trim they must balance the terms on the right side at each point along the trajectory. Note that in equation 6.4 the aerodynamic and propulsion moments and forces on the left side consist of both base vehicle plus effector moments and forces.

$$
\begin{bmatrix} M\ (aero) \\ F\ (aero) \end{bmatrix} + \begin{bmatrix} M\ (thrust) \\ F\ (thrust) \end{bmatrix} + \begin{bmatrix} M\ (disturb) \\ F\ (disturb) \end{bmatrix} = \begin{bmatrix} Inertia \times \dot{\omega}\ (traject) \\ Mass \times Acceleration\ (traject) \end{bmatrix}
$$

$$(6.4)$$

In order to solve this equation we must separate the base terms from the effector terms on the left side of equation 6.4. The base terms cannot be modified by the effectors because they are caused by the aerodynamic forces due to the angles of attack and sideslip and also by the constant propulsion forces of the engine thrusts applied on the base body. The only moments and forces that can be adjusted on the left side of this equation in order to match the terms on the right side and balance equation 6.4 are the contributions from the effector deflections and the thrust variations. They will determine the trim positions of the control surfaces or TVC engine deflections from installation (zero) positions and the thrust variations from nominal thrusts. The effectors should not only be capable of providing the control authority to balance the equation but they must also have sufficient controllability left for maneuvering and overcoming unexpected disturbances, such as wind shear. In a typical trajectory derived from a point mass simulation the angular accelerations in the trajectory are zero because it assumes that the vehicle moments are perfectly balanced and the effector trimming boils down to zeroing the moments on the left side of equation 6.4. When the rotational accelerations are available, however, either from a 6-DOF simulation or test data, they can be included to provide a more efficient trimming. In equation 6.4 the translational accelerations from the trajectory on the RHS are multiplied with the vehicle mass to calculate the total force on the vehicle along x, y, and z. The linear accelerations (A_x and A_z) in a typical trajectory are not zero, and in some cases we may want to use the effectors in order to trim along those directions and to match the translational accelerations, because the axial acceleration affects the range and the normal acceleration affects the altitude. When trimming along the translational directions we may have to use additional effector activity such as propulsion, TVC, and aerosurfaces in order to balance the forces.

In this section we will present a trimming algorithm for a flight vehicle with multiple types of effectors. The number of effectors must be greater than or equal to the number of DOF that must be balanced, plus all directions to be trimmed and eventually controlled should be accessible by at least one effector. When the number of effectors exceeds the number of DOFs or trim directions the solution is overdetermined. The more the effectors the better the controllability because they can be combined more efficiently to control the directions they can influence. Having an abundance of effectors is also good for redundancy. The trimming algorithm uses pseudo-inversion that gives preference to the effectors that are more effective in the demanded directions by allocating them bigger control authority and hence increased activity, than the effectors which are less capable along those directions and are, therefore, less active. All four types of effectors: gimbaling engines (pitch and yaw), throttling engines, RCS jets, and aerosurfaces, are combined by the trim algorithm and used as a system.

The trimming algorithm requires aerodynamic increment coefficients for each aerosurface and the orientation of each engine with respect to the vehicle. The direction of the engine thrust is defined by two angles (elevation and azimuth or simply pitch and yaw) which are measured relative to the $-x$ axis. An engine is either mounted at a fixed position relative to the vehicle or it can be gimbaled in the pitch and yaw directions with respect to its mounting position and capable to provide TVC. The thrust is either constant or it can be modulated up and down relative to the nominal thrust to provide throttle control. An engine may also be able to gimbal and throttle simultaneously. Aircraft engines, for example, are modeled as throttling engines having a nominal thrust (T_e), with a certain amount of thrust variation about T_e. Reaction control jets (RCS) are also considered to be throttling engines. They are mounted at fixed angles relative to the vehicle axes and their thrust can vary between zero and $\pm T_{max}$. For trimming purposes the RCS jets are not considered to be "on/off" devices, but they are continuous (analog) thrusters and negative thrusting is allowed. After all, the purpose of trimming is to size the thrusters and not to perform a dynamic simulation. A single thruster can be used to model a pair of back-to-back firing jets producing positive or negative forces as a function of the throttle control input. In order to solve the trimming equation numerically we must rewrite equation 6.4 by separating the moments and forces produced by each effector as consisting of two parts: a fixed part and an adjustable part. In the equations that follow we will write in detail the moments and forces produced by the engines, jets, and the aerosurfaces and separate them into two parts: (a) the steady-state part that is produced when the effector is at trim position or nominal thrust, and (b) the adjustable part due to the deflection or throttle that is used for controlling and trimming the vehicle.

6.2.1   Aerodynamic Moments and Forces

The aerodynamic moments and forces on the vehicle consist of two parts: (a) the aero moments and forces on the base body resolved along the body axes, assuming that the control surface positions are at zero, equation 6.5 and, (b) the moment and force increments produced by the control surface deflections, equation 6.6.

The base moments and forces are

$$
\begin{aligned}
L_{BX} &= \overline{Q}S_{ref}bC_l\,(M,\alpha,\beta) & F_{BX} &= -\overline{Q}S_{ref}C_A\,(M,\alpha,\beta)\\
M_{BY} &= \overline{Q}S_{ref}\overline{c}C_m\,(M,\alpha,\beta) & F_{BY} &= \overline{Q}S_{ref}C_Y\,(M,\alpha,\beta) & (6.5)\\
N_{BZ} &= \overline{Q}S_{ref}bC_n\,(M,\alpha,\beta) & F_{BZ} &= \overline{Q}S_{ref}C_Z\,(M,\alpha,\beta)
\end{aligned}
$$

where

$\overline{Q}$ is the dynamic pressure,
S_{ref} is the reference area,
$\overline{c}$ is the reference length or mean aerochord, and
b is the wing span.

The six basic aero coefficients (three moments and three forces) are nonlinear functions of Mach number M, and the angles of attack α and the sideslip β.

The control surfaces generate additional moment and force increments that are used to balance the moments and forces on the base vehicle along the trajectory. The aerodynamic moments and forces are functions of the aerosurface increment coefficients, the dynamic pressure, and the reference length, as shown in equation 6.6.

The aerosurface increment coefficients are nonlinear functions of four variables, for example, $C_m(\alpha, \beta, M, \Delta_{asi})$:

- The surface deflection from zero position (Δ_{asi}),
- The angles of attack and sideslip (α and β), and
- The Mach number (M),

$$
\begin{aligned}
L_{XSi} &= \overline{Q}S_{ref}\, b\left\{C_l\left(\alpha, \beta, M, \Delta_{asi}\right)\right\} & F_{XSi} &= -\overline{Q}S_{ref}\left\{C_A\left(\alpha, \beta, M, \Delta_{asi}\right)\right\} \\
M_{YSi} &= \overline{Q}S_{ref}\,\overline{c}\left\{C_m\left(\alpha, \beta, M, \Delta_{asi}\right)\right\} & F_{YSi} &= \overline{Q}S_{ref}\left\{C_Y\left(\alpha, \beta, M, \Delta_{asi}\right)\right\} \quad (6.6) \\
N_{ZSi} &= \overline{Q}S_{ref}\,b\left\{C_n\left(\alpha, \beta, M, \Delta_{asi}\right)\right\} & F_{ZSi} &= \overline{Q}S_{ref}\left\{C_Z\left(\alpha, \beta, M, \Delta_{asi}\right)\right\}
\end{aligned}
$$

In order to trim we must solve equation 6.6 for the surface deflections Δ_{asi} which are needed to balance the base moments and forces. It is not easy, however, to solve directly for the surface deflections Δ_{asi} because the equations are nonlinear and not explicitly available, but they are usually defined by wind-tunnel data. The equations are solved numerically by linearizing them at fixed (α, β, Δ_{asi}, M) for each control surface, and using the control surface derivatives $C_{m\delta asi}$, etc. to propagate the solution toward a deflection that will balance the vehicle moments and forces. The derivatives are calculated at each iteration, and are also functions of (α, β, Δ_{asi}, and M).

In addition to the base aero moments and forces described in equation 6.5, the moment/force increment from each aerosurface is solved by separating equation 6.6 into two parts, as shown in equation 6.7: (a) a steady-state term M_{S0i} representing the moments and forces at a fixed surface deflection Δ_{S0i}, and (b) a linear term representing additional moments and forces due to a small deflection increment δ_{asi} relative to the deflection Δ_{S0i}. In the aerodata base, the surface coefficients are measured or calculated at fixed discrete angles, and Δ_{S0i} represents the deflection of surface (i) that is nearest to the expected trim angle Δ_{asi}. The second term in the moment/force equation 6.7 is a linear derivative term that calculates an increment relative to the first term. It is scaled by dividing the deflection increment δ_{asi} with the max surface deflection (δ_{asiMax}). We must also multiply the coefficients in the second term with the max deflection (δ_{asiMax}). The input ($\delta_{asi}/\delta_{asiMax}$) in the normalized second term in equation 6.7 becomes nondimensional and it can only vary between zero and ± 1. It is this increment that we must solve for, during each iteration, in order to calculate the new surface deflection Δ_{S0i} that will eventually converge to the trim angle Δ_{asi}.

$$
M_{Si} = M_{S0i} + \left[DM_{si}\right]\left(\delta_{asi}\big/\delta_{asi_{MAX}}\right)
$$

$$
M_{Si} = \overline{Q}S_{ref}
\begin{bmatrix}
bC_l\left(m, \alpha, \beta, \Delta_{S0i}\right) \\
\overline{c}C_m\left(m, \alpha, \beta, \Delta_{S0i}\right) \\
bC_n\left(m, \alpha, \beta, \Delta_{S0i}\right) \\
-C_A\left(m, \alpha, \beta, \Delta_{S0i}\right) \\
C_Y\left(m, \alpha, \beta, \Delta_{S0i}\right) \\
C_Z\left(m, \alpha, \beta, \Delta_{S0i}\right)
\end{bmatrix}_{(i)}
+ \overline{Q}S_{ref}\,\delta_{asi_{MAX}}
\begin{bmatrix}
bC_{l\delta as} \\
\overline{c}C_{m\delta as} \\
bC_{n\delta as} \\
-C_{A\delta as} \\
C_{Y\delta as} \\
C_{Z\delta as}
\end{bmatrix}_{(i)}
\left(\delta_{asi}\big/\delta_{asi_{MAX}}\right)
$$

$$(6.7)$$

6.2.2   Moments and Forces from an Engine Gimbaling in Pitch and Yaw

The moments and forces on the vehicle generated by a single engine (i) are also nonlinear functions of the pitch and yaw deflection angles and they also depend on the thrust value. We will linearize the force equation produced by the ith engine and separate it into three parts:

1. The nominal moments and forces created by the nominal engine thrust T_{ei} and at fixed deflection angles (Δ_{Ei} and Δ_{Zi}) that correspond to the engine mounting or trim positions, as defined in Figure 5.3.
2. The moment and force increments generated due to small engine deflections in pitch and yaw (δ_{Yi}, δ_{Zi}) relative to the engine mounting positions, and
3. The additional moment and force increments generated by the variation in engine thrust $D_{Thr(i)}$ relative to its nominal value.

Each term in equation 6.8 is a six-dimensional vector consisting of three moments and three forces. The pitch and yaw engine deflections (δ_{Yi}, δ_{Zi}) in equation 6.8 are normalized by dividing with the maximum engine deflections. Similarly, the thrust variation inputs $D_{Thr(i)}$ are normalized by dividing with the maximum thrust variation of each engine $D_{ThrMax(i)}$. The controls for each engine are, therefore, scaled to vary between zero and ± 1, and we must solve for these increments at each iteration, in order to calculate the new engine trim positions and thrust value.

$$M_{E(i)} = M_{E0(i)} + \left[DM_{E(i)}\right] \left\{ \begin{matrix} \delta_{Y(i)}/\delta_{YMAX(i)} \\ \delta_{Z(i)}/\delta_{ZMAX(i)} \end{matrix} \right\} + \left[DM_{T(i)}\right] \left\{ \frac{D_{Thr(i)}}{D_{ThrMax(i)}} \right\} \quad (6.8)$$

Equation 6.9 calculates the forces at the gimbal along the vehicle body axes generated by a single engine, where Δ_{Ei} is the pitch (elevation angle with respect to the x–y plane), and Δ_{Zi} is the yaw (azimuth angle about the body z-axis), see Figure 5.3.

$$\begin{aligned} F_{Xe(i)} &= T_{e(i)} \cos(\Delta_{Ei}) \cos(\Delta_{Zi}) \\ F_{Ye(i)} &= T_{e(i)} \cos(\Delta_{Ei}) \sin(\Delta_{Zi}) \\ F_{Ze(i)} &= -T_{e(i)} \sin(\Delta_{Ei}) \end{aligned} \quad (6.9)$$

The moment arms distances between the ith engine gimbal and the vehicle CG are:

$$\begin{aligned} l_{Xei} &= X_{ei} - X_{CG} \\ l_{Yei} &= Y_{ei} - Y_{CG} \\ l_{Zei} &= Z_{ei} - Z_{CG} \end{aligned} \quad (6.10)$$

Equation 6.11 calculates the nominal moments and forces, excluding variations, generated by a single engine at its nominal trim deflection angles (Δ_{Ei} and Δ_{Zi}).

It is the first term in equation 6.8 represented by $M_{EO(i)}$. This term assumes that the engine thrust is at its nominal value $T_{e(i)}$, and it does not include the small variations.

$$M_{EO(i)} = \left\{ \begin{bmatrix} 0 & -l_{zei} & l_{yei} \\ l_{zei} & 0 & -l_{xei} \\ -l_{yei} & l_{xei} & 0 \end{bmatrix} \begin{bmatrix} F_{Xe} \\ F_{Ye} \\ F_{Ze} \end{bmatrix}_{(i)} \right. \\ \left. \begin{bmatrix} F_{Xe} \\ F_{Ye} \\ F_{Ze} \end{bmatrix}_{(i)} \right\} \tag{6.11}$$

The additional moments and forces on the vehicle generated by the small pitch and yaw angle deflections $\delta_{Y(i)}$ and $\delta_{Z(i)}$ of the ith engine relative to its trim positions (Δ_{Ei} and Δ_{Zi}), are shown in equation 6.12. The deflection inputs to the equation are normalized by dividing with the maximum pitch and yaw engine deflection capability ($\delta_{YMAX(i)}$ and $\delta_{ZMAX(i)}$). This normalization makes the inputs to equation 6.12 vary between zero and ± 1. The elements inside the matrix are also scaled accordingly. Equation 6.12 represents the second term in equation 6.8.

$$[DM_{E(i)}] \left\{ \begin{matrix} \delta_{Y(i)}/\delta_{YMAX(i)} \\ \delta_{Z(i)}/\delta_{ZMAX(i)} \end{matrix} \right\} =$$

$$T_e(i) \left\{ \begin{bmatrix} 0 & -l_{zei} & l_{yei} \\ l_{zei} & 0 & -l_{xei} \\ -l_{yei} & l_{xei} & 0 \end{bmatrix} \begin{bmatrix} -\cos(\Delta_Z)\sin(\Delta_E)\delta_{YMAX} & -\cos(\Delta_E)\sin(\Delta_Z)\delta_{ZMAX} \\ -\sin(\Delta_Z)\sin(\Delta_E)\delta_{YMAX} & \cos(\Delta_E)\cos(\Delta_Z)\delta_{ZMAX} \\ -\cos(\Delta_E)\delta_{YMAX} & 0 \end{bmatrix}_{(i)} \right.$$

$$\left. \begin{bmatrix} -\cos(\Delta_Z)\sin(\Delta_E)\delta_{YMAX} & -\cos(\Delta_E)\sin(\Delta_Z)\delta_{ZMAX} \\ -\sin(\Delta_Z)\sin(\Delta_E)\delta_{YMAX} & \cos(\Delta_E)\cos(\Delta_Z)\delta_{ZMAX} \\ -\cos(\Delta_E)\delta_{YMAX} & 0 \end{bmatrix}_{(i)} \right\}$$

$$\times \left\{ \begin{matrix} \delta_{Y(i)}/\delta_{YMAX(i)} \\ \delta_{Z(i)}/\delta_{ZMAX(i)} \end{matrix} \right\} \tag{6.12}$$

Similarly, equation 6.13 calculates the moments and force increments on the vehicle generated by thrust variations $D_{Thr(i)}$ of the ith engine. The throttle control input $D_{Thr(i)}$ has no units and it can be made to vary between zero and ± 1 maximum. The actual engine thrust is defined as: $T(i) = T_e(i)\{1 + D_{Thr(i)}\}$ where $T_e(i)$ is the nominal engine thrust. It means that the engine thrust can be made to vary between zero and a maximum value of $2T_e(i)$. However, the maximum throttling capability of an engine is usually less than one. The throttle parameter $D_{ThrMax(i)}$ of a throttling engine is used to define the max thrust variation from nominal, and it is typically less than one. For example, if the value of the maximum throttling parameter

$D_{ThrMax} = 0.3$, it means that the engine thrust can only vary up to $\pm 30\%$ from nominal T_e. In equation 6.13 we must normalize the throttle control input the same way we normalized the deflection inputs of the aerosurfaces and the gimbaling engines. We scale the input by dividing it with the maximum throttling parameter $D_{ThrMax(i)}$ and the normalized throttle input now varies between zero and ± 1, representing a thrust variation $\pm 30\%$ T_e. Equation 6.13 is the third term in equation 6.8 and calculates the moment and force variations due to throttling.

$$
[DM_{T(i)}]\left\{\frac{D_{Thr(i)}}{D_{ThrMax}}\right\} =
$$

$$
T_e(i)D_{ThrMax}\left\{
\begin{bmatrix}
0 & -l_{zei} & l_{yei} \\
l_{zei} & 0 & -l_{xei} \\
-l_{yei} & l_{xei} & 0
\end{bmatrix}
\begin{bmatrix}
\cos(\Delta_E)\cos(\Delta_Z) \\
\cos(\Delta_E)\sin(\Delta_Z) \\
-\sin(\Delta_E)
\end{bmatrix}_{(i)} \\
\begin{bmatrix}
\cos(\Delta_E)\cos(\Delta_Z) \\
\cos(\Delta_E)\sin(\Delta_Z) \\
-\sin(\Delta_E)
\end{bmatrix}_{(i)}
\right\}\left\{\frac{D_{Thr(i)}}{D_{ThrMax}}\right\} \quad (6.13)
$$

By scaling the control inputs when solving the trimming equation numerically, it adjusts the trim angles proportionally according to the control capability of each effector. For example, when a vehicle has three engines the combined moments and forces due to gimbaling and throttling is given by equation 6.14.

$$
M_{3TVC} = \sum_{i=1}^{Neng=3} M_{EO(i)} + \begin{bmatrix} DM_{E1} & DM_{E2} & DM_{E3} \end{bmatrix}
\begin{bmatrix}
\delta_{y1}/\delta_{y1Max} \\
\delta_{z1}/\delta_{z1Max} \\
\delta_{y2}/\delta_{y2Max} \\
\delta_{z2}/\delta_{z2Max} \\
\delta_{y3}/\delta_{y3Max} \\
\delta_{z3}/\delta_{z3Max}
\end{bmatrix}
$$

$$
+ \begin{bmatrix} DM_{T1} & DM_{T2} & DM_{T3} \end{bmatrix}
\begin{bmatrix}
D_{Thr1}/D_{Thr1Max} \\
D_{Thr2}/D_{Thr2Max} \\
D_{Thr3}/D_{Thr3Max}
\end{bmatrix} \quad (6.14)
$$

6.2.3 Numerical Solution for Calculating the Effector Trim Deflections and Throttles

Having obtained the equations for calculating the moments and forces on the vehicle generated from each individual effector separately, which are: aerosurfaces, TVC, and engine throttling, we may now combine all the effectors together with the base vehicle moments and forces in a single moment/force balance equation, as shown in 6.15.

$$M\ (basic) + \sum_{i=1}^{Neng} M_{EOi} + \sum_{i=1}^{Nsurf} M_{SOi} + M\ (disturb) - M_V \times acceleration$$
$$= -M\ (residual) = -\delta M\ (aero\ surface) - \delta M\ (gimbaling) - \delta M\ (throttling)$$

(6.15)

Equation 6.15 is nonlinear and it must be solved numerically at each point along the trajectory. The residual terms on the left side of equation 6.15 must converge to zero at the completion of the iterations. The left-hand side (LHS) consists of the base vehicle moments and forces, which are eventually balanced by the effectors as they are converging toward their trim positions. The effector increments are calculated from the RHS, after each iteration, and the increments are added to the trim estimate from the previous iteration, and the iterations continue until the trim estimate converges to the trim angle. There is also a term included on the LHS for adding external disturbances. It helps in analyzing the vehicle capability to trim against known disturbances. If the vehicle is capable to trim perfectly without requiring any assistance from the effectors, the forces and moments on the left side of the equation 6.15 would perfectly match the (M × acceleration) term which is also on the left side of the equation without any additional assistance from the (δM) control terms which are on the RHS of equation 6.15. But this is rarely the case. The (δM) terms are the contributions from the three types of control effectors, which are normalized as already described, and calculated after each iteration, that is:

- Control surface deflections (δ_{asi}) relative to the trim positions (Δ_{SOi}),
- Pitch and yaw TVC engine deflections ($\delta_{Yei}, \delta_{Zei}$) relative to their trim angles ($\Delta_{Eei}, \Delta_{Zei}$),
- Moments and forces due to engine thrust variations $D_{Thr(i)}$ from their nominal thrust T_{ei}.

During each iteration, if the moments and forces on the left side of the equation do not balance with the M × acceleration terms, and the M(residual) term is not equal to zero, then we must solve for the (δM) terms on the RHS to calculate how much additional deflections or thrust variations are needed in order to balance the LHS of equation 6.15. The deflection increments are then added to the deflections from the previous iteration to calculate the new trim positions and to adjust the residual terms. This is repeated until the M(residual) term converges to zero. The matrix equation 6.15 is solved for the unknown effector increments on the RHS which are stacked together in a single column vector, as shown in the two aerosurfaces/two engines illustration in equation 6.16. This shapes the equation 6.15 to a matrix equation form, as shown in equation 6.16, which is solved numerically for the effector trim positions vector (δ_T) which is needed to balance the residual forces and moments M(residual), assuming of course that the matrix [DM] is pseudo-invertible. This happens when the rank of [DM] is greater than or equal to the number of vehicle directions (DOFs) that must be trimmed.

$$M(resid) = \begin{bmatrix} DM_{S1} & DM_{S2} & | & DM_{E1} & DM_{E2} & | & DM_{T1} & DM_{T2} \end{bmatrix}$$

$$\times \begin{bmatrix} \delta_{as1}/\delta_{as1Max} \\ \delta_{as2}/\delta_{as2Max} \\ ---- \\ \delta_{y1}/\delta_{y1Max} \\ \delta_{z1}/\delta_{z1Max} \\ \delta_{y2}/\delta_{y2Max} \\ \delta_{z2}/\delta_{z2Max} \\ ---- \\ D_{Thr1}/D_{Thr1Max} \\ D_{Thr2}/D_{Thr2Max} \end{bmatrix} \qquad (6.16)$$

In Matrix Form: $M(resid) = [DM] \left(\dfrac{\delta_T}{\delta_{MAX}} \right)$

Notice that the normalized deflections obtained by the pseudo-inversion are multiplied by the max deflections because the inputs of equation 6.16 are already divided by the max deflections and the matrix DM was properly scaled. This scaling allows the effectors that have greater control authority in certain directions to deflect at greater amplitudes than other effectors which are less capable in those directions.

$$\delta_T = diag(\delta_{MAX}) \times Pseudo\ Inverse\ [DM] \times M(resid) \qquad (6.17)$$

The equation 6.17 is solved numerically at each trajectory point as follows:

1. Starting with the first trajectory time point at time = T(0). The control surface positions are initialized at zero or at some bias angles (δ_{S0i}), the engines are initialized at the pitch and yaw mounting angles (Δ_{Eeoi}, Δ_{Zeoi}), and the thrusts at nominal $T_e(i)$. Then we calculate the initial matrix $[DM]^0$ of equation 6.16 and the residual moment/force vector M^0(residual) from the left side of equation 6.15 using the Mach number, the angles of attack and sideslip, and the engine positions.
2. Solve the pseudo-inverse equation 6.17 for the trim angle and throttle increments (first trajectory point, first iteration).

$$\delta_T^1 = diag(\delta_{MAX}) \times Pseudo\ Inverse\ [DM]^0 \times M^0(residual)$$

3. Calculate new values for the control surface deflections, engine gimbal deflections, and thrusts by adding the effector increments and throttle values

(obtained from the first iteration step 2) to the corresponding previous trim values.

$$\Delta^1_{E(i)} = \Delta^0_{E(i)} + \delta^1_{y(i)} \qquad \Delta^1_{Z(i)} = \Delta^0_{Z(i)} + \delta^1_{z(i)}$$
$$\Delta^1_{S(i)} = \Delta^0_{S(i)} + \delta^1_{as(i)} \qquad T^1_{(i)} = T^0_{(i)}\left(1 + D^1_{Thr(i)}\right)$$

4. Obtain new values for matrices $[DM]^1$ and M^1(residual) from equations 6.16 and 6.15, repeat step 2 for the same trajectory time point and solve for the new trim variables using equation 6.17, and repeat the iterations in steps 2 and 3, still for the same trajectory point until the M(residual) term converges close to zero and the trim angles towards steady-state values.

$$\delta^2_T = diag(\delta_{MAX}) \times Pseudo\_Inverse\,[DM]^1 \times M^1(residual)$$

5. Select the next trajectory point at time $= T(1)$ and repeat the same iterative process described in steps 1–4. Initialize using the trim angles from the previous trajectory time point and solve for the trim angles and throttle values at this point. Continue this process with the remaining trajectory points, all the way to the final point $T = T(n)$, and obtain a time history of the effector trim angles and throttle values as a function of trajectory time.

6.2.4   Adjusting the Trim Profile along the Trajectory

The effector trimming algorithm calculates the effector trim angles and throttle values as a function of the trajectory time by adjusting the effector deflections and thrusts as necessary to balance the moments and forces on the vehicle. The utilization of each effector depends on its control authority along the trim directions which also depends on its maximum deflection or throttling capability. The algorithm calculates the trim positions by taking into consideration the effector's control authority as we have already described. The analyst has the capability of trading-off between effector utilization. Obviously, when there is a multiplicity of effectors, the better the flexibility of trading-off among them. The effector trimming is an iterative process and adjustments in the trimming parameters are often used in order to improve the trimming condition. A previous effector trim history versus trajectory time can also be used for initializing a new trim under different trimming conditions. The initialization trim history can also be manually adjusted as needed in order to bias the next trim results. For example, the designer may wish to reduce the reaction jet usage against allowing bigger aerosurface deflections in order to save fuel. In this case, the designer can adjust the trimming conditions by constraining the jet throttle values and opening up the max deflections on the aerosurfaces. Initializing the jet throttle values at lower magnitudes also helps reducing their activity after retrimming. This of course is only possible when the vehicle has other effectors that can provide sufficient authority to to trim along the required directions in the entire trajectory.

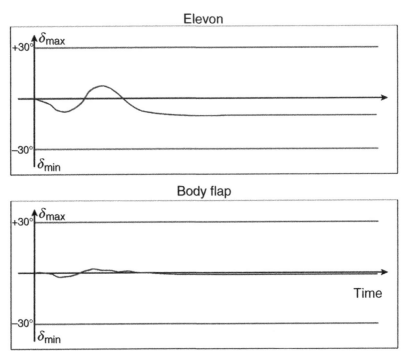

FIGURE 6.1   Initial Pitch Trimming of a Vehicle Using an Elevon and a Body Flap

The initial trimming is performed based on the maximum effector deflections provided without imposing any additional limitations on the effector deflections or throttles. Then adjustments can be made on the effector initialization trim profile to constrain usage of some effectors at the expense of increasing the contributions from others. This trade-off can be performed graphically by adjusting the initial effector positions and their maximum deflections, as a function of time, and retrimming. Consider, for example, a vehicle that may have an elevon, a body flap, a speed brake, and thruster engines. It may be possible to eliminate or to reduce activity in some of the effectors during trimming, like for example, the body flap, by keeping it at a fixed position or scheduling its deflection versus time, and allow the other effectors to be adjusted by trimming. If the vehicle configuration does not have sufficient or it has barely enough effectors to trim with the limitations imposed, the algorithm will not be able to converge and modify its default trimming positions or it will allow very small amounts of effector adjustments from the original trim.

We can illustrate this trimming trade-off between effectors using the following example. Figure 6.1 shows the trim angles versus time of a vehicle's elevon and body flap along a trajectory. The initial trim positions are obtained based on the max surface deflections, which are both set to $\pm30°$, without applying any additional trimming constraints. The maximum deflections determine the amount of aerosurface utilization at each trajectory point during trim. Notice that the body flap trim angle is very close to zero. The elevon, however, is biased in the negative direction during

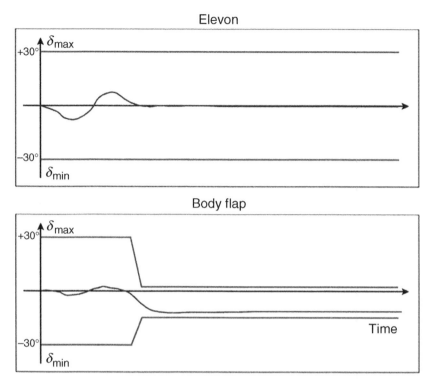

FIGURE 6.2   Retrimming of the Elevon and Body Flap after Applying the BF Constraints

most of the time in order to balance the pitching moment. However, in this particular vehicle the elevon is used for flight control, but not the body flap, which is supposed to be only a trimming device. We would rather prefer to deflect the body flap in order to trim the pitching moment and keep the elevon closer to zero in order to provide greater deflection capability for maneuvering.

Figure 6.2 shows the retrimming results after applying constraints on the body flap deflection. Before retrimming the body flap position versus time is now initialized at constant negative value $-12°$. We would like to keep it at $-12°$ by reducing the upper and lower limits on the body flap during the next trim. Lowering the deflection limits reduces its capability to change from the $-12°$ initialization value. The limits are reduced a few seconds later to allow for the dynamic pressure to increase. The elevon max deflections were not reduced prior to retrimming in order to allow it more authority to adjust. After retrimming the body flap remains close to the $-12°$ initialization position that was set (because it was constrained by the reduced limits). The unconstrained elevon position, however, allowed it to adjust from its initial position and it is now trimming very close to zero as we would like it to be.

The reader will find more trimming examples in Section 10.2 where we design a launch vehicle that is controlled by multiple TVC and throttling engines, and by aerosurfaces in both rotational and translational directions. We also examine a trimming situation that involves thrust failure in one of the engines.

7

STATIC PERFORMANCE ANALYSIS ALONG A FLIGHT TRAJECTORY

Before analyzing the dynamic characteristics of a flight vehicle the designer must first evaluate if the airframe satisfies certain performance characteristics along the mission trajectory. Low airspeed and gusts place the greatest demands on control authority. In addition, agile maneuvers accomplished by frequent excursions into high angle of attack regimes and high roll performance can result in critical control power conditions, including adverse coupling effects. To achieve a successful design, it is important to assess the control authority of a proposed design concept against the performance requirements early in the conceptual stage. The static and dynamic performance of the flight vehicle is captured in its data and its flying performance must be evaluated along the expected trajectory prior to any control analysis and simulations. The proposed flight trajectory captures most of the mission requirements and the vehicle control authority, stability, and maneuverability characteristics depend on the environment, the vehicle configuration, and the effector capability.

In this section we will define some important parameters that help the analyst evaluate in a static sense the overall performance quality of a generic flight vehicle by processing the flight vehicle data along the trajectory as a function of time. This evaluation is not only for aircraft but it includes all types of flight vehicles which are controlled by aerosurfaces, thrust vector control (TVC) and throttling engines, and RCS jets. The performance parameters are calculated at each trajectory point as a function of the trajectory data, mass properties, aerodynamic coefficients for the vehicle and the control surfaces, hinge moment coefficients, engine data, reaction control jets (RCS), vehicle geometry, and the control effector combination logic.

Performance Evaluation and Design of Flight Vehicle Control Systems, First Edition. Eric T. Falangas.
© 2016 by Eric T. Falangas. Published 2016 by John Wiley & Sons, Inc.

The mixing logic defines the control allocation among the effectors. That is, how the acceleration demands from the flight control system are converted to effector deflection and throttle commands and, therefore, it plays an important role in the evaluation of performance. The aerosurface and engine trim angles and the throttle values are also needed in the performance calculations. The effectors must, therefore, be trimmed, as described in Chapter 6, prior to evaluating the vehicle performance.

The parameters used in the performance evaluation are: static stability (percent), center of pressure, aerodynamic center (along the x-axis), time to double amplitude in (sec), short-period and Dutch-roll frequencies in (rad/sec), Cnβ-dynamic, the control authority of the effectors as a system to maneuver the vehicle against expected wind disturbances, the lateral control departure parameter (LCDP) which affects roll controllability, inertial coupling effects that occur between axes due to fast maneuvering, hinge moments at the control surfaces which are needed for sizing the actuators, the bank angle and the sideslip angle β generated due to cross wind near landing, and also the maximum control accelerations along each axis provided by the effectors system. Dispersions and biases in some of the input data are also introduced in the performance analysis process, where the performance is reevaluated using modified values in the mass properties, trajectory parameters, CG location, angles of attack and sideslip, aerocoefficients, etc. This allows us to evaluate the system robustness to variations and disturbances.

7.1   TRANSFORMING THE AEROMOMENT COEFFICIENTS

The basic aerodynamic moment coefficients C_L, C_M, and C_N, the aerosurface coefficients, and their derivatives which are measured or estimated from the aerogroup are not necessarily calculated about the vehicle CG because the mass properties and the CG position vary during flight as the fuel is depleted. The moment coefficients are calculated instead with respect to a fixed point on the vehicle called the moment reference center (MRC). The aerodynamic coefficients must, therefore, be transformed from the MRC to the instantaneous vehicle CG at each trajectory point, and the performance parameters are calculated relative to the CG. The location of the MRC (X_{MRC}, Y_{MRC}, Z_{MRC}) in vehicle coordinates is usually included in the basic aerodata. Equation 3.23 is used to transform the moment coefficients from the MRC to the vehicle CG, where l_{ch} and l_{sp} are the reference length and span. The aerosurface coefficients and the aeroderivatives are transformed similarly.

7.2   CONTROL DEMANDS PARTIAL MATRIX (C_T)

The derivation of the mixing logic matrix was described in Chapter 5, and it is part of the control software that connects between the flight control system outputs and the vehicle effectors. A typical vehicle is controlled by multiple types of effectors and the mixing logic matrix combines the effectors together as a system. It converts the FCS acceleration demands to effector deflection or throttle commands. The FCS outputs are mainly roll, pitch, and yaw rotational acceleration demands and may possibly include some linear acceleration demands (A_x, A_y, A_z). The effector commands are: engine and control surface deflections, main engine, and RCS jet thrust variations. In the event of an effector failure it is the mixing logic matrix that must be changed and not the flight control gains. If we consider a vehicle that is controlled by multiple types of effectors, such as gimbaling engines, throttling engines, RCS jets, and aerosurfaces, it is evident that having a multiplicity of controls the vehicle would be better controllable and it would be able to accelerate in many directions, both rotational and translational. Launch vehicles, for example, use TVC engines to provide control forces and moments and sometimes in combination with engine throttling and RCS jets for linear acceleration control. In most aircraft the FCS controls three rotational axes using aileron, elevator, and rudder. Velocity along the x-axis is also controlled by varying the engine thrust or by modulating drag using the speed brake. The z acceleration is usually controlled indirectly through pitching but some aircrafts have the capability to control normal acceleration directly and independently from pitching, by using flaps or jets. It is also possible, although not as frequent, to control lateral translation along the y-axis independently from other axes using jets.

Figure 7.1 shows the main participants in the flight control loop. The flight control system generates the acceleration demands δ_{FCS} in three to six directions. The mixing

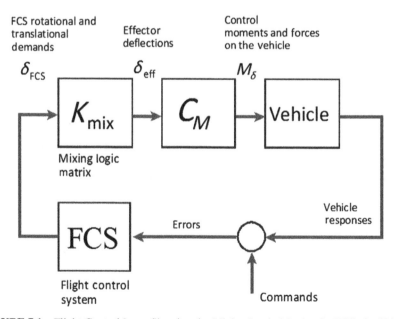

FIGURE 7.1   Flight Control Loop Showing the Mixing Logic Matrix, the FCS, the Vehicle Dynamics, and the Matrix C_M

logic matrix (K_{mix}) converts the demands vector to effector deflections δ_{eff} vector (including throttle commands). The matrix (C_M) converts the effector deflections to forces and moments on the vehicle (M_δ). In our performance calculations, in addition to the effector mixing matrix (K_{mix}), we also need to derive the matrix (C_M) along the trajectory. The combined matrix $(C_T = C_M K_{mix})$ is used in the calculation of the performance parameters because its elements consist of moment and force partials with respect to the FCS demands. The following section describes the derivation of the demands partial matrix C_T.

7.2.1   Vehicle Moments and Forces Generated from a Double-Gimbaling Engine

Equation 7.1 calculates the three moments and three forces acting on the vehicle from a single engine (i) that is gimbaling in pitch (δy) and yaw (δz) directions relative to the trim positions (Δ_E, Δ_Z), see Figure 5.3. This equation is also written in matrix form at the bottom.

$$
\begin{bmatrix} L_{G2} \\ M_{G2} \\ N_{G2} \\ F_{XG2} \\ F_{YG2} \\ F_{ZG2} \end{bmatrix}_{(i)} = T_e(i) \left\{ \begin{bmatrix} 0 & -l_{zei} & l_{yei} \\ l_{zei} & 0 & -l_{xei} \\ -l_{yei} & l_{xei} & 0 \end{bmatrix} \begin{bmatrix} -\cos(\Delta_Z)\sin(\Delta_E) & -\cos(\Delta_E)\sin(\Delta_Z) \\ -\sin(\Delta_Z)\sin(\Delta_E) & \cos(\Delta_E)\cos(\Delta_Z) \\ -\cos(\Delta_E) & 0 \end{bmatrix}_{(i)} \right.
$$

$$
\left. \begin{bmatrix} -\cos(\Delta_Z)\sin(\Delta_E) & -\cos(\Delta_E)\sin(\Delta_Z) \\ -\sin(\Delta_Z)\sin(\Delta_E) & \cos(\Delta_E)\cos(\Delta_Z) \\ -\cos(\Delta_E) & 0 \end{bmatrix}_{(i)} \right\} \begin{Bmatrix} \delta_{Y(i)} \\ \delta_{Z(i)} \end{Bmatrix}
$$

$$
= \{ V_{G2Y} \quad V_{G2Z} \} \begin{Bmatrix} \delta_{Y(i)} \\ \delta_{Z(i)} \end{Bmatrix} \tag{7.1}
$$

The total moments and forces on the vehicle generated by N number of engines $(N = 3$ in this case) which are gimbaling in pitch and yaw directions $(\delta_{Zi}, \delta_{Yi})$ can be written in the compact matrix form as shown in equation 7.2 with the six deflection inputs combined in a single vector (three pitch and three yaw). Where V_{G2Yi} and V_{G2Zi} are column vectors for for the ith engine obtained from equation 7.1.

$$
\begin{bmatrix} L_{G2} \\ M_{G2} \\ N_{G2} \\ F_{XG2} \\ F_{YG2} \\ F_{ZG2} \end{bmatrix}_{(T)} = \begin{bmatrix} \uparrow & \uparrow & \uparrow & \uparrow & \uparrow & \uparrow \\ V_{G2Y1} & V_{G2Z1} & V_{G2Y2} & V_{G2Z2} & V_{G2Y3} & V_{G2Z3} \\ \downarrow & \downarrow & \downarrow & \downarrow & \downarrow & \downarrow \end{bmatrix} \begin{Bmatrix} \delta_{Y1} \\ \delta_{Z1} \\ \delta_{Y2} \\ \delta_{Z2} \\ \delta_{Y3} \\ \delta_{Z3} \end{Bmatrix} \tag{7.2}
$$

7.2.2 Vehicle Moments and Forces Generated by an Engine Gimbaling in Single Direction

We can also calculate the moments and forces vector for a vehicle with TVC engines that gimbal only in a single skewed direction (γ_{ei}). The moments and forces on the vehicle generated by an engine (i) that is gimbaling in a single direction (γ_{ei}), and deflects at an angle ($\delta_{\gamma ei}$) along (γ_{ei}), are obtained from the following equation 7.3.

$$
\begin{bmatrix} L_{SG} \\ M_{SG} \\ N_{SG} \\ F_{XSG} \\ F_{YSG} \\ F_{ZSG} \end{bmatrix}_{(i)} = T_e(i) \left\{ \begin{bmatrix} 0 & -l_{zei} & l_{yei} \\ l_{zei} & 0 & -l_{xei} \\ -l_{yei} & l_{xei} & 0 \end{bmatrix} \begin{bmatrix} -\cos(\Delta_Z)\sin(\Delta_E)\cos(\gamma_i) - \cos(\Delta_E)\sin(\Delta_Z)\sin(\gamma_i) \\ -\sin(\Delta_Z)\sin(\Delta_E)\cos(\gamma_i) + \cos(\Delta_E)\cos(\Delta_Z)\sin(\gamma_i) \\ -\cos(D_E)\cos(\gamma_i) \end{bmatrix}_{(i)} \\ \begin{bmatrix} -\cos(\Delta_Z)\sin(\Delta_E)\cos(\gamma_i) - \cos(\Delta_E)\sin(\Delta_Z)\sin(\gamma_i) \\ -\sin(\Delta_Z)\sin(\Delta_E)\cos(\gamma_i) + \cos(\Delta_E)\cos(\Delta_Z)\sin(\gamma_i) \\ -\cos(\Delta_E)\cos(\gamma_i) \end{bmatrix}_{(i)} \right\} \{\delta_{\gamma(i)}\}
$$

$$(7.3)$$

The total moments and forces on the vehicle generated by ($N = 3$) single gimbaling engines which are gimbaling along some skewed direction (γ_{ei}) and are deflecting at an angle ($\delta_{\gamma ei}$) can be obtained from equation 7.4, where V_{SGi} are the column vectors for each engine (i), as defined in equation 7.3 above.

$$
\begin{bmatrix} L_{SG} \\ M_{SG} \\ N_{SG} \\ F_{XSG} \\ F_{YSG} \\ F_{ZSG} \end{bmatrix}_{(T)} = \begin{bmatrix} V_{SG1} & V_{SG2} & V_{SG3} \end{bmatrix} \begin{Bmatrix} \delta_{\gamma 1} \\ \delta_{\gamma 2} \\ \delta_{\gamma 3} \end{Bmatrix}
$$

$$(7.4)$$

7.2.3 Moment and Force Variations Generated by a Throttling Engine

The variations in moment and force on the vehicle generated by the thrust variation of a single throttling engine (i) are:

$$
\begin{bmatrix} L_{TH} \\ M_{TH} \\ N_{TH} \\ F_{XTH} \\ F_{YTH} \\ F_{ZTH} \end{bmatrix}_{(i)} = T_e(i) \left\{ \begin{bmatrix} 0 & -l_{zei} & l_{yei} \\ l_{zei} & 0 & -l_{xei} \\ -l_{yei} & l_{xei} & 0 \end{bmatrix} \begin{bmatrix} \cos(\Delta_E)\cos(\Delta_Z) \\ \cos(\Delta_E)\sin(\Delta_Z) \\ -\sin(\Delta_E) \end{bmatrix}_{(i)} \\ \begin{bmatrix} \cos(\Delta_E)\cos(\Delta_Z) \\ \cos(\Delta_E)\sin(\Delta_Z) \\ -\sin(\Delta_E) \end{bmatrix}_{(i)} \right\} \{\delta_{Thr(i)}\} \quad (7.5)
$$

where

$\delta_{Thr(i)}$ is the throttle control that varies from 0 to $\pm\delta_{ThrMax}$, (where $\delta_{ThrMax} < 1$)

$T_{e(i)}$ is the nominal thrust for the ith engine.

The actual thrust of engine (i) $T_{(i)}$ is equal to $T_{e(i)}(1 + \delta_{Thr(i)})$. The force and moment variations are due to the thrust variation $T_{e(i)} \delta_{Thr(i)}$. For an RCS jet the force variation equation is the same, except that, in this case $T_{e(i)}$ represents the maximum jet thrust because its nominal thrust is zero. The moment and force variations on the vehicle generated by $(N = 3)$ throttling engines are shown in equation 7.6.

$$
\begin{bmatrix} L_{TH} \\ M_{TH} \\ N_{TH} \\ F_{XTH} \\ F_{YTH} \\ F_{ZTH} \end{bmatrix}_{(T)} = \begin{bmatrix} \uparrow & \uparrow & \uparrow \\ V_{TH1} & V_{TH2} & V_{TH3} \\ \downarrow & \downarrow & \downarrow \end{bmatrix} \begin{Bmatrix} \delta_{Thr1} \\ \delta_{Thr2} \\ \delta_{Thr3} \end{Bmatrix} \tag{7.6}
$$

where

δ_{Thri} is the throttle control of engine (i) and

V_{THi} are column vectors for engine (i) obtained from equation 7.5.

7.2.4 Vehicle Moments and Forces Generated by Control Surfaces

Similarly, we can calculate the moment and force variations from each aerosurface. We are assuming that the coefficients for each surface correspond to separate rotating panels and they do not represent a combination of surfaces, such as, an aileron, for example, which is defined to be the differential rotation of two surfaces. An individual aerosurface often excites multiple directions, and by defining the aero-coefficients for each surface separately it allows us to combine the surfaces more efficiently by means of surface combination logic. The control moments and forces on the vehicle generated by a single aerosurface panel (i) rotating at an angle (δ_{asi}) is shown in equation 7.7a. Equation 7.7b shows the control moments and forces in matrix form generated by multiple surfaces.

$$
\begin{bmatrix} L_{AS} \\ M_{AS} \\ N_{AS} \\ F_{XAS} \\ F_{YAS} \\ F_{ZAS} \end{bmatrix}_{\delta(i)} = \overline{Q}S_{ref} \begin{bmatrix} bC_{l\delta i} \\ cC_{m\delta i} \\ bC_{n\delta i} \\ -C_{A\delta i} \\ C_{Y\delta i} \\ C_{Z\delta i} \end{bmatrix} \{\delta_{asi}\} ; \qquad \begin{bmatrix} L_{AS} \\ M_{AS} \\ N_{AS} \\ F_{XAS} \\ F_{YAS} \\ F_{ZAS} \end{bmatrix}_{(T)} = \begin{bmatrix} \uparrow & \uparrow & \uparrow \\ V_{AS1} & V_{AS2} & V_{AS3} \\ \downarrow & \downarrow & \downarrow \end{bmatrix} \begin{Bmatrix} \delta_{as1} \\ \delta_{as2} \\ \delta_{as3} \end{Bmatrix}
$$

$$\text{(7.7 a \& b)}$$

7.2.5 Total Vehicle Moments and Forces due to All Effectors Combined

We will now combine the equations 7.2, 7.6, and 7.7 together to calculate the moments and forces on the vehicle due to multiple type of effectors consisting of double or single gimbaling engines, throttling engines (or RCS), and aerosurfaces. Equation 7.8

is an example showing how the moment and force vectors from individual effector deflections are stacked together in matrix form to calculate the total moment/force vector. The control inputs are lined up as a column vector on the right side. The equation is also written below in compact matrix form.

$$
\begin{bmatrix} L_X \\ M_Y \\ N_Z \\ F_X \\ F_Y \\ F_Z \end{bmatrix}_{(Tot)} = \begin{bmatrix} \uparrow & \uparrow & \uparrow & \uparrow & \uparrow & \uparrow & \uparrow & \uparrow & \uparrow & \uparrow \\ V_{GY1} & V_{GZ1} & V_{GY2} & V_{GZ2} & V_{G\gamma1} & V_{G\gamma2} & V_{TH1} & V_{TH2} & V_{AS1} & V_{AS2} \\ \downarrow & \downarrow & \downarrow & \downarrow & \downarrow & \downarrow & \downarrow & \downarrow & \downarrow & \downarrow \end{bmatrix} \begin{Bmatrix} \delta_{y1} \\ \delta_{z1} \\ \delta_{y2} \\ \delta_{z2} \\ \delta_{\gamma1} \\ \delta_{\gamma2} \\ \delta_{Th1} \\ \delta_{Th2} \\ \delta_{as1} \\ \delta_{as2} \end{Bmatrix}
$$

or $\quad \underline{M_\delta} = (C_M)\,\underline{\delta_{\text{eff}}} \qquad$ where, $\qquad \underline{\delta}_{\text{eff}} = (K_{\text{mix}})\,\underline{\delta}_{\text{FCS}}$

$$(7.8)$$

The deflections vector $(\underline{\delta}_{\text{eff}})$ on the right side of equation 7.8, generally consist of all three types of effectors. It is the product of the FCS output vector $(\underline{\delta}_{\text{FCS}})$ multiplied with the mixing logic matrix (K_{mix}), see Figure 7.1. The FCS output demands vector $(\underline{\delta}_{\text{FCS}})$ consists of control acceleration demands along the controlled directions, which are a minimum of three rotations (roll, pitch, and yaw) plus some optional translations along the x-, y-, and z-axes. By multiplying the two matrices together we obtain the demands partial matrix $(C_T = C_M\,K_{\text{mix}})$, shown in equation 7.9, which converts the FCS output demands $(\underline{\delta}_{\text{FCS}})$ to vehicle control moments and forces (M_δ). The equation 7.9 is shown with all six directional demands included on the right hand side.

$M_\delta = C_T \delta_{\text{FCS}} \qquad$ where,

$$
\begin{bmatrix} L_X \\ M_Y \\ N_Z \\ F_X \\ F_Y \\ F_Z \end{bmatrix}_{(Tot)} = \begin{bmatrix} L_{\delta P} & 0 & L_{\delta R} & 0 & L_{\delta Y} & 0 \\ 0 & M_{\delta Q} & 0 & M_{\delta X} & 0 & M_{\delta Z} \\ N_{\delta P} & 0 & N_{\delta R} & 0 & N_{\delta Y} & 0 \\ 0 & F_{X\delta Q} & 0 & F_{X\delta X} & 0 & F_{X\delta Z} \\ F_{Y\delta P} & 0 & F_{Y\delta R} & 0 & F_{Y\delta Y} & 0 \\ 0 & F_{Z\delta Q} & 0 & F_{Z\delta X} & 0 & F_{Z\delta Z} \end{bmatrix} \begin{bmatrix} \delta_P \\ \delta_Q \\ \delta_R \\ \delta_X \\ \delta_Y \\ \delta_Z \end{bmatrix}_{\text{FCS}} \qquad (7.9)
$$

By introducing the demands partial matrix representation C_T we are essentially replacing the classical $C_{m\delta_{\text{elevon}}}$, $C_{l\delta_{\text{aileron}}}$, and $C_{n\delta_{\text{rudder}}}$ partials (which apply mainly to aircraft using classical controls) with a more generic form that is applicable to multiple types of controls and in more than three directions. The rotational flight controls $(\delta_{P_{\text{FCS}}}, \delta_{Q_{\text{FCS}}}, \delta_{R_{\text{FCS}}})$ are the FCS acceleration demands in roll, pitch, yaw, rather than the classical aileron, elevon, and rudder deflections. The translational x-acceleration demand is $(\delta_{X\text{FCS}})$ rather than throttle or speed brake command, etc.

The individual effector deflections (or thrust variations) become transparent as they are combined together by the mixing logic matrix toward a common goal, which is to provide the required acceleration in the direction demanded by the FCS. The mixing matrix takes care of the control distribution among the effectors. It is calculated from the mass properties, geometry, engines, and aerodata, which depend on the flight conditions.

The elements of matrix C_T in equation 7.9 consist of moment and force partials with respect to the FCS demands which are useful in the calculations of the performance parameters, as we shall see in Section 7.3. For example, the (2,2) element $M_{\delta Q}$ is the pitch moment partial with respect to pitch FCS demand $(\delta_{Q_{FCS}})$. The size of matrix C_T is $(6 \times N_{dof})$, where $N_{dof} \leq 6$ is the number of the controlled directions (mainly the three rotations plus some translations). The diagonal elements of C_T measure the effector system capability to maneuver the vehicle along the controllable directions. The noncontrollable directions are ignored by removing the corresponding rows in equation 7.9. If the mixing logic matrix is properly designed, the product matrix $(C_T = C_M K_{mix})$ will be diagonal or at least diagonally dominant. The first three inputs $\delta_{P_{FCS}}, \delta_{Q_{FCS}}, \delta_{R_{FCS}}$ on the RHS of the matrix equation 7.9 are the rotational acceleration demands coming from the flight control system that control vehicle rotations. The last three inputs $\delta_{X_{FCS}}, \delta_{Y_{FCS}}, \delta_{Z_{FCS}}$ are the optional translational acceleration demands.

Translational demands in the FCS and in the mixing logic should be included only when the effector system has the translational control capability along those demanded directions without degrading the authority of the moments, such as for example, throttling capability, reaction control jets, body flap, or speed brake, and a requirement to provide direct control along those directions. Otherwise, translations are often performed indirectly by means of rotational maneuvering. The most commonly used translation demands are: $\delta_{X_{FCS}}$ and $\delta_{Z_{FCS}}$ for controlling the axial and normal accelerations. The y direction $\delta_{Y_{FCS}}$ is rarely controlled directly by the FCS. The Harrier aircraft has such a feature using reaction control thrusters when hovering.

7.3 PERFORMANCE PARAMETERS

The following parameters are used for analyzing the vehicle performance in a static sense along a predefined trajectory.

7.3.1 Aerodynamic Center

The aerodynamic center is defined to be the point where the partial of pitch moment with respect to angle of attack is zero, that is: $\left(\partial C_m / \partial \alpha\right) = 0$

$$\frac{\partial Cm_{ac}}{\partial \alpha} = \frac{\partial Cm_{MRC}}{\partial \alpha} + \frac{\partial Cz}{\partial \alpha}\left[\frac{x_{ac} - x_{MRC}}{l_{ch}}\right] = 0$$

$$x_{ac} = x_{MRC} - \frac{Cm_{\alpha_{MRC}}}{Cz_\alpha} l_{ch}$$

(7.10)

where l_{ch} is the vehicle reference length or chord length. The location of the x_{AC} with respect to the x_{CG} determines the static stability of the vehicle. If the CG is ahead of the AC then the vehicle is stable and it has a negative C_{ma}.

7.3.2 Static Margin

Static margin is defined to be the ratio of:

$$SM = \left[\frac{x_{CG} - x_{AC}}{\text{Vehicle length}} \right] 100 \qquad (7.11)$$

7.3.3 Center of Pressure

The center of pressure is defined to be the point about which the pitch aerodynamic moment is zero.

$$Cm_{CP} = Cm_{MRC} + Cz \left[\frac{x_{CP} - x_{MRC}}{l_{ch}} \right] = 0$$

$$x_{CP} = x_{MRC} - \frac{Cm_{MRC}}{Cz} l_{ch} \qquad (7.12)$$

The location of the CP with respect to the x_{CG} determines how much effector deflection is required to trim the vehicle. If the CP and the CG are colocated then the vehicle can be trimmed with zero surface deflection.

7.3.4 Pitch Static Stability/Time to Double Amplitude Parameter (T2)

Static stability refers to the tendency of a flight vehicle under static conditions to return to its trimmed condition. The transfer function in equation 7.13 is an approximate relationship between a pitch axis FCS demand δ_{QFCS} and the vehicle angle of attack, damping is ignored. The coefficient $M_{\delta Q}$ is the pitch control moment partial due to pitch demand (δ_{QFCS}). It is obtained from matrix $C_T(2,2)$ in equation 7.9.

$$\frac{\alpha(s)}{\delta_{Q_{FCS}}(s)} = \left\{ \frac{M_{\delta Q}/I_{YY}}{s^2 + \omega_P^2} \right\} \quad \text{where,} \quad \omega_P^2 = \left[-\frac{M_\alpha}{I_{YY}} \right] \quad \text{and} \quad M_\alpha = \overline{Q} S_{ref} \overline{c} C_{ma}$$

$$(7.13)$$

The static stability of a flight vehicle is determined by the sign of the pitch moment derivative coefficient C_{ma}. When C_{ma} is negative, $\omega_P^2 > 0$, the vehicle is statically stable and it has a short-period resonance ω_P (rad/sec). When C_{ma} is positive and $\omega_P^2 < 0$, the solution of the transfer function is divergent, and the vehicle is open-loop unstable. We typically like to have statically stable vehicles, but not too stable, because if the vehicle is too stable it becomes less maneuverable and bigger control surfaces and also deflections are required to control it. If on the other hand C_{ma} is a little positive (slightly unstable) the vehicle becomes more maneuverable and it can be trimmed and controlled with smaller effectors. The flight control system is usually able to tolerate a certain amount of instability, but not too much. The actuators must be fast enough to respond when the angle of attack diverges due to instability and latency. The amount of static instability in an open-loop unstable vehicle is measured by the time it takes to double its amplitude or α. The time to double amplitude (T2) is obtained from the equation 7.14, and as a rule of thumb it should not be less than 0.45 sec, depending on the vehicle size and its actuator bandwidth.

$$T2 = \ln(2)/|\omega_P| \qquad (7.14)$$

The T2 requirement on a vehicle is determined from the speed capability of its actuators (maximum rate, acceleration, delays, and bandwidth) and the maximum deflections capability. Obviously, the T2 requirement on a flight vehicle must be greater than the time it takes for its effectors to travel from zero to maximum deflection, plus some additional margin to account for any latency in the system, such as computational, etc. When the vehicle is statically stable, on the other hand, and C_{ma} is negative, static stability is measured by the frequency of the short-period resonance (ω_P). A useful parameter for evaluating the static stability or instability of a flight vehicle along the trajectory is to plot the inverse of T2 versus time when the vehicle is unstable, and the frequency of the short-period resonance with a negative sign, ($-\omega_P$) in (rad/sec) in the regions where it is stable. This makes it easy to differentiate between stable and unstable regions along the trajectory and to prevent T2 from diverging toward infinity when static stability fluctuates between stable and unstable regions. In fact, the vehicle is well behaved when the "T2-inverse/short-period" parameter is close to zero (neutrally stable), either a little positive or a little negative. When the T2-inverse parameter is positive, the vehicle is unstable and, typically, it should not exceed 2.2 sec^{-1}. When it is negative the vehicle is stable, and as a rule of thumb (ω_P) should not exceed 5 (rad/sec), otherwise, it becomes difficult to maneuver and requires a wider control bandwidth and bigger effectors.

7.3.5 Derivation of Time to Double Amplitude

Assume that the pitch moment equation has the following simple form that is excited by aerodynamics due to α plus an external excitation δ_e, and $\ddot{\alpha} \cong \ddot{\theta}$.

$$I_Y \ddot{\alpha} = M_\alpha \alpha + \delta_e$$

Assuming that the aerodynamic moment M_α is unstable and the transfer function relating the disturbance to α is in the following form:

$$\frac{\alpha(s)}{\delta_e(s)} = \frac{k}{s^2 - b^2} = \frac{k}{(s+b)(s-b)}$$

where

$$b = \sqrt{\frac{M_\alpha}{I_Y}}$$

When the input is zero, this system has an exponentially diverging solution of this form:

$$\alpha(t) = k(e^{bt} + e^{-bt})$$

Ignoring the stable term because it does not contribute to the divergence, and let us assume that at time t_1 the amplitude is α_1, and at time t_2 the amplitude is $2\alpha_1$.

$$\alpha_1 = k e^{bt_1} \ and \ 2\alpha_1 = k e^{bt_2}$$

Dividing the two equations with each other we obtain:

$$\frac{e^{bt_2}}{e^{bt_1}} = 2$$

The time to double amplitude is defined to be $T2 = t_2 - t_1$, and after taking natural logarithms on both sides,

$$b(t_2 - t_1) = \ln(2); \qquad T2 = \ln(2)\sqrt{\frac{I_Y}{M_\alpha}}; \qquad where, \ M_a > 0$$

7.3.6   Directional Stability ($C_{n\beta}$-dynamic)

The $C_{n\beta}$-dynamic parameter, in equation 7.15, is useful in predicting directional stability at high angles of attack. The vehicle will be directionally stable when the parameter $C_{n\beta}$-dynamic is greater than zero. Otherwise, it may experience yaw departure and Dutch-roll instability. In general, aircrafts that have $C_{n\beta}$-dynamic greater than (0.004) deg^{-1} tend to exhibit very little tendency toward yaw departure.

$$C_{n\beta}\text{-dynamic} = C_{n\beta} \cos \alpha_o - \frac{I_Z}{I_X} C_{l\beta} \sin \alpha_o \qquad (7.15)$$

Where α_0 is the trim angle of attack. Increasing the wing dihedral improves against yaw departure. Too much dihedral, however, makes the vehicle sensitive to gusts and lateral control inputs. When the aircraft is directionally stable, the "Dutch-roll" resonance (ω_D) is obtained from the following equation:

$$\omega_D^2 = \overline{Q} S_{ref} b \left[C_{n\beta} \frac{\cos \alpha_o}{I_{ZZ}} - C_{l\beta} \frac{\sin \alpha_o}{I_{XX}} \right] \qquad (7.16)$$

7.3.7   Lateral Static Stability/Time to Double Amplitude Parameter (T2)

When the yaw axis is unstable (which happens when $C_{n\beta}$-dynamic < 0), the time to double sideslip (β) amplitude should, in general, be greater than 0.5 sec. The time to double amplitude in the lateral direction is defined by an equation similar to the pitch T2.

$$T2 = \frac{\ln(2)}{|\omega_D|} \tag{7.17}$$

The lateral stability parameter is defined similar to the pitch stability, by plotting the T2-inverse when the vehicle is directionally unstable. Otherwise, when the vehicle is directionally stable we plot the Dutch-roll frequency with a negative sign ($-\omega_D$). When the vehicle is directionally unstable, having positive lateral T2-inverse stability parameter, the T2-inverse parameter should, in general, not exceed 2 sec^{-1}, and when the vehicle is stable, the Dutch-roll frequency (ω_D) should not exceed 5 (rad/sec).

7.3.8   Authority of the Control Effectors

We will now derive equations for evaluating the capability of the control effectors to maneuver the vehicle in the demanded control directions. The demands vector δ_{FCS} from the flight control system in equation 7.9 consists of up to six elements as already described. The control authority of the effector system in a certain direction is evaluated by its capability to maneuver the vehicle in the commanded direction. The purpose of the mixing logic matrix is to optimize the control allocation among the effectors system in order to maximize controllability, to provide the expected accelerations in the demanded directions, and to minimize cross-coupling between the control directions. An FCS acceleration demand during a maneuver will typically command several effectors. Each effector has a different saturation limit and for the maneuver to be performed efficiently, none of the effector deflections or throttles should reach their saturation limits.

The control authority of the effector system can be evaluated by the maximum magnitudes of angles of attack and sideslip that can be achieved (or tolerated) before saturating at least one of the effectors. These maximum dispersion angles ($\pm\alpha_{max}$ and $\pm\beta_{max}$) from trim conditions (α_0 and β_0) are either due to wind disturbances or due to maneuvering. The control authority can also be evaluated by the maximum accelerations that can be achieved along the control directions when the FCS demands are maximized. So far we have not discussed about hard limits in the flight control demands vector (δ_{FCS}). We know that this is a software input signal to the mixing logic and the mixing logic generates position and throttle commands driving the effectors which have physical limits. The effector displacements (or throttle variations), however, are transparent in this analysis and we need to figure out a way to translate the effector physical limits back to (δ_{FCS}) software demand limits. So the question is; what is the maximum FCS demand (δ_{FCS_Max}) that the mixing logic matrix will be able to accept along the (3–6) controlled directions before it saturates at least one of the vehicle effectors. The mixing logic matrix (K_{mix}) plays an important role in affecting the control authority because ideally it should be distributing the control capability as evenly as possible among the effectors so that they should all be reaching toward saturation simultaneously when the demand begins to exceed the capability of the effector system. The control distribution should be allocated optimally according to the capability of each individual effector, and thus avoiding the weaker guy in the team from dropping the ball when others can play better. The size of the effector mixing matrix is ($N_{eff} \times N_{dof}$), where N_{eff} and N_{dof} are the number of effectors and the number of controllable degrees of freedom, respectively.

Now let us try to place some limits on the FCS demands that reflect the hardware capabilities. The mixing logic matrix equation can be normalized by dividing each row with the maximum deflection (δ_{imax}) of the corresponding effector, as shown in equation 7.18a that relates the FCS demands (δ_{FCS}) to the normalized effector deflections (δ_i/δ_{imax}). The magnitudes of the normalized effector deflection outputs on the LHS must never exceed 1 in order to prevent the effectors from reaching their hard limits.

$$
\begin{pmatrix} \delta_1/\delta_{1\mathrm{Max}} \\ \delta_2/\delta_{2\mathrm{Max}} \\ \vdots \\ \delta_n/\delta_{n\mathrm{Max}} \end{pmatrix} = \begin{bmatrix} \uparrow & \uparrow & \uparrow & \uparrow & \uparrow & \uparrow \\ U_P & U_Q & U_R & U_X & U_Y & U_Z \\ \downarrow & \downarrow & \downarrow & \downarrow & \downarrow & \downarrow \end{bmatrix} \begin{bmatrix} \delta_P \\ \delta_Q \\ \delta_R \\ \delta_X \\ \delta_Y \\ \delta_Z \end{bmatrix}_{FCS}
\tag{7.18a}
$$

Each of the column vectors in equation 7.18, $\{\underline{U}_P, \underline{U}_Q, \underline{U}_R, \underline{U}_X, \underline{U}_Y, \underline{U}_Z\}$ defines the normalized deflections of the effectors vector resulting from 1 unit of FCS acceleration demand along one of the control directions: $\{\delta_{P_{FCS}}, \delta_{Q_{FCS}}, \delta_{R_{FCS}}, \delta_{X_{FCS}}, \delta_{Y_{FCS}}, \delta_{Z_{FCS}}\}$. The largest magnitude element of each of these normalized vectors are: ($U_{Pmax}, U_{Qmax}, U_{Rmax}, U_{Xmax}, U_{Ymax}, U_{Zmax}$), respectively, and they determine which effector (i) produces the largest deflection from one unit of FCS demand in the corresponding direction: $\delta_{P_{FCS}}, \delta_{Q_{FCS}}, \delta_{R_{FCS}}, \delta_{X_{FCS}}, \delta_{Y_{FCS}}$, or $\delta_{Z_{FCS}}$. In pitch, for example, the deflection of the most active effector (i) due to a pitch demand $\delta_{Q_{FCS}}$ is:

$$
\delta_i = U_{Qmax} \, \delta_{i\mathrm{Max}} \, \delta_{Q_{FCS}}
\tag{7.19}
$$

Let us assume that the most active effector (i) produces the largest normalized deflection magnitude $\delta_i/\delta_{i\mathrm{Max}}$ coming from a pitch demand δ_{QFCS}. The element in vector $\underline{U}_Q$ that corresponds to the largest deflection is U_{Qmax}. From equation 7.19 we conclude that the most active effector in pitch reaches its limit when the FCS command $\delta_{Q_{FCS}}$ is maximized at:

$$
\delta_{Q_{FCS}\,MAX} = \frac{1}{U_{Q\,Max}}
\tag{7.20}
$$

Now that we have calculated the maximum pitch demand that can be tolerated by the effector system let us calculate the maximum steady-state angle of attack $\alpha_{ss\,max}$ displacement from trim α_0 that can be achieved (or tolerated) before saturating the effectors. Consider the transfer function in equation 7.13 which is an approximate relationship between the pitch FCS demand (δ_{QFCS}) and the vehicle angle of attack response. When the aircraft is statically stable the parameter M_α is negative which implies that ω_p^2 is positive. The partial derivative $M_{\delta Q}$ represents the pitch moment on the vehicle due to a pitch FCS demand (δ_{QFCS}) and it is obtained from equation 7.9. The pitch control authority can be measured by the amount of alpha variation $\alpha_{ss\,Max}$ from trim α_0 that can be achieved (or tolerated) by maximizing the control demand ($\delta_{Q_{FCS\,Max}}$), as calculated in equation 7.21. It occurs when the most active effector in the pitch direction reaches its limit, that is, ($\delta_i = \delta_{i\,Max}$), due to maximizing the pitch demand.

$$\alpha_{ss\,MAX} = \frac{M_{\delta Q}/I_{YY}}{\omega_p^2}\delta_{Q_{FCSMAX}} = -\frac{M_{\delta Q}}{U_{QMax}\,M_\alpha} \tag{7.21}$$

where U_{QMax} is the largest element in the column vector U_Q in the normalized mixing logic equation 7.18a.

7.3.9 Biased Effectors

In equation 7.18a we have assumed that the effector zero positions are centered with equal max peak deflections in both directions resulting from either positive or negative FCS demands (δ_{FCS}), but this is not always true. Some aerosurfaces may be biased or trimmed closer to the positive or to the negative limits and the peak displacement ($\delta_{i\,Max}$) due to a positive max demand (δ_{+FCS_Max}) may be different than the peak displacement ($\delta_{i\,Min}$) due to a negative peak demand (δ_{-FCS_Max}) in the opposite direction. If we assume that equation 7.18a corresponds to positive FCS demands, we can also write a similar equation 7.18b for negative FCS demands.

$$\begin{pmatrix} \delta_1/\delta_{1Min} \\ \delta_2/\delta_{2Min} \\ \vdots \\ \delta_n/\delta_{nMin} \end{pmatrix} = \begin{bmatrix} \uparrow & \uparrow & \uparrow & \uparrow & \uparrow & \uparrow \\ U'_P & U'_Q & U'_R & U'_X & U'_Y & U'_Z \\ \downarrow & \downarrow & \downarrow & \downarrow & \downarrow & \downarrow \end{bmatrix} \begin{bmatrix} -\delta_P \\ -\delta_Q \\ -\delta_R \\ -\delta_X \\ -\delta_Y \\ -\delta_Z \end{bmatrix}_{FCS} \tag{7.18b}$$

where $\delta_{i\,Min}$ are now the peak deflections of the effectors due to negative demands (δ_{-FCS_Max}). They are in the opposite direction of $\delta_{i\,Max}$ (not necessarily negative) and different in magnitude from $\delta_{i\,Max}$. The column vectors $\{U'_P,\ U'_Q,\ U'_R,\ U'_X,\ U'_Y,\ U'_Z\}$ are now scaled differently than in 3.19 because their corresponding rows are divided by $\delta_{i\,Min}$. The largest magnitude element of each of these normalized vectors are also different: ($U_{Pmin}, U_{Qmin}, U_{Rmin}, U_{Xmin}, U_{Ymin}, U_{Zmin}$) and they do not necessarily correspond to the same effector element as $(U_{P\max}, U_{Q\max}, U_{R\max}, U_{X\max}, U_{Y\max}, U_{Z\max})$. Based on this result we can rewrite equation 7.20 to calculate the maximum FCS demand in the negative direction, and equation 7.21b to calculate the peak alpha dispersion as a result of this negative peak demand δ_{-QFCS_Max}.

$$\delta_{-Q FCSMAX} = \frac{1}{U_{Q\,\text{Min}}}$$

$$\alpha_{-ss\,\text{Max}} = -\frac{M_{\delta Q}}{U_{Q\,\text{Min}}\,M_\alpha} \qquad (7.21b)$$

7.3.10   Control to Disturbance Moments Ratio (M_α/M_δ)

A flight vehicle must be designed to be capable of counteracting a certain amount of wind shear, gusts and other disturbances by using its controls. The vehicle controllability in the aerodynamic environment is typically evaluated in terms of dispersions in α and β which are introduced in simulations as gusts and wind shear. The gusts are short-period disturbances used in dynamic models for the purpose of evaluating the dynamic response of the flight vehicle. In static analysis the aerodynamic disturbance is defined as a steady wind shear. A wind shear causes steady-state variations in the angles of attack and sideslip from the trim conditions. This α and β variation creates additional forces and moments to be applied on the aircraft that must be counteracted by further deflecting the control surfaces or the engines. The ratio of moment per alpha excitation, divided by the control moment per control deflection (M_α/M_δ) is a parameter that is typically used for evaluating controllability, and in general this parameter should be less than 1. This M_α/M_δ controllability parameter, however, does not take into account the maximum value of the angle of attack (or sideslip in the lateral direction) and the maximum authority of the control system, and we shall derive a different criteria.

Note, when the vehicle is statically unstable (i.e., CG is behind the CP) the smallest control system bandwidth for achieving a minimum of 6 dB gain margin occurs when the attitude feedback gain is: $K_P = 2M_\alpha/M_{\delta Q}$, assuming a PD type controller $\delta_{QFCS} = -K_p\theta - K_r\dot\theta$. With this controller gain the control system bandwidth is $\omega_b = \sqrt{M_\alpha/I_{YY}}$. Increasing the attitude feedback gain K_p and bandwidth further improves rigid-body stability margin and also the system performance to attitude commands. However, you cannot increase it too much without exciting the structure flexibility. In general there should be a separation factor of 10 between the control system bandwidth and the first structural mode.

7.3.11   Pitch Control Authority Against an Angle of Attack $\alpha_{\max}$ Dispersion

Let us now define another parameter for measuring the control authority in pitch and we shall name it the "pitch control effort". The control authority can be defined by the amount of moment it uses to achieve the required angle of attack variation from trim in comparison with the maximum available moment. That is, an aircraft should have enough control authority to tolerate a certain amount of α-variation, let us say $\alpha_{max} = \pm 5°$ due to a wind-shear disturbance, and to be able to counteract against it without affecting its attitude and flight direction, by using its effectors without saturating its pitch control demand (δ_{QFCS}) which is limited to ($\delta_{Q_{FCSMax}}$).

For good controllability in pitch we would expect the pitching moment produced when the pitch demand is maximized to be considerably greater (at least twice as big) than the moment produced at (α_{max}).

$$\left| C_{m\delta Q}\, \delta_{Q_{FCSMax}} \right| > 2 \left| C_{m\alpha}\, \alpha_{max} \right|$$

We introduced a factor of 2 to allow some flight control authority for maneuvering and for controlling gust disturbances, etc. If we take into consideration that the max control demand is: $\delta_{QFCS\,max} = 1/U_{QMax}$, the pitch control authority criterion becomes:

$$\text{Pitch Control Effort} = \left(\frac{\delta_{QFCS}}{\delta_{Q_{FCSMax}}} \right) = \left| \frac{C_{m\alpha}}{C_{m\delta Q}} \right| U_{QMax}\, \alpha_{max} < 0.5 \qquad (7.23a)$$

In other words, the ratio of the pitch control used against a dispersion α_{max} over the max pitch control availability should be less than one, or even better, less than half to be conservative.

When the vehicle has multiple control effectors to provide translational control we may extend this control authority criterion to translations, mainly along x and z. Let us say that we would like the vehicle to have the control authority to counteract translational accelerations in the x and z directions in the presence of winds. This may obviously require the use of throttle control or speed brake in the x direction or RCS jets to provide forces along those directions and to regulate its speed in the presence of winds. The pitch control authority definition in equation 7.23a can be extended to translational control and to evaluate the effector system's control authority against α_{max} disturbances. That is, evaluate its capability to generate the accelerations required to counteract the disturbance accelerations in the x and z directions, due to $\pm \alpha_{max}$.

$$\{Z\ \text{Accel. Control Effort}\} = \left(\frac{\delta_{Z_{FCS}}}{\delta_{Z\,FCSMax}} \right) = \left| \frac{C_{Z\alpha}}{C_{Z\delta Z}} \right| U_{Z\,Max}\, \alpha_{max} < 0.5$$

$$(7.23b)$$

$$\{X\ \text{Accel. Control Effort}\} = \left(\frac{\delta_{X_{FCS}}}{\delta_{X\,FCSMax}} \right) = \left| \frac{C_{A\alpha}}{C_{X\delta X}} \right| U_{X\,Max}\, \alpha_{max} < 0.5$$

The coefficients $C_{Z\delta Z}$ and $C_{X\delta X}$ are the nondimensional force partials with respect to the FCS acceleration demands in the z and x directions, respectively. They are obtained directly from the demands partial matrix C_T in equation 7.9. They are not only due to aerosurfaces, but they also include contributions from all vehicle effectors which are combined together by the effector mixing logic. The parameters $U_{Q\text{Max}}$, $U_{Z\text{Max}}$, and $U_{X\text{Max}}$, are the biggest elements in the column vectors $\underline{U}_Q$, $\underline{U}_Z$, and $\underline{U}_X$ in the normalized mixing matrix, equation 7.18a or 7.18b.

7.3.12 Lateral Control Authority Against an Angle of Sideslip β_{max} Disturbance

An important requirement for the lateral design is to have sufficient control authority to perform a steady sideslip maneuver during turning or to be able to react against cross-winds near landing. When an aircraft is near landing with a cross-wind, the side force creates a sideslip angle and the aircraft has to bank toward the wind. For an aircraft to maintain a steady angle of sideslip β, the net side force, and the rolling and yawing moments must be equal to zero. Both, the roll and yaw FCS controls (δ_{PFCS} and δ_{RFCS}) must be used in order to maintain a constant sideslip. The designer must evaluate if the effector deflections due to the FCS demands (δ_{PFCS} and δ_{RFCS}) are within range and if the vehicle bank angle (ϕ) is acceptable. From the roll and yaw moment balance equations the amount of roll and yaw FCS control required to achieve a constant angle of sideslip (β_{ss}) are calculated from equation 7.24.

$$
\begin{aligned}
\delta_{P_{FCS}} &= \frac{\left(L_\beta N_{\delta R} - N_\beta L_{\delta R}\right)}{\left(L_{\delta R} N_{\delta P} - N_{\delta R} L_{\delta P}\right)} \beta_{ss} \\
\delta_{R_{FCS}} &= \frac{\left(N_\beta L_{\delta P} - L_\beta N_{\delta P}\right)}{\left(L_{\delta R} N_{\delta P} - N_{\delta R} L_{\delta P}\right)} \beta_{ss}
\end{aligned}
\tag{7.24}
$$

The partial derivatives $N_{\delta R}$ and $L_{\delta R}$ represent the yaw and roll moments on the vehicle due to a yaw flight control demand ($\delta_{R_{FCS}}$). Similarly, the partial derivatives $N_{\delta P}$ and $L_{\delta P}$ represent the yaw and roll moments on the vehicle per roll demand ($\delta_{P_{FCS}}$). They are obtained directly from matrix C_T in equation 7.9 which includes the effector combination matrix. The FCS authority in roll and yaw is evaluated by the amount of roll and yaw effort that is produced in order to generate the expected maximum steady-state angle of sideslip β_{max}. In general, when the rudder is at full deflection it should be able to provide at least five degrees of sideslip β. After dividing equation 7.24 with the max roll and yaw FCS demands, $\delta_{P_{FCS\_Max}}$ and $\delta_{R_{FCS\_Max}}$, we obtain equation 7.25 that define the roll and yaw control authority or control effort.

$$
\begin{aligned}
\{\text{Roll Control Effort}\} &= \left(\frac{\delta_{P_{FCS}}}{\delta_{P_{FCS\,Max}}}\right) = \left|\frac{\left(L_\beta N_{\delta R} - N_\beta L_{\delta R}\right)}{\left(L_{\delta R} N_{\delta P} - N_{\delta R} L_{\delta P}\right)}\right| U_{P_{\text{Max}}} \beta_{\text{max}} < 0.5 \\
\{\text{Yaw Control Effort}\} &= \left(\frac{\delta_{R_{FCS}}}{\delta_{R_{FCS\,Max}}}\right) = \left|\frac{\left(N_\beta L_{\delta P} - L_\beta N_{\delta P}\right)}{\left(L_{\delta R} N_{\delta P} - N_{\delta R} L_{\delta P}\right)}\right| U_{R_{\text{Max}}} \beta_{\text{max}} < 0.5
\end{aligned}
\tag{7.25a}
$$

The ratios $(\delta P_{FCS}/\delta P_{FCSMax})$ and $(\delta R_{FCS}/\delta R_{FCSMax})$ represent the roll and yaw control authority (or control effort) and they should obviously be less than 1. In fact, half is a better number that would allow for dynamic control. They represent the fraction of effort required from the roll and yaw controls to produce (or tolerate) a certain amount of sideslip β_{max}, such as, from a lateral wind disturbance. Similarly, we can extend the controllability definition in the y direction to evaluate if the vehicle has sufficient control authority (side acceleration) to counteract a steady-state (β_{max}) disturbance. We assume of course that the vehicle must have lateral jets to provide controllability in the y direction.

$$\{Y \text{ acceler Control Effort}\} = \left(\frac{\delta_{Y_{FCS}}}{\delta_{Y\,FCS\,Max}}\right) = \left|\frac{C_{Y\beta}}{C_{Y\,\delta Y}}\right| U_{Y\,Max}\beta_{max} < 0.5 \quad (7.25b)$$

where U_{PMax}, U_{RMax}, and U_{YMax} are the largest elements in the column vectors $\underline{U}_P$, $\underline{U}_R$, and $\underline{U}_Y$ in the normalized mixing logic equation 7.18a or 7.18b.

7.3.13   Normal and Lateral Loads

The normal and lateral load parameters (Q_α and Q_β) are often used as indicators of structural loading due to the angles of attack and sideslip. The vehicle must be capable of withstanding a certain amount of normal and lateral loading which ranges between 3000 and 4000 (lb-deg/ft^2). The loading increases during periods of high dynamic pressure and for this reason some vehicles have a load relief system which attempts to zero-out the angles of attack and sideslip during high pressures. It is, therefore, useful to evaluate normal and lateral loading along the trajectory, including dispersions.

$$Q_\alpha = \overline{Q} \times \alpha_0; \qquad Q_\beta = \overline{Q} \times \beta_0$$

7.3.14   Bank Angle and Side Force During a Steady Sideslip

Another set of parameters which are important to calculate when the aircraft is under constant sideslip, is the side force and the bank angle. This is particularly important near landing. The side force due to a steady cross-wind from the starboard side is balanced with a gravity component by banking the aircraft toward the wind at a small angle φ, maintaining a zero side force, as shown in Figure 7.2. After solving for δ_{PFCS} and δ_{RFCS} from equation 7.24, which are required for maintaining a steady sideslip, let us say $\beta_{ss} = 5$ (deg), the bank angle required to balance the side force is obtained from equation 7.26. Typically, near landing the bank angle φ should be less than 5°. In level flight and near landing the normal acceleration A_z is approximately 1g. The partials $F_{Y\delta R}$ and $F_{Y\delta P}$ are the forces along the y-axis per yaw and roll FCS demands, and they are calculated from equation 7.9. In general, it is sufficient to

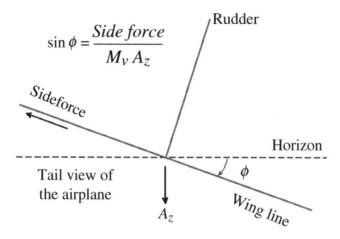

$$\sin \phi = \frac{Side\ force}{M_V A_Z}$$

Rudder

Sideforce

Horizon

Tail view of
the airplane

A_Z

ϕ

Wing line

FIGURE 7.2   Aircraft Banking with Cross-Wind near Landing

demonstrate that no more than 75% of the roll and yaw control authority be devoted to maintaining a steady sideslip. Typically the bank angle must be less than 5°.

$$\text{Side force} = \overline{Q}S_{\text{ref}}C_{Y\beta}\beta_{\text{ss}} + F_{Y\delta R}\delta_{R_{FCS}} + F_{Y\delta P}\delta_{P_{FCS}}$$

$$\text{Bank angle:}\quad \sin\varphi = \frac{\text{Side force}}{M_V A_Z \cos\gamma} \tag{7.27}$$

7.3.15   Engine-Out or Y_{cg} Offset Situations

The lateral control authority analysis can be extended to include asymmetric situations due to engine-out, thrust mismatch, or Y_{cg} offset. The roll and yaw effectors must also have sufficient control authority to cope with asymmetric forces and moments generated due to asymmetric propulsion in the lateral direction or Y_{cg} offset due to an asymmetric payload. This requirement becomes most demanding when operating at very low speeds. To maintain steady straight flight, the roll and yaw effectors must counter the effect of an asymmetric thrust or payload, and to produce zero side force and zero rolling and yawing moments. The following three equations in 7.27 must be solved simultaneously.

$$\overline{Q}S_{\text{ref}}C_{Y\beta}\beta + F_{Y\delta R}\delta_{R_{FCS}} + F_{Y\delta P}\delta_{P_{FCS}} + M_V A_Z \cos\gamma \sin\varphi + \Delta F_Y = 0$$

$$L_\beta\beta + L_{\delta R}\delta_{R_{FCS}} + L_{\delta P}\delta_{P_{FCS}} + \Delta L = 0 \tag{7.27}$$

$$N_\beta\beta + N_{\delta R}\delta_{R_{FCS}} + N_{\delta P}\delta_{P_{FCS}} + \Delta N = 0$$

where F_Y, L, and N are the side force, rolling, and yawing moments generated due to the thrust mismatch and the Y_{cg} offset. The Y_{cg} offset will only produce moments and no side force.

7.3.16 Lateral Control Departure Parameter

At high angles of attack the effectiveness of the control surfaces in roll may sometimes be reduced or lost, plus at high alpha the roll effectors may produce significant adverse yaw that may lead to the so-called "roll-reversal" phenomenon. As a result, the aircraft rolls in opposite direction to the roll command input. The LCDP ratio is a good indicator of roll axis controllability and is defined in equation 7.28. It is the ratio of the lateral departure parameter (ω_0^2) divided by the Dutch-roll frequency (ω_D^2).

$$
\text{LCDP} = \left(\frac{\omega_0^2}{\omega_D^2} \right) = \frac{N_\beta - L_\beta \frac{N_{\delta P}}{L_{\delta P}}}{N_\beta - L_\beta \frac{I_{ZZ}}{I_{XX}} \tan \alpha_0}
\tag{7.28}
$$

This parameter determines the susceptibility of the flight vehicle to departure during high angle of attack operation. It is usually positive but it changes sign as a function of the angle of attack. Its departure from unity is an indicator of coupling between roll and yaw axes. Values greater than unity are favorable in yaw but when they are too large it has a tendency to induce beta oscillations. When the LCDP ratio is near zero the aircraft response to roll commands is sluggish. When it becomes negative it causes reversal in the roll (aileron) control which can be catastrophic if unaccounted. The flight control system should be able to detect this situation and to take an appropriate action when it happens.

For a reentry vehicle, such as a Space Shuttle entering the atmosphere at high angles of attack, there are periods where directional stability is poor because the vertical stabilizer is ineffective since the wake from the wings is preventing it from operating in a clean airflow. During this period the rudder flight control gain is phased out and the aileron is used for roll and directional control and often in combination with yaw RCS jets. The vehicle must, therefore, have an acceptable LCDP in order to be stabilizable with the ailerons. When the angle of attack is reduced the rudder becomes more effective to provide directional stability and the gain in the rudder control loop is increased appropriately. At high angles of attack the LCDP ratio may be negative which implies that the roll gain must be reversed. This is not a problem, however, as long as the LCDP is not too close to zero which implies uncertain and sluggish roll controllability. As the angle of attack decreases further there is a period during which the LCDP is transitioning from a negative to positive LCDP and its magnitude becomes very small before it changes sign to positive. During this period the roll controllability becomes unreliable and ineffective for lateral control, the yaw control loop alone may not be sufficient, and the vehicle has to rely fully on the RCS jets.

Equation 7.29 is a transfer function that approximates the roll acceleration due to a roll flight control command. The terms N_β and L_β represent the vehicle yaw and roll moment partials per sideslip angle β. The partials $N_{\delta p}$ and $L_{\delta p}$ represent the yaw and roll moment partials due to variations in roll FCS demand δ_{PFCS}, and they are obtained from matrix C_T. Notice that ω_D^2 is only a function of the vehicle aerodynamics and inertias, but ω_0^2 also depends on the mixing logic matrix (K_{mix}), which is a software. This makes the LCDP ratio to depend on the effector combination matrix and on the roll/yaw effectors.

$$\frac{\dot{p}(s)}{\delta_{P_{FCS}}(s)} = \frac{L_{\delta p}}{I_{XX}} \left\{ \frac{s^2 + \omega_0^2}{s^2 + \omega_D^2} \right\} \quad \text{where,}$$

$$\omega_0^2 = \frac{\cos \alpha_o}{I_{ZZ} L_{\delta p}} \left[N_\beta L_{\delta p} - L_\beta N_{\delta p} \right] \quad \text{and} \quad \omega_D^2 = \left[\frac{N_\beta \cos \alpha_o}{I_{ZZ}} - \frac{L_\beta \sin \alpha_o}{I_{XX}} \right]$$

$$N_\beta = \overline{Q} S_{ref} b C_{n\beta} \quad L_\beta = \overline{Q} S_{ref} b C_{l\beta}$$

$$N_{\delta p} = \frac{\partial N}{\partial \delta_{P_{FCS}}} \quad L_{\delta p} = \frac{\partial L_X}{\partial \delta_{P_{FCS}}} \tag{7.29}$$

Notice that the transfer function has two terms, a gain term that includes the roll effector partial $L_{\delta p}$, and a dynamic term consisting of a pair of poles and a pair of zeros. The dynamic term is a function of the aeromoments. When the aerodynamics is weak and the vehicle is controlled by TVC or RCS jets the dynamic term in the brackets becomes small in comparison with the gain term and the transfer function simplifies to:

$$\frac{\dot{p}(s)}{\delta_{P_{FCS}}(s)} = \frac{L_{\delta p}}{I_{XX}} \tag{7.30}$$

The term in parenthesis introduces a low frequency transient in the transfer function response. The LCDP to Dutch-roll frequency ratio becomes important to analyze roll controllability when the $L_{\delta p}$ partial is weak due to loss of TVC or RCS control. In the situation where the vehicle is controlled by aerosurfaces alone the dynamic term in the parenthesis begins to dominate. You would like to have directional stability ($\omega_D^2 > 0$), and the (ω_0^2/ω_D^2) ratio sufficiently far from zero, otherwise it will not respond to roll commands. Notice that pole/zero cancellations also occur when the LCDP ratio is equal to one. In this case the vehicle response to a roll command becomes ideal, producing a perfectly coordinated turn with zero β. An acceptable range for the LCDP ratio is: $\{15 > |LCDP| > 0.2\}$. This range provides a satisfactory compromise between performance and robustness against aerodynamic uncertainties. Negative LCDP ratios are also acceptable as long as they are not very close to zero. They cause, however, reversal in roll control gains because it changes the sign of the transfer function 7.29 and the roll control system must be able to predict the reversal and to compensate for it.

When the magnitude of the LCDP ratio becomes small or negative at high angles of attack, the mixing matrix may be adjusted to include contributions from the yaw effectors. This can improve the value of the LCDP at high alphas, sometimes at the expense of other properties such as control authority. The weakness in the LCDP ratio, however, can be tolerated if the $L_{\delta p}$ term in equation 7.29 dominates through the use of propulsion. Good reliability and robustness against lateral closed-loop instability due to the aerodynamic uncertainties require that the magnitude of the LCDP ratio be sufficiently greater than zero (either positive or negative), especially, during periods of poor yaw controllability. Reaction control jets are often required to control the vehicle in roll and yaw during LCDP transitioning. It is not sufficient to control only the roll axis, but both, roll and yaw, are commanded by the FCS because the vehicle is maneuvering about the velocity vector at high angles of attack. This type of maneuvering minimizes sideslip and lateral loads. During the transitioning period the RCS torques must be properly sized in order to be able to overcome the uncertain aerodynamic moments and it should be able to provide at least 1.5 deg/sec^2 angular acceleration in roll and yaw.

Figure 7.3 is a plot of $\omega_0{}^2$ in the vertical scale versus $\omega_D{}^2$ in the horizontal scale. It shows the acceptable regions of the LCDP ratio. At high angles of attack a Shuttle type of reentry vehicle operates in the regions of negative $\omega_0{}^2$ where the roll control gain is reversed. As the angle of attack decreases the operation shifts towards the positive $\omega_0{}^2$ region. The aileron and in general the roll control is unreliable when the LCDP ratio is too small and near the horizontal shaded region of Figure 7.3. It should not be very large either because it causes large sideslip transients. The region where $\omega_D{}^2 < 0$ is statically unstable, and it is not recommended to operate there without a powerful RCS. The ideal value of the LCDP ratio is 1, which is along the dashed line

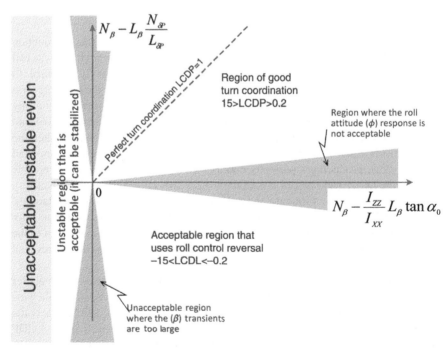

FIGURE 7.3   Plot of $\omega_0{}^2$ versus $\omega_D{}^2$ Showing the Acceptable Regions of the LCDP Ratio

in Figure 7.3. This value produces a perfectly coordinated turn in response to a roll command with zero angle of sideslip and this happens when:

$$\frac{N_{\delta P}}{L_{\delta P}} = \frac{I_{ZZ}}{I_{XX}} \tan \alpha$$

In general, it is desirable to have the value of LCDP ratio as close to 1.0 as possible and to avoid sign reversals but this is impossible to achieve. With strong RCS, however, it is possible to operate in the marginally stable and mildly unstable regions and also where the LCDP ratio is small.

7.3.17   Examples Showing the Effects of LCDP Sign Reversal on Stability

The effect of the LCDP and $C_{n\beta}$-dynamic variations on lateral stability and on the control feedback gains can be illustrated with an example. Let us consider a lateral vehicle model that is controlled by two inputs, an aileron and a rudder, and its states consisting of: roll and yaw rates (p, r), roll attitude ϕ, and the angle of sideslip β. This system is controlled by a (2×4) state-feedback matrix (KSpr1) and the roll angle is controlled by a command ϕ_{cmd}. This system is open-loop statically stable with a Dutch-roll resonance of 2 (rad/sec), $C_{n\beta}$-dynamic > 0, and it has a negative LCDP ratio, $\omega_0^2/\omega_D^2 = -0.334$. It corresponds to the region in Figure 7.3 where $(\omega_0{}^2 < 0$; reverse roll) and $(\omega_D{}^2 > 0$; statically stable). This system is shown in Figure 7.4, where it responds to a $\phi_{cmd} = 1$ (rad). The state-feedback matrix is also included in the figure.

A parameter of this system was modified which causes the LCDP to change sign. Figure 7.5 is almost identical to the previous system but it has a different $C_{n\delta\text{-aileron}}$. The yawing moment due to aileron was changed from negative to a positive value. The new vehicle system is also open-loop stable with the same Dutch-roll resonance, but the LCDP ratio is now +0.302, almost the opposite of the previous system. It

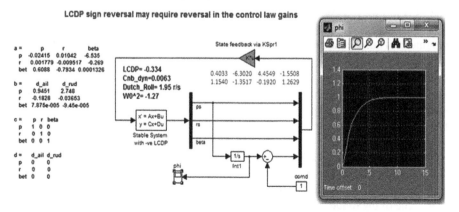

FIGURE 7.4   Closed-Loop System with Negative LCDP Ratio and $C_{n\beta}$-Dynamic > 0

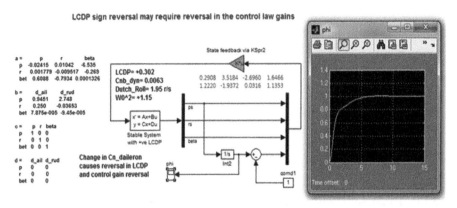

FIGURE 7.5 Closed-Loop System with Positive LCDP Ratio and $C_{n\beta}$-Dynamic > 0

corresponds to the region in Figure 7.3 where ($\omega_0^2 > 0$) and ($\omega_D^2 > 0$; statically stable). This system is no longer stabilizable with the previous state-feedback matrix. A different state-feedback (KSpr2) was designed in order to stabilize the system and to achieve a similarity to the previous performance in response to a $\phi_{cmd} = 1$ (rad). Notice how the signs in most of the elements are reversed in this new state-feedback gain matrix.

This example demonstrates that an unexpected reversal in the sign of the LCDP ratio can have destabilizing effect on the vehicle, even though the open-loop stability was almost identical in both vehicle cases. In general, the exact timing of the sign reversal in the aerodynamic parameters is not known due to the uncertainties in the coefficients, so there is a transitioning period where we cannot rely on the aileron for control. The magnitude of the LCDP ratio should not be allowed to drop below 0.02. A solution when the LCDP ratio is small and is transitioning between negative and positive regions is to introduce some rudder control in the roll loop. It often increases the magnitude of the LCDP, either positive or negative. An even better solution is to use RCS jets for roll control during this transitioning period.

LCDP ratio reversal can also be caused by a change in the sign of $C_{n\beta}$-dynamic parameter. This can happen when the vehicle is transitioning from a statically stable to a statically unstable region where the $C_{n\beta}$-dynamic < 0. This corresponds to the statically unstable region in Figure 7.3, where ($\omega_D^2 < 0$). This change in open-loop stability also requires a change in the state-feedback control law in order to properly stabilize it, as shown in Figure 7.6. In this example, which is similar to Figure 7.4, the sign of $C_{n\beta}$-dynamic was changed to negative by a modification in the aero coefficients $C_{n\beta}$ and $C_{l\beta}$. A new state-feedback control law (KSpr3) is designed to stabilize the modified system and enable it to respond to a $\phi_{cmd} = 1$ (rad) by achieving similar performance. The new state-feedback is also significantly different from the original state-feedback matrix (KSpr1), which is no longer able to stabilize it. Notice how some of the rudder gains in the second row are reversed. Notice also how the ϕ response is not monotonic, as it was in the previous two cases, but it responds first

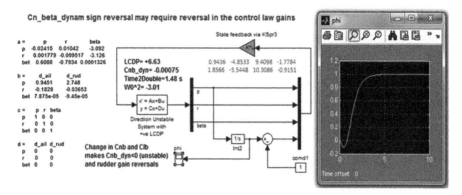

FIGURE 7.6   Closed-Loop System with Positive LCDP Ratio and $C_{n\beta}$-Dynamic < 0 (Unstable)

toward the opposite direction and then reverses. This is also an unreliable situation and one cannot depend on the aerosurfaces for lateral control during this type of transitioning and, therefore, RCS is the only reliable solution during periods of uncertainty in the LCDP.

7.3.18   Effector Capability to Provide Rotational Accelerations

The capability of the control effectors system to provide rotational acceleration in roll, pitch, and yaw can be evaluated in terms of the largest rotational accelerations achieved in the corresponding direction before saturating at least one of the effectors. From the moment equations we can calculate the accelerations as follows:

$$\dot{p} = \frac{L_{\delta P}}{I_{XX}}\delta_{P_{FCS}} \qquad \dot{q} = \frac{M_{\delta Q}}{I_{YY}}\delta_{Q_{FCS}} \qquad \dot{r} = \frac{N_{\delta R}}{I_{ZZ}}\delta_{R_{FCS}} \qquad (7.31)$$

where $\delta_{P_{FCS}}, \delta_{Q_{FCS}}, \delta_{R_{FCS}}$ are the rotational acceleration demands originating from the FCS. The rotational accelerations in roll, pitch, and yaw are maximized when the demands in the corresponding directions are also maximized. This happens when the deflection of the most active effector (i) reaches its peak limit $\delta_{i\text{Max}}$. By substituting equation 7.20 for the maximum pitch demand we obtain the maximum pitch acceleration in (rad/sec^2) in equation 7.32. Similarly for the roll and yaw accelerations.

$$\dot{p}_{max} = \frac{L_{\delta P}}{I_{XX}U_{P\,Max}} \qquad \dot{q}_{max} = \frac{M_{\delta Q}}{I_{YY}U_{Q\,Max}} \qquad \dot{r}_{max} = \frac{N_{\delta R}}{I_{ZZ}U_{R\,Max}} \qquad (7.32)$$

where U_{QMax} is the element that has the largest magnitude in vector ($\underline{U}_Q$). The column vector ($\underline{U}_Q$) corresponds to the pitch demand $\delta_{Q_{FCS}}$ in the normalized mixing-logic equation 7.18a, and its dimension is equal to the number of control effectors.

7.3.19 Effector Capability to Provide Translational Accelerations

Similarly, the capability of the effectors system to perform translation maneuvers along the vehicle x-, y-, and z-axes can be evaluated in terms of the largest translational accelerations that can be achieved along those directions, before at least one of the effectors reaches its saturation limit. The acceleration along a linear direction is maximized when the deflection of the most active effector (i) reaches its saturation limit $\delta_{i\text{Max}}$. By applying a similar argument as in equation 7.32, the maximum accelerations along x, y, and z are shown in equation 7.33

$$\dot{u}_{\max} = \frac{F_{X\delta X}}{M_V U_{X\,\text{Max}}} \quad \dot{v}_{\max} = \frac{F_{Y\delta Y}}{M_V U_{Y\,\text{Max}}} \quad \dot{w}_{\max} = \frac{F_{Z\delta Z}}{M_V U_{Z\,\text{Max}}} \quad (7.33)$$

where $U_{X\text{max}}$ is the element that has the largest magnitude in vector ($\underline{U}_X$). The column vector ($\underline{U}_X$) corresponds to the FCS demand $\delta_{X_{FCS}}$ in equation 7.18. Similarly for $U_{Y\text{max}}$ and $U_{Z\text{max}}$.

7.3.20 Steady Pull-Up Maneuverability

Another important parameter which is applicable mainly to aircraft is the amount of pitch FCS command and the angle of attack required to hold the vehicle in a steady pull-up maneuver with a load factor n_z. When the vehicle performs a circular pull-up maneuver at constant speed V, see Figure 7.7, the net normal force vertically upward

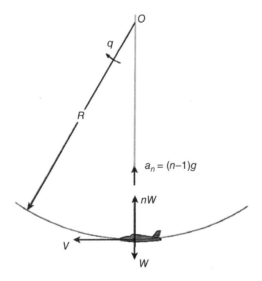

FIGURE 7.7   Steady Pull-Up Maneuver

at the point where the flight path tangent is horizontal, is $L - W = (n_z - 1)W$. The normal acceleration, therefore, is $(n_z - 1)g$, and the angular velocity of the vehicle is:

$$q = \frac{(n_z - 1)\, g}{V} \tag{7.34}$$

The increment in lift above what is required for 1g level flight is $\Delta L = (n_z - 1)\, W$ or in coefficient form it is:

$$\Delta C_L = \frac{\Delta L}{\overline{Q}S} = \frac{(n_z - 1)\, W}{\overline{Q}S} \tag{7.35}$$

The increment in lift and pitch moment $\{\Delta C_L \text{ and } \Delta C_m\}$ caused by increments in pitch command $(\Delta\delta_{Q_{FCS}})$ and in angle of attack $(\Delta\alpha)$ during the pull-up maneuver are:

$$\Delta C_L = C_{L_\alpha}\Delta\alpha + C_{L_q}\hat{q} + C_{L_{\delta Q}}\Delta\delta_{Q_{FCS}}$$

$$\Delta C_m = C_{m_\alpha}\Delta\alpha + C_{m_q}\hat{q} + C_{m_{\delta Q}}\Delta\delta_{Q_{FCS}} = 0;$$

$$\text{where } \hat{q} = \frac{(n_z - 1)\, \bar{c}g}{2V^2} \tag{7.36}$$

This is reduced to two equations with two unknowns: change in alpha $(\Delta\alpha)$ and change in pitch control $(\Delta\delta_{Q_{FCS}})$, which must be solved to obtain the required load factor:

$$C_{L_\alpha}\Delta\alpha + C_{L_{\delta Q}}\Delta\delta_{Q_{FCS}} = (n_z - 1)\left[\frac{W}{\overline{Q}S} - C_{L_q}\frac{g\bar{c}}{2V^2}\right]$$

$$C_{m_\alpha}\Delta\alpha + C_{m_{\delta Q}}\Delta\delta_{Q_{FCS}} = -(n_z - 1)\, C_{m_q}\frac{g\bar{c}}{2V^2} \tag{7.37}$$

If the required load factor is one, which is equivalent to a steady level flight, the RHS of equation 7.37 becomes zero, and the increments in alpha and pitch control $(\delta_{Q_{FCS}})$ also become zero. From equation 7.37 we can solve for the change in pitch control per g, and the change in angle of attack per g.

$$\left[\frac{\Delta\delta_{Q_{FCS}}}{n_z - 1}\right] = -\frac{1}{D}\left\{\left(C_{m_q}\frac{g\bar{c}}{2V^2}\right)C_{L_\alpha} + \left(\frac{W}{\overline{Q}S} - C_{L_q}\frac{g\bar{c}}{2V^2}\right)C_{m_\alpha}\right\} \tag{7.38a}$$

$$\left[\frac{\Delta\alpha}{n_z - 1}\right] = \frac{1}{D}\left\{\left(C_{m_q}\frac{g\bar{c}}{2V^2}\right)C_{L_{\delta Q}} + \left(\frac{W}{\overline{Q}S} - C_{L_q}\frac{g\bar{c}}{2V^2}\right)C_{m_{\delta Q}}\right\} \tag{7.38b}$$

where $D = C_{L\alpha}C_{m_{\delta Q}} - C_{m\alpha}C_{L_{\delta Q}}$ and $\frac{W}{\overline{Q}S} = C_L$ which is the lift coefficient at level flight.

The issue here is to evaluate the flight vehicle maneuverability and if the vehicle has enough control authority to achieve its maximum loading specifications during a pull-up maneuver. Another way of solving these equations is to calculate what would the increase in the pitch FCS command $\Delta \delta_{Q_{FCS}}$ be to raise the load factor $(n_z - 1)$ to the max specification limit, or a better measure of maneuverability would be to evaluate the ratio $(\Delta \delta_{Q_{FCS}} / \Delta \delta_{Q_{FCS\,Max}})$, where $\Delta \delta_{Q_{FCS\,Max}}$ is the max FCS demand before it saturates the effectors. From equation 7.20 the maximum pitch command increment is:

$$\Delta \delta_{Q_{FCS\,Max}} = \frac{1}{U_{Q\,Max}} \tag{7.39}$$

From equation 7.38a, and assuming that the angle of attack is small, we can calculate the ratio of increment to achieve max load over max increment to saturation, assuming a constant speed V and pitch rate q. This ratio should be less than 1, but we should be aiming toward a smaller number ($f = 0.7$), because the aircraft is already using some pitch control to trim at level flight.

$$\left[\frac{\Delta \delta_{Q_{FCS}}}{\Delta \delta_{Q_{FCS\,Max}}} \right] = (n_z - 1) \frac{U_{Q\,Max}}{D} \left\{ \left(C_{m_q} \frac{g\bar{c}}{2V^2} \right) C_{Z_\alpha} + \left(C_{Z0} - C_{Z_q} \frac{g\bar{c}}{2V^2} \right) C_{m_\alpha} \right\} < f \tag{7.40}$$

From equation 3.38b we calculate the change in angle of attack that results from maximizing the load,

$$\left[\frac{\Delta \alpha}{n_z - 1} \right] = -\frac{1}{D} \left\{ \left(C_{m_q} \frac{g\bar{c}}{2V^2} \right) C_{Z_{\delta Q}} + \left(C_{Z0} - C_{Z_q} \frac{g\bar{c}}{2V^2} \right) C_{m_{\delta Q}} \right\} \tag{7.41}$$

where $D = -C_{Z_\alpha} C_{m_{\delta Q}} + C_{m_\alpha} C_{Z_{\delta Q}}$

The pitch control moment derivative is: $C_{m_{\delta Q}} = M_{\delta Q} / \bar{Q} S \bar{c}$, and the normal control force derivative is $C_{Z_{\delta Q}} = Z_{\delta Q} / \bar{Q} S$, where the partials $M_{\delta Q}$ and $Z_{\delta Q}$ are the pitch moment and the normal force (along $+z$) partials with respect to the pitch FCS command ($\delta_{Q_{FCS}}$), obtained from equation 7.9.

7.3.21 Pitch Inertial Coupling Due to Stability Roll

A flight vehicle must possess sufficient nose down pitch authority to overcome the nose up moment as a result of inertial cross-coupling during high angle of attack, stability axis, roll maneuvers. That is, roll maneuvers about the velocity vector V, assuming a constant flight path without sideslip. The pitching moment due to roll and yaw body rates is obtained from the following equation:

$$M = I_Y \dot{q} - I_{XZ} \left(r^2 - p^2 \right) - \left(I_Z - I_X \right) r p \tag{7.42}$$

The roll rate in stability axis (p_{stab}), which is pure roll about the velocity vector, is a function of the body rates (p and r). After some equation manipulations, the pitching moment (M_{IC}) due to inertial coupling with p_{stab} can be written as shown in equation 7.43.

$$M_{IC} = \left[I_{XZ} \cos(2\alpha) - 0.5 \left(I_Z - I_X \right) \sin(2\alpha) \right] p_{stab}^2 \qquad (7.43)$$

which illustrates how the induced pitching moment M_{IC} is dramatically increased with roll rate (p_{stab}). This nose up moment due to roll coupling reaches its maximum at $\alpha = 45°$, so the vehicle must have enough pitch control authority to overcome this coupling moment. By combining equations 7.43 with 7.9 and 7.20, the max roll rate must be less than

$$p_{stab}^2 < \frac{M_{\delta Q}}{\left[I_{XZ} \cos(2\alpha) - 0.5 \left(I_Z - I_X \right) \sin(2\alpha) \right] U_{QMax}} \qquad (7.44)$$

Actually, it should be significantly less than what is shown in equation 7.44 and we should allow a margin factor because some additional control power is needed for normal flight path control. Some pitch authority is needed only to maintain attitude at zero roll rate.

7.3.22   Yaw Inertial Coupling Due to Loaded Roll

The yaw control system must possess sufficient authority to overcome the yaw inertial coupling moments during a rolling pullout maneuver (which produces simultaneously roll and pitch rates). The adverse yawing moment during a rolling pullout maneuver is:

$$N_{IC} = \left(I_X - I_Y \right) pq \cos(\alpha) \qquad (7.45)$$

The max pitch rate q is determined by the bank angle and the normal load factor applied to the airframe. The adverse yawing moment is most severe, because it results in highest pitch rate, when the loading occurs while the airplane is inverted (due to additional contribution from gravity). The pitch rate of the aircraft while inverted at max loading is:

$$q = \frac{\left(n_z + 1 \right) g}{V} \qquad (7.46)$$

The yaw FCS control ($\delta_{R_{FCS}}$) needed to counteract this adverse yawing moment is:

$$N_{\delta R} \delta_{R_{FCS}} = \left(I_Y - I_X \right) p \cos(\alpha) \frac{\left(n_z + 1 \right) g}{V} \qquad (7.47)$$

After taking into consideration the maximum control authority in yaw

$$\delta_{R_{FCS\,Max}} = \frac{1}{U_{R\,Max}}$$

We can solve for the max roll rate (at max pitch rate) that the yaw FCS can tolerate before it saturates the yaw effectors.

$$p_{max} < \frac{V N_{\delta R}}{U_{R\,Max} \left(I_Y - I_X\right) \left(n_Z + 1\right) g \cos \alpha} \qquad (7.48)$$

7.3.23   Moments at the Hinges of the Control Surfaces

The moments at the hinges of the control surfaces are used for sizing the actuator torques. Control surfaces rotate about a hinge line which is parallel to the wings and tails and powered by electromechanical or hydraulic actuators. The torque supplied by the actuators must be greater than the torques generated by the aerodynamic moments at the surface hinges.

Equation 7.49 calculates the moment at the hinge of the ith control surface. It is a function of the dynamic pressure, the reference area of the aerosurface (S_{Rsi}), the surface chord which is the distance between the surface center of pressure and the hinge line, and the hinge moment coefficient which varies as a function of: α, β, Mach number, and the surface increment (δ_{so}). There is a fixed number of surface increments in the database. The first term in equation 7.49 calculates the hinge moment at the nearest surface increment. The equation includes a correction term to smooth out the hinge moment calculations. It uses the derivative of the hinge moment coefficient per surface deflection to interpolate the hinge moments between the surface increments using the difference ($\delta_{si} - \delta_{so}$), where δ_{si} is the actual surface deflection.

$$M_{HSi} = \overline{Q} S_{Rsi} \bar{c}_{hsi} \left\{ C_{hmc}\left(i, M, \alpha, \beta, \delta_{so}\right) + \left. \left(\frac{\partial C_{hmc}\left(i, M, \alpha, \beta, \delta_{so}\right)}{\partial \delta_{si}} \right) \right|_{so} \left\{ \delta_{si} - \delta_{so} \right\} \right\}$$

$$(7.49)$$

where

δ_{si}	is the control surface (i) deflection	(deg)
δ_{so}	is the nearest deflection with available data	(deg)
C_{hsi}	is the chord of the control surface (i)	(ft)
$\overline{Q}$	is the aerodynamic pressure	(PSF)
S_{Rsi}	is the reference area of the control surface (i)	(ft^2)
C_{hmc} $(i, M, \alpha, \beta, \delta)$	is the hinge moment coefficient of surface (i)	(−)

The hinge moment coefficients of a surface (i), C_{hmc} $(M, \alpha, \beta, \delta)$, and the derivatives are generated by aerodynamics specialists. They are functions of four variables: the Mach number M, the aerodynamic angles (α, β), and the control surface increment δ_{so}.

7.4 NOTES ON SPIN DEPARTURE

By Aditya A. Paranjape

A spin is an uncontrolled motion of an aircraft at high angles of attack accompanied by large angular rates and a near-vertical flight path. An important point to bear in mind is that spin is not necessarily an instability, in fact, spins are usually quite stable as equilibria or as limit cycles. The onset of spin in an aircraft is related quite closely to stall. Imagine a pilot pulling back on the control column steadily, and assume that the aircraft is flying wings-level. When the angle of attack gets past the stalling value, both wings stall and the roll damping of the aircraft typically becomes negative. The ensuing roll instability leads even to a divergence in the roll and yaw rates in the event of a small disturbance or pilot input, a phenomenon termed as autorotation.

The mechanism described above leads to spin only if spin solutions exist, that is, if the forces and moments balance each other to produce equilibrium spin or oscillatory spin. Therefore, from the perspective of aircraft design, it is important to understand two distinct points: (1) the characteristics of an aircraft which make spin solutions likely to exist, and (2) the flight regime in which the aircraft is likely to depart into a spin. The second problem is a classic stability problem. The former problem is one of determining the existence of equilibrium or periodic solutions. This viewpoint is supported by the three stages into which a generic spin may be split (see [1,2]):

- Incipient spin: This is the phase described above, starting with stall, going through autorotation, and ending with the axis of rotation and flight path becoming nearly vertical.
- Developed spin: The rotational motion of the aircraft is stabilized about an equilibrium condition or a period oscillation. The corresponding spins are called equilibrium and oscillatory spins, respectively.
- Recovery: This is usually a result of pilot inputs or automatic control inputs exerted against the spinning motion to annul the rotation, followed by a pitch-down to safe angles of attack.

The onset of incipient spin corresponds to a loss of stability. Thereafter, a developed spin occurs only if the corresponding equilibrium or periodic solutions exist. Of particular interest during the design phase are criteria which predict incipient spin, of which several have been proposed over the years.

7.4.1 Stability-Based Criteria

One of the earliest criteria for spin susceptibility assessment was due to Moul and Paulson who proposed the use of $C_{n\beta}$-dynamic and LCDP together.

$$\text{LCDP} = C_{n\beta} - \left(\frac{C_{n\delta a}}{C_{l\delta a}}\right) C_{l\beta} \qquad C_{n\beta}\text{-dynamic} = C_{n\beta}\cos\alpha - \frac{I_Z}{I_X}C_{l\beta}\sin\alpha$$

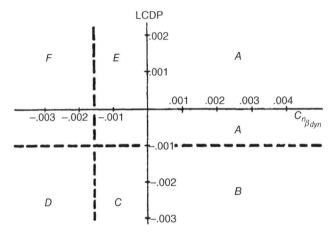

FIGURE 7.8   STI-Modified Weissman's Criterion (From [1,6]), The Regions of the Plot Are Explained in Table 7.1.

Since positive values of these parameters would likely rule out lateral-directional departures and control reversal, it can be assumed that spin is ruled out as well. Moul and Paulson's criterion was subsequently modified to improve the estimate of the departure-free range of parameters and additionally qualify the problems arising out of one or both parameters lying outside the safe region (see [2] for more details). One such criterion, called the STI-modified Weissman's criterion, is shown in Figure 7.8, and tabulated in Table 7.1. Notice that spin susceptibility correlates well with the nature of the rolling and yaw departures.

Figure 7.9 shows a plot of $C_{n\beta}$-dynamic and LCDP for the F-16 data. The variation of $C_{n\beta}$-dynamic with α is also helping to interpret the first plot. Since $C_{n\beta}$-dynamic < 0 for $0.6 < \alpha < 1$ rad and LCDP is negative in the same range, we conclude that the aircraft is susceptible to spin in this regime. Moreover, from Figure 7.8, we infer that the aircraft is highly susceptible to spin for $\alpha > 0.6$ rad. A more complete picture of spin susceptibility is presented by Kalviste's criterion which is found by solving the coupled longitudinal-lateral rotational equations of motion. Since it does not look

TABLE 7.1   Regions in Figure 7.8

Region	Departures	Spin susceptibility
A	No departures	None
B	Mild roll departures	Low
C	Moderate roll departures	Moderate
D	Strong roll departures	High
E	Random yaw departures	Moderate
F	Severe yaw departures	High

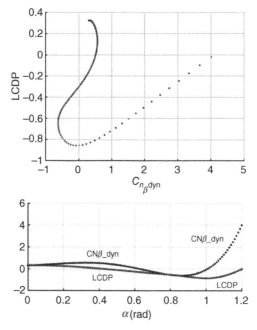

FIGURE 7.9   A Plot of Moul and Paulson's Criterion Applied to the F-16 Data

at lateral-directional dynamics in isolation, it may be viewed as an extension of the $C_{n\beta}$-dynamic based criteria.

While a complete derivation is beyond the scope of this chapter, a few salient points are worth noting:

- Kalviste's criterion identifies regions of the flight envelope (in terms of angle of attack and sideslip) where the aircraft is susceptible to coupled longitudinal divergence, lateral-directional divergence and an oscillatory instability involving α and β.
- Spin susceptibility is determined using the size of the unstable regions, when plotted on a contour plot with α and β as the axes.

While Kalviste's criterion represents a generalization of Moul and Paulson's criterion, it requires additional aerodynamic data which is typically available only in later stages of preliminary design. On the other hand, Moul and Paulson's criterion can be used even in the conceptual design phase, but will miss onset mechanisms arising from coupled longitudinal-lateral instabilities. A common feature of the stability-based criteria is that they are probabilistic in nature—they yield the likelihood of spin rather than an unambiguous, deterministic assessment. This owes to the fact that they use very little information about the dynamics of the aircraft and look for generic instabilities rather than any intrinsic property of spin.

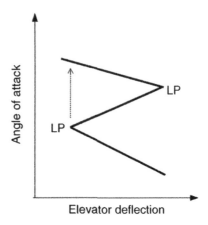

FIGURE 7.10 Elevator versus Angle of Attack at Trim, Showing Equilibrium Branches and Limit Points

7.4.2 Solution-Based Criteria

The limitations of stability-based criteria were addressed partially by criteria which attempted to solve for steady spin solutions rather than look for directional instabilities. The earliest of these criteria was due to Bihrle and Barnhart which computes equilibrium spin solutions together with associated control inputs. This criterion is meant primarily for use with rotary balance data from wind tunnels and is thus useful in later stages of design. It can help design spin-resistant airframes (no spin solutions with neutral control input) or those with desirable spin characteristics. A hybrid trim stability, but procedurally simpler, approach to determining the spin susceptibility was proposed by Paranjape and Ananthkrishnan [5]. Their criterion is based on two hypotheses:

- Incipient as well as developed spin equilibria occur on a common equilibrium branch in the state-control space provided it turns around twice, as shown in Figure 7.10. The points at which the branch turns around are referred to as limit points (each is denoted by LP in Figure 7.10).
- It suffices to determine whether or not LPs exist in order to infer spin susceptibility.

Figure 7.10 shows the variation of α trim versus elevon deflection δ_e. It depicts the limit points and the jump of α in the state-control space from a lower-α branch to an upper-α branch. In this figure, we see a single solution branch of angle of attack equilibrium as a function of δ_e, while keeping other control inputs fixed. The line joining the two LPs is also an equilibrium branch, but not the same equilibrium. It has different stability than the parent low-α branch. The LPs occur where the slope $\partial \delta_e / \partial \alpha = 0$, or where the solution branch turns around. LPs are neutrally stable equilibria and they typically occur in pairs. Solutions exist to only one side of the LP and there are no trim solutions in the vicinity of the LP. The systems jump toward the closest equilibrium for that value of the control input.

When the elevator up-deflection crosses the first LP, there is no equilibrium solution locally and the angle of attack jumps to a value on the upper solid branch. The arrow denotes the jump to the high-α branch associated with spin. The upper branch may be a spin solution, depending on the associated angular rates. When the state is on the upper branch, if you keep increasing the elevator toward the left, the angle of attack keeps increasing. If you lower the elevator toward the right instead, you could reach the LP on the right. Further down-elevator would be expected to return the aircraft to the lower branch via another jump, leading thereby to recovery from spin. In most cases, however, there are other solutions (e.g., limit cycles corresponding to oscillatory spin) that block the second jump and the aircraft ends up in those solutions. LPs also correspond to a qualitative change in stability, but this property is not explicitly used in the derivations. It can be argued that $\beta \approx 0$ and $q \approx 0$ at LPs since they are incipient spin solutions in their own right.

The first step in locating the incipient spins is to solve the rotational equations for equilibria as functions of elevator input as the independent control input. This procedure can be performed with the other control inputs with no change whatsoever in the final form of the criterion. The trim equations are given by equation 7.4.1.

$$
\begin{aligned}
\dot{p} &= L_0 + L_p p + L_r r = 0 \\
\dot{q} &= A\,p\,r + M(\alpha) + M_{\delta e}\,\delta_e = 0; \ \ \text{where}, A = \frac{I_Z - I_X}{I_Y} \\
\dot{r} &= N_0 + N_p p + N_r r = 0
\end{aligned}
\tag{7.4.1}
$$

The term $M(\alpha)$ includes contributions from the wing, fuselage, and the undeflected tail. The terms N_0 and L_0 are assumed to be essentially independent of α. They may depend on nonzero aileron and rudder inputs and all other geometric and control asymmetries, such as thrust vectored inputs, reaction jets, etc. They may also arise due to asymmetric vortex shedding from the fore-body. The trim values of the roll and yaw rates are given by equation 7.4.2.

$$
p = \frac{L_r N_0 - N_r L_0}{\Delta}; \ \ r = \frac{N_p L_0 - L_p N_0}{\Delta}; \ \ \text{where}: \ \Delta = L_p N_r - L_r N_p \tag{7.4.2}
$$

With these trim values, we get a single consolidated trim equation for δ_e as a function of α:

$$
\delta_e = -\frac{1}{\Delta^2 M_{\delta e}} \left[A\left(L_r N_0 - N_r L_0\right)\left(N_p L_0 - L_p N_0\right)\right] - \frac{M(\alpha)}{M_{\delta e}} \tag{7.4.3}
$$

The LPs are found by imposing the turnaround condition: $\partial \delta_e / \partial \alpha = 0$ together with the implicit assumption that the variation of $M_{\delta e}$ with α is not significant. The resulting expression makes for a simpler presentation after resubstituting for N_0 and

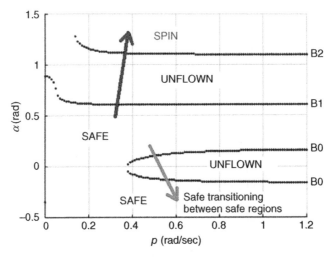

FIGURE 7.11   Trace of Limit Points in the α - p Space for the F-16 Model

L_0 in terms of p and r, together with the approximation $r = p \tan \alpha$:

$$Ap^2 \left[\left(\frac{\partial L_p}{\partial \alpha} N_p - \frac{\partial N_p}{\partial \alpha} L_p \right) + \left(\frac{\partial N_r}{\partial \alpha} L_r - \frac{\partial L_r}{\partial \alpha} N_r \right) - \frac{\partial \Delta}{\partial \alpha} \tan \alpha \right] + \Delta \frac{\partial M(\alpha)}{\partial \alpha} = 0$$

$$(7.4.4)$$

Equation 7.4.4 yields the loci of LPs in the α - p space. They are not equilibrium branches in the same sense as Figure 7.10, which are obtained by varying a single control parameter independently. Rather, the loci are the LPs that you would see on all such branches for the various (α, p) combinations. It is best understood through a numerical example. Figure 7.11 shows the trace of LPs, found by solving equation 7.4.4, for the F-16 data in the appendix. The aircraft can transition between various (α, p) trim states depending on the controls that it receives. Only positive values of roll rate are considered since the model is symmetric in p. There are three branches of LPs, two of which correspond to high alphas have been labelled as B1 and B2, and a pair of low alpha branches is labeled B0.

- The region below the branch B1, marked "Safe," denotes the equilibria in the flight regime at which the aircraft is safe from departures and spin. The aircraft can move between points in the safe region without undergoing jumps and with a quasi-static variation of control inputs. When the aircraft is given control inputs corresponding to an (α, p) combination that reaches and exceeds the B1 boundary, it departs into a spin.
- The region that lies between the branches B1 and B2, and in the region enclosed by the two B0 branches, marked "Unflown," corresponds to the segments

TABLE 7.2   Summary of Spin Prediction Criteria

Criterion	Nature	Derivatives required	Utility
Moul and Paulson	Stability	Static β and δ_α	Early design
Kalviste	Stability	Static and damping	Late preliminary design
Bihrle and Barnhart	Trim	Static and damping	Late design, rotary tests
Paranjape and Ananthkrishnan	Trim and stability hybrid	Damping	Late design, operations

between successive LPs on an equilibrium branch. These are rarely ever flown under usual circumstances or even in exigencies. They exist, nevertheless.

- The B0 branches, at low α, marks the boundary of the highest alpha/roll-rate combination that may be attained safely without undergoing a departure. A jump from the safe region to the unflown region will not lead to an equilibrium point in the unflown region itself, but in the region immediately beyond the unflown region, as shown by the lower arrow in Figure 7.11. The transition across the two B0 branches takes place between two points in the safe region itself. This means that there is a control combination for transitioning between those two points without a jump.

- The B2 branch represents the second set of limit points. The jump from below the B1 branch to a region beyond the B2 branch (the region is marked "Spin") indicates a spin departure, see the upper jump arrow in Figure 7.11. The developed spin is either an equilibrium spin found by extrapolating beyond the branch B2, or a limit cycle about some equilibrium. The existence of branch B2 allows us to predict spin, but not whether it would be an equilibrium spin or oscillatory spin. A loose rule of thumb is as follows: if the angle of attack on B2 is under 45°, we should expect a steep spin (equilibrium or oscillatory). At alphas higher than 45°, it would be a moderately flat spin, and at alphas over 65° we should expect a thorough flat spin, (see [2]).

In this context, it is interesting to note that the F-16 in our example is equipped with an α-limiter which prevents the commanded angle of attack from exceeding 30°. This is to prevent the aircraft from entering into a deep stall and an accompanying spin. The conclusions obtained from Figure 7.11 correlate well with those obtained from Figure 7.9. It also indicates the possibility of departure at low α (the unlabeled branch) which is not suggested by Figure 7.9. Interestingly enough, Paranjape and Ananthkrishnan's criterion ignores the β derivatives completely and therefore has no overlap with the previous criteria despite yielding the correct spin susceptibility assessment. The criteria examined here have been summarized in Table 7.2.

A simple way to solve equation 7.4.4 is to perform a manual raster scan of the (α, p) envelope: fix α, vary p, and calculate the criterion. The LPs occur wherever the criterion changes sign. This method is simple and reliable. Numerical continuation is an alternate, perhaps more elegant, approach. However, it requires several runs to ascertain that all the LP branches are computed.

7.5 APPENDIX

F-16 data used for computations consists of two parts. The inertial and geometric properties are reproduced here, while the aerodynamic coefficients were used from [4]. Wing span and mean chord: 9 m, 3 m; wing area: 27 m^2; principal moments of inertia, I_x; I_y; I_z: 12875; 75674; 85552 kg $\times$ m^2.

REFERENCES

[1] Bihrle Jr., W. and Barnhart, B. (1983) "Spin Prediction Techniques," *Journal of Aircraft*, 20 (2), pp. 97–101.

[2] Burk Jr, S.M., Bowman Jr, J.S., and White, W.L. (1977) "Spin-Tunnel Investigation of the Spinning Characteristics of Typical Single-Engine General Aviation Designs, I-Low Wing Model A: Effects of Tail Configurations," *NASA* TP 1009, .

[3] Chambers, J.R. and Grafton, S.B. (1977) "Aerodynamic Characteristics of Airplanes at High Angles of Attack," *NASA* TM 74097.

[4] Morelli, E.A. (1998) "Global Nonlinear Parametric Modelling With Application to F-16 Aerodynamics," *Proceedings of American Control Conference*, June 1998.

[5] Paranjape, A. and Ananthkrishnan, N. (2010) "Analytical Criterion for Aircraft Spin Susceptibility," *Journal of Aircraft*, 47(5), pp. 1804–1807.

[6] Seltzer, R.M. and Rhodeside, G.R. (1988) "Fundamentals and Methods of High Angle of Attack Flying Qualities Research," *Office of Naval Technology Report*, NADC-88020-60.

8

GRAPHICAL PERFORMANCE ANALYSIS

This chapter describes some graphical tools which are used for analyzing flight vehicle performance either along a trajectory or at fixed flight conditions. They provide a visual interpretation of vehicle performance and stability in the Mach versus alpha operational range, maneuverability in different directions, and they assess the control authority of the effector system against wind-shear disturbances. They are implemented in the Flixan program which is documented in Flixan.com.

8.1 CONTOUR PLOTS OF PERFORMANCE PARAMETERS VERSUS (MACH AND ALPHA)

In Chapter 7 we discussed how to calculate some important parameters that characterize vehicle stability and performance, as a function of trajectory time. The performance parameters, such as static stability, time to double amplitude, lateral control departure (LCDP), control authority, etc. are functions of the trajectory variables, aerodynamic coefficients, and also the trim angles at each point along the trajectory. Now let us suppose that some of the performance parameters do not meet our required criteria and we would like to modify the trajectory in order to improve the vehicle stability, maneuverability, etc. This would be difficult to accomplish because we would not know how to modify the trajectory in order to improve the performance. It would be nice to have an analytic tool that would give us a more global viewpoint that could help us toward reshaping the trajectory and to avoid regions of unacceptable performance. Furthermore, the performance requirements in most

Performance Evaluation and Design of Flight Vehicle Control Systems, First Edition. Eric T. Falangas.
© 2016 by Eric T. Falangas. Published 2016 by John Wiley & Sons, Inc.

aircrafts are not defined along a trajectory but over a range of speeds (Mach numbers) and angles of attack. So there is a need for an analytic method that would expand our capability to analyze vehicle performance in a wider range of Mach and alphas, rather than restricting it in the vicinity of a trajectory. The trajectory variables that strongly affect vehicle performance are the Mach number and the angle of attack because the aerodynamic data strongly depend on them. So the Mach versus alpha combination is a good choice of variables along which to examine performance.

Contour plots are three-dimensional surface plots that provide a wider depiction on how a vehicle performance parameter varies over the entire Mach versus alpha envelope. The Mach number is plotted in the horizontal *x*-axis, the angle of attack (deg) in the *y*-axis, and the performance parameter under evaluation is plotted in the normal to the paper *z*-axis, except that instead of using a three-dimensional display the value of the performance parameter in the *z*-axis is color coded. The trajectory is also shown as a black line traveling across the Mach versus alpha field. This type of presentation helps the analysts to identify any potentially unacceptable regions to be avoided across the Mach versus alpha range, and how to reshape the trajectory trail, or modify the effectors, as needed, in order to attain an acceptable performance. Trajectory reshaping, however, is not always easy because there are other factors to consider in a trajectory, such as payload maximization, structural loading, aeroheating.

Figure 8.1 shows the contour plot of the pitch stability parameter (T2-inverse) for an unpowered descending vehicle from Mach 4.5 to landing at Mach 0.25. This parameter is described in equation 7.14, and it is plotted versus Mach and angle

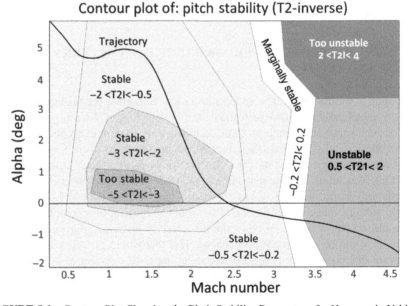

FIGURE 8.1 Contour Plot Showing the Pitch Stability Parameter of a Hypersonic Vehicle Along the Trajectory as a Function of Mach and Alpha; The Trajectory Begins From a Region of Instability and Becomes Stable Below Mach 3.3

of attack in (deg). The trajectory begins in the lower right-hand corner where the vehicle is slightly unstable at ($\alpha = -1.6°$, Mach = 4.5). It briefly passes through a neutrally stable region (white) when Mach is 3.3 and alpha slightly negative, and it crosses into the stable region as the angle of attack increases toward positive values. It remains in the stable region all the way to landing in the upper left-hand corner, where ($\alpha = 5.5°$, Mach = 0.25). The shading of the regions in the surface plot shows the approximate range of the pitch stability parameter in that region. Initially, at high Mach the trajectory passes through a statically unstable (divergent) region with a time to double amplitude T2 = 0.67 (sec). Then it crosses into the stable region that has short-period frequencies peaking at around 3 (rad/sec). At Mach 2 the trajectory passes near a region where the stability parameter would have been a little "too stable" but it does not cross that region and there is no reason to modify it. The trajectory ends up slightly stable and close to neutral stability near landing at ($\alpha = 5.5°$, Mach = 0.25). A similar plot can be drawn for the lateral T2-inverse stability parameter.

Figure 8.2 shows the contour plot for the LCDP ratio parameter for the same hypersonic vehicle. The LCDP ratio was described in equation 7.28. The LCDP values are drawn in different shades as a function of Mach and alpha with the same trajectory line across. White represents the region corresponding to LCDP values near one where the roll/yaw coordination is perfect. Values of LCDP magnitude which are less than 0.2 perform poorly in roll because the aileron to roll-rate transfer function (7.29) becomes weak, and the aileron would be unreliable for roll control, as described in Chapter 7. Low LCDP values are also likely to reverse sign due to aero uncertainties and would require changes in the sign of the roll control gain. In

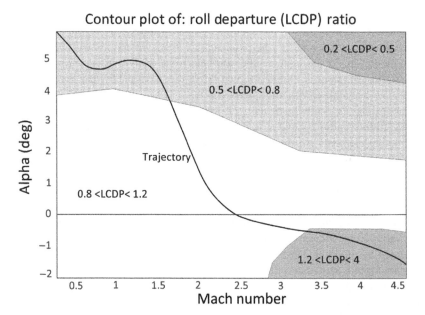

FIGURE 8.2   LCDP Ratio of a Hypersonic Vehicle Shows Excellent Roll Performance Across the Entire Mach versus Alpha Region

the example of Figure 8.2, however, the LCDP ratio contour plot is almost perfect because there are no bad regions and no sign reversals across the Mach versus alpha range. Contour plots are also used to plot other performance parameters against Mach and alpha, such as, control authority in each direction and the parameters described in Section 7.3.

8.2   VECTOR DIAGRAM ANALYSIS

Vector diagrams are two-dimensional plots that analyze vehicle controllability in two directions and compare the vehicle control authority against moments and forces produced by aerodynamic disturbances in the same directions. They compare the control moments and forces generated by the effectors system in two directions, for example, in roll and yaw, or in pitch and normal acceleration, against the moments and forces produced by a steady wind-shear disturbance. The disturbance is defined by the maximum alpha and beta dispersion angles ($\pm\alpha_{max}$ and $\pm\beta_{max}$) generated by the expected wind-shear relative to the trim alpha and beta (α_0, β_0) defined in the trajectory. Vector diagrams examine if the vehicle has the control authority required to counter the disturbance along the control directions. They also measure maneuverability because the α_{max} and β_{max} dispersion angles may be due to maneuvering. Vector diagrams compare not only magnitudes but they also allow us to examine the directions of the controls versus disturbance. They helps us evaluate the orthogonality of the control system by comparing the moments, forces, or accelerations of the controls against those generated by the wind disturbance and to determine if the controls are powerful enough and pointing in the proper directions to counteract the disturbances along the control directions. Vector diagrams also show the effects of aerodynamic uncertainties in the coefficients. Uncertainties are drawn as rectangles around the vector tips and they help the analyst to decide about the efficiency of the the aerodata. That is, to request either more accurate data or airframe modifications. Since vector diagrams are limited to two directions we typically use several plots to analyze control authority in multiple directions.

Vector diagram is obviously an open-loop static analysis of the airframe alone at specific flight conditions and it is not related to closed-loop control analysis. The contributing elements are shown in Figure 8.3. On the left side we have the control demands δ_{FCS} which are generated by the flight control system (FCS) demanding vehicle acceleration in certain directions, mainly in roll, pitch, and yaw, and possibly in some translational directions. A mixing-logic matrix translates the acceleration demands to effector displacements (or throttles) relative to their trim positions which are converted to control moment and force variations M_δ. The control moment and force variation vectors generate the vehicle accelerations, hopefully, in the directions demanded by the FCS. The disturbances are introduced at the bottom of Figure 8.3 by the maximum anticipated variations ($\pm\alpha_{max}$ and $\pm\beta_{max}$) from trim. They generate the disturbance moments and forces vector M_α and disturbance accelerations. The vehicle is expected to have sufficient control authority M_δ to counter M_α at any given flight condition. In the following sections we will describe four different types of vector diagrams used to address different aspects of static controllability.

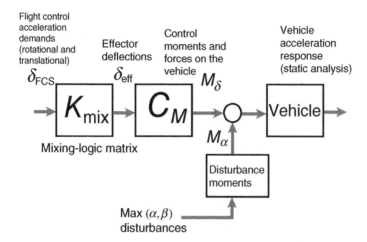

FIGURE 8.3   Elements Used for Analyzing Vehicle Controllability by Vector Diagrams

8.2.1   Maximum Moment and Force Vector Diagrams

The first type of vector diagrams to examine is the maximum moment/force diagrams that plot the maximum control moments and forces generated by the effectors system in two directions against those generated by the alpha and beta dispersions. The control moments and forces are calculated by maximizing the flight control demands to their saturation limits in both the positive and in the negative directions relative to trim. The FCS demand in one of the control directions is maximized when at least one of the effectors saturates. Equation 8.1 denotes maximum moments and forces produced when the controls saturate in two opposite directions. That is, when the FCS demand is maximized either in the positive ($\underline{\delta}_{+\text{FCS Max}}$) or in the negative ($\underline{\delta}_{-\text{FCS Max}}$) directions.

$$M_\delta = C_M K_{\text{mix}}\delta_{\text{FCS}}$$

$$
\begin{bmatrix} L_X \\ M_Y \\ N_Z \\ F_X \\ F_Y \\ F_Z \end{bmatrix}_{+\text{Max}} =
\begin{bmatrix}
L_{\delta P} & 0 & L_{\delta R} & 0 & L_{\delta Y} & 0 \\
0 & M_{\delta Q} & 0 & M_{\delta X} & 0 & M_{\delta Z} \\
N_{\delta P} & 0 & N_{\delta R} & 0 & N_{\delta Y} & 0 \\
0 & F_{X\delta Q} & 0 & F_{X\delta X} & 0 & F_{X\delta Z} \\
F_{Y\delta P} & 0 & F_{Y\delta R} & 0 & F_{Y\delta Y} & 0 \\
0 & F_{Z\delta Q} & 0 & F_{Z\delta X} & 0 & F_{Z\delta Z}
\end{bmatrix}
\begin{bmatrix} \delta_P \\ \delta_Q \\ \delta_R \\ \delta_X \\ \delta_Y \\ \delta_Z \end{bmatrix}_{+\text{FCS Max}}
\quad
\begin{array}{l}\text{for max}\\\text{positive}\\\text{demands}\end{array}
\qquad (8.1)
$$

$$
\begin{bmatrix} L_X \\ M_Y \\ N_Z \\ F_X \\ F_Y \\ F_Z \end{bmatrix}_{-\text{Max}} =
\begin{bmatrix}
L_{\delta P} & 0 & L_{\delta R} & 0 & L_{\delta Y} & 0 \\
0 & M_{\delta Q} & 0 & M_{\delta X} & 0 & M_{\delta Z} \\
N_{\delta P} & 0 & N_{\delta R} & 0 & N_{\delta Y} & 0 \\
0 & F_{X\delta Q} & 0 & F_{X\delta X} & 0 & F_{X\delta Z} \\
F_{Y\delta P} & 0 & F_{Y\delta R} & 0 & F_{Y\delta Y} & 0 \\
0 & F_{Z\delta Q} & 0 & F_{Z\delta X} & 0 & F_{Z\delta Z}
\end{bmatrix}
\begin{bmatrix} \delta_P \\ \delta_Q \\ \delta_R \\ \delta_X \\ \delta_Y \\ \delta_Z \end{bmatrix}_{-\text{FCS Max}}
\quad
\begin{array}{l}\text{for max}\\\text{negative}\\\text{demands}\end{array}
$$

Maximum moment and force diagrams compare the maximum control authority of the effectors system against the maximum disturbance moments and forces generated by the $\pm\alpha_{max}$ and $\pm\beta_{max}$ dispersions from trim. It is not sufficient to show only the maximum moment/force generated from the max positive FCS demand because the effectors are not always trimmed exactly at the midpoint between the two saturation limits, plus their effectiveness is not always symmetric between positive and negative directions, so in this type of vector diagrams we plot the max moments and forces produced by maximizing the controls in both positive and negative directions. The dispersions play the role of design requirements that the vehicle should be able to tolerate and we show the max moments and forces produced due to $\pm\alpha_{max}$ and $\pm\beta_{max}$. The control vectors are contrasted against the disturbance vectors to demonstrate the vehicle controllability or maneuverability and its capability to react against the worst case disturbances. We will illustrate them with two examples, where the units of moments and forces are nondimensionalized by the usual scaling.

The vector diagram in Figure 8.4a analyzes the longitudinal controllability of an accelerating vehicle that possesses pitch and axial acceleration control against the aerodynamic disturbances generated by dispersions in the angle of attack. Figure 8.4a shows the max pitch moment (C_m) and max axial force (C_X) control vectors produced by maximizing the pitch control demand in the positive ($\delta_{+QFCS\ Max}$) direction (solid line) and also in the negative ($\delta_{-QFCS\ Max}$) direction (dashed line). Notice that the pitch control vectors affect mostly in the $\pm$pitch direction and are pointing toward their intended directions with a small amount of cross-coupling in the axial force direction. The figure also shows the moment and force vectors produced by throttling when the axial acceleration demand is maximized in the positive ($\delta_{+XFCS\ Max}$) and also in the negative ($\delta_{-XFCS\ Max}$) directions from trim. Axial control is achieved by varying the thrust, as shown by the two control vectors pointing up and down along the axial force axis, in the $+C_X$ (solid line) and in the $-C_X$ (dashed line) directions, with a small amount of cross-coupling in the pitch direction. Since the vehicle in this flight condition is accelerating under constant thrust, at trim the axial force is biased at $C_{X_0} = 0.128$ and by throttling C_X can be varied between 0.085 and 0.19. Notice that the controls are not symmetric relative to trim (origin). The peak negative pitching moment is greater than the peak positive pitch moment, and the variation in acceleration due to throttling is also asymmetric. Figure 8.4a also shows the pitching moment and axial force generated due to the variations in the angle of attack $\pm\alpha_{max}$ from trim. An increase in alpha causes an increase in pitching moment and a reduction in axial force (drag increase).

Figure 8.4b shows the pitch moment C_m versus the normal force C_Z because the pitch control also couples in the C_Z direction which is not directly controlled by the effectors system. It shows how the pitch moment C_m and normal force C_Z vary by maximizing the pitch controls ($\delta_{\pm QFCS\ Max}$) in both directions. Control authority, however, is not evaluated along C_Z because C_Z is not directly controllable. So we do not care if the pitch control couples into C_Z and also the $\pm\alpha_{max}$ variations affect the C_Z direction almost as much as the pitch control affects it in z. Notice also that the vehicle is trimmed with a negative normal force (up), $C_{Z_0} = -0.4$, obviously because the trim angle of attack is positive.

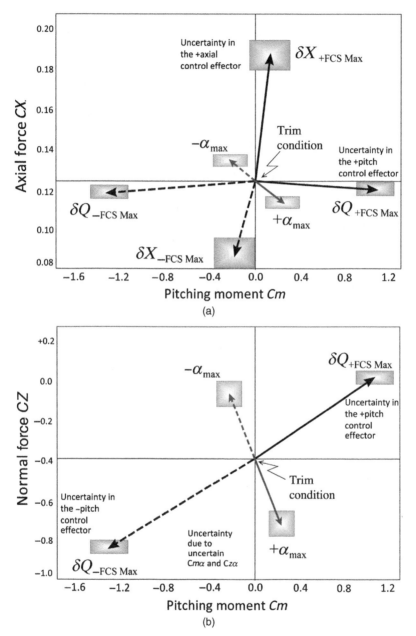

FIGURE 8.4 (a) Maximum Pitch Moment and Axial Force Produced by Maximizing the Pitch and Axial Acceleration Controls Plotted Against the Moment and Force Produced Due to Maximum Alpha Dispersions $\pm\alpha_{max}$; (b) Maximum Pitch Moment and Normal Force Produced by Maximizing the Pitch Control, It is Compared Against the Moment and Force Generated by the Maximum Alpha Dispersion $\pm\alpha_{max}$

The square rectangles at the tips of the vectors represent the possible variation of the vectors as a result of aerodynamic uncertainties. The bigger the uncertainties the larger the rectangles. The uncertainties in the aero-coefficients are estimated by the aerodynamics group. In this example the two longitudinal control vectors obviously dominate against the disturbance vectors due to the $\pm\alpha_{max}$ variations, including uncertainties.

The vector diagram in Figure 8.5 shows similar results in the lateral, roll, and yaw directions, where we plot the nondimensional moments, C_l versus C_n. The two controls are roll and yaw FCS demands. The two horizontal vectors are the moments produced by the max positive yaw demand (δR_{+FCS_Max}) and the max negative yaw demand (δR_{-FCS_Max}). The two nearly vertical vectors are the moments generated by maximizing the roll FCS demands in two opposite directions δP_{+FCS_Max} and δP_{-FCS_Max}. All control vectors are pointing in the proper directions with small cross-coupling into each other's direction. The two disturbance vectors represent the roll and yaw moments generated by the max dispersions in the angle of sideslip $\pm\beta_{max}$. The dispersion angles define the worst case wind-shear or maneuverability conditions, and they typically vary between $\pm 2°$ and $\pm 5°$ from trim, depending on the flight condition. The $+\beta_{max}$ vector shows an increase in yawing moment (and negative rolling moment) when the angle of sideslip is increased from trim to ($\beta_0 + \beta_{max}$). It proves that the vehicle is statically stable in yaw because it turns toward the airflow. The $-\beta_{max}$ vector shows the reverse effect due to negative sideslip.

Notice that the yaw control vector slightly couples into roll but the roll control has a stronger component into yaw. This is intentionally caused by the mixing logic

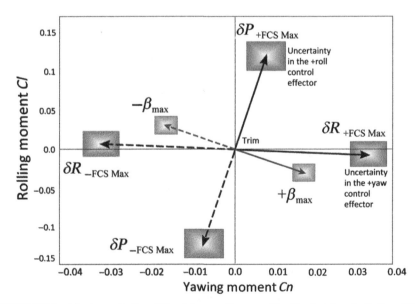

FIGURE 8.5   Maximum Roll and Yaw Moment Generated by Maximizing the Roll and Yaw Control Demands Against the Moments Produced by Maximum Sideslip Dispersion ($\pm\beta_{max}$)

which provides lateral decoupling to compensate against the vehicle I_{xz} cross product of inertia. In vector diagrams we typically expect to see the two control vectors almost perpendicular to each other and their magnitudes exceeding those of the disturbance vectors, such as in this case.

The square rectangles at the tip of the disturbance vectors represent the possible variations of this vector as a result of the aerodynamic uncertainties. They are due to the uncertainty in the aero-coefficients (C_l and C_n). The rectangles at the tips of the yaw control vectors represent possible variations due to the uncertainties in the control surface derivatives, which are mainly in the rudder ($C_{n\delta_{rudd}}$), but also the aileron ($C_{n\delta_{ailr}}$) is contributing because they both combine for yaw control. Similarly, the rectangles at the tips of the roll control vectors represent variations in the roll vector caused by the uncertainties in the aerosurface derivatives, mainly in aileron ($C_{l\delta_{ailr}}$), but also in the rudder ($C_{l\delta_{rudd}}$), because they are both contributing in roll control. The uncertainty rectangles should be sufficiently small to preclude the possibility that the disturbance moments may become stronger than the control moments in either direction.

8.2.2   Maximum Acceleration Vector Diagrams

In some applications it may be more appropriate to plot accelerations rather than moments and forces. The maximum acceleration vector diagrams are very similar to the maximum moments and force diagrams. The only difference is that the moment vectors from equation 8.1 are scaled by dividing them with the moments of inertia dyadic matrix and the force vectors are divided by the vehicle mass to be converted to angular and linear accelerations, as shown in equation 8.2.

$$
\begin{bmatrix} \dot{p} \\ \dot{q} \\ \dot{r} \\ \ddot{x} \\ \ddot{y} \\ \ddot{z} \end{bmatrix} = \begin{bmatrix} I_{XX} & -I_{XY} & -I_{XZ} & 0 & 0 & 0 \\ -I_{XY} & I_{YY} & -I_{YZ} & 0 & 0 & 0 \\ -I_{XZ} & -I_{YZ} & I_{ZZ} & 0 & 0 & 0 \\ 0 & 0 & 0 & M_V & 0 & 0 \\ 0 & 0 & 0 & 0 & M_V & 0 \\ 0 & 0 & 0 & 0 & 0 & M_V \end{bmatrix}^{-1} \begin{pmatrix} L \\ M \\ N \\ F_X \\ F_Y \\ F_Z \end{pmatrix} \tag{8.2}
$$

In the acceleration vector diagrams we would ideally expect the peak control accelerations generated by the max control demands (either positive or negative) to be in the directions demanded by the FCS with minimal coupling into other control directions. For example, the roll control ($\delta_{\pm PFCS\,Max}$) should affect mostly the $\pm$roll acceleration and not yaw, and the yaw control ($\delta_{\pm RFCS\,Max}$) should affect the $\pm$yaw acceleration and not roll. This, of course, is not always the case and it depends on the efficiency of the mixing-logic matrix. It is difficult to have perfect decoupling between axes and a certain amount of cross-coupling between axes is unavoidable. Uncertainty rectangles are also used at the tips of the max acceleration vectors, similar to Section 8.2.1, to represent possible variations in the acceleration (angular or linear) due to the aerodynamic uncertainties. They are calculated from the uncertainty moments and forces and scaled as in equation 8.2.

The maximum acceleration diagrams are also used in analyzing the effectiveness of the mixing-logic matrix against disturbances in multiple directions. For example, the vehicle effectors system may have accessibility to span multiple directions including translational, and in theory, a mixing-logic matrix can be derived to provide control in multiple directions. The practicality, however, of this design is evaluated by the effectors capability to maneuver and to counter disturbances caused by variations in $\pm\alpha_{max}$ and $\pm\beta_{max}$. When the effector system attempts to spread and control multiple directions, in general, it loses its control authority against aero disturbances. The effectors system is in general more efficient for countering disturbances when it controls fewer directions.

8.2.3 Moment and Force Partials Vector Diagrams

The two previous vector diagrams analyze the control authority of an effector system in terms of its maximum control capability against maximum steady-state wind-shear disturbances defined in terms of maximum aerodynamic angles. The partial vector diagrams measure gains. They compare the open-loop system controllability gain versus disturbability gain (or sensitivity to disturbance). The controllability partial is measured in terms of moment or force produced per acceleration demand in a certain direction. The disturbance partial measures the ratio of moment or force produced per alpha in pitch or per beta in the lateral directions. They compare the moment partials per flight control acceleration demands: $(C_{m\delta Q}, C_{n\delta R}, C_{l\delta P})_{FCS}$ per (rad/sec^2) or the force partials per translational acceleration demands: $(C_{X\delta X}, C_{Y\delta Y}, C_{Z\delta Z})_{FCS}$ per (g), against the base vehicle moment and force partials with respect to alpha and beta angles $(C_{m\alpha}, C_{n\beta}, C_{l\beta}, C_{z\alpha}, C_{y\beta})$ per degree. The control and disturbance partials are plotted together in two-dimensional vector diagrams because the relative magnitudes and directions of these vectors are important for control design.

Since the flight control demands $(\delta_P, \delta_Q, \delta_R, \delta_X, \delta_Y, \delta_Z)_{FCS}$ are defined in vehicle body frame, instead of the classical $(\delta_{elevon}, \delta_{aireron}, \delta_{rudder})$ definition, for good controllability we would like to see the control moment and force vector partials to be greater in magnitude along the control directions, than the moment and force partials per alpha and beta variations. These partials, however, cannot be compared directly because they are different in nature and units. They must be properly scaled in order to be associated in the same diagram. For example, let us consider the pitch moment equation: $I\ddot{\theta} = M_{\alpha}\alpha + M_{\delta}\delta$. A good test for pitch controllability is to compare the magnitudes of the partials M_{α} and M_{δ}. However, they cannot be compared directly because they are of different substance. The first one is moment per degree of alpha and the second one is moment per angular acceleration demand in (rad/sec^2). For the vehicle to have good controllability we would like the $M_{\delta}\delta$ term to be considerably greater than the $M_{\alpha}\alpha$ term. This is possible when the control contribution due to δ_{max} is greater than the aero disturbance due to α_{max}. One way to make the control versus alpha partials comparable is to scale the magnitudes of the control vectors by multiplying them with $(\delta_{max}/\alpha_{max})$. This scaling allows us to compare the two vector partials and to quantify controllability versus disturbability by their relative size and direction. Obviously, the scaled control vector partials should be greater than the moment or force partials due to α or β. One might argue that the rates of torque build up rather than the max torques are more important in this comparison, especially when you have a divergent vehicle, in which case we would like the $M_\delta \dot{\delta}_{max}$ term to be greater than the $M_\alpha \dot{\alpha}_{max}$ term, and the scaling factor becomes: $\left(\dot{\delta}_{max} / \dot{\alpha}_{max} \right)$, where the max actuator rate is defined in the actuator specifications and the max vehicle rate can also be estimated.

Let us consider, for example, the partials in two longitudinal directions for a vehicle that has pitch control and also axial force control by means of thrust variation or a speed brake. The $(C_{ma}, C_{X\alpha})$ vector in Figure 8.6 corresponds to the pitch moment and x-force partials per alpha. The two partials are calculated at the trim condition. The vehicle has a negative C_{ma} which indicates that it is statically stable, and also a negative $C_{X\alpha}$ that causes an increase in aft force due to α. The nearly horizontal pitch control vector $(C'_{m\delta Q}, C'_{X\delta Q})$ represents the pitch moment and axial force partials per pitch control acceleration demand (δQ_{FCS}). Both terms in this vector are scaled by multiplying them with $(\delta_{Qmax} / \alpha_{max})$ so that they can be made comparable with (C_{ma}) and plotted on the same scale. Similarly, the vertical axial force control vector $(C'_{m\delta X}, C'_{X\delta X})$, represents the pitch moment and axial force partials per axial acceleration demand (δX_{FCS}). Both terms are scaled by multiplying them with $(\delta_{Xmax} / \alpha_{max})$ so that they can be made comparable with $C_{X\alpha}$. They are scaled as shown in equations 8.3 and 8.4. For good controllability in both axes, the directions of the control moment and force partials should be pointing toward their intended directions and their scaled magnitudes should exceed those of the moment and force partials with respect to alpha. It means that the scaled pitch moment per pitch demand

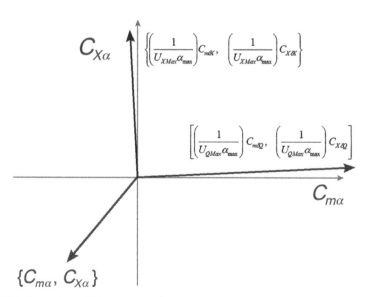

FIGURE 8.6   Partials of Pitch Moment and Axial Force Vector Diagram per Pitch and Axial Control, also Partials per Alpha

$C_{m\delta Q_{FCS}}$ should be bigger than the pitch partial $C_{m\alpha}$, and the scaled x-force per x-acceleration demand $C_{X\delta X_{FCS}}$ should be bigger than the x-force partial C_{X_α}.

$$|C_{m\alpha}| < \left(\frac{\delta_{Q\max}}{\alpha_{\max}}\right)|C_{m\delta Q}| \quad \text{or} \quad |C_{m\alpha}| < \left(\frac{1}{U_{Q\text{Max}}\,\alpha_{\max}}\right)|C_{m\delta Q}| \quad \text{and}$$

$$\left\{C'_{m\delta Q} \quad C'_{X\delta Q}\right\} = \left\{\left(\frac{1}{U_{Q\text{Max}}\,\alpha_{\max}}\right)C_{m\delta Q} \quad \left(\frac{1}{U_{Q\text{Max}}\,\alpha_{\max}}\right)C_{X\delta Q}\right\}$$

$$(8.3)$$

$$|C_{A\alpha}| < \left(\frac{\delta_{X\max}}{\alpha_{\max}}\right)|C_{X\delta X}| \quad \text{and} \quad |C_{A\alpha}| < \left(\frac{1}{U_{X\text{Max}}\,\alpha_{\max}}\right)|C_{X\delta X}|$$

$$\left\{C'_{m\delta X} \quad C'_{X\delta X}\right\} = \left\{\left(\frac{1}{U_{X\text{Max}}\,\alpha_{\max}}\right)C_{m\delta X} \quad \left(\frac{1}{U_{X\text{Max}}\,\alpha_{\max}}\right)C_{X\delta X}\right\}$$

$$(8.4)$$

A similar moment partials vector diagram is shown in Figure 8.7 for the lateral directions. It shows the roll and yaw moment partials per sideslip angle, $(C_{l\beta}, C_{n\beta})$ and the two roll and yaw control vector partials. The dashed vector represents the roll/yaw moment partials per yaw control demand $(C'_{l\delta R}, C'_{n\delta R})$. Both components of this vector are scaled by multiplying them with $(\delta_{R\max}/\beta_{\max})$ in order to make $C'_{n\delta R}$ comparable with $C_{n\beta}$, as shown in equation 8.6. This vector is nearly horizontal indicating that the yaw control demand δR_{FCS} affects mainly the yaw direction with very little coupling in roll. Similarly, the vertical vector represents the moment partials per roll control demand $(C'_{l\delta P}, C'_{n\delta P})$, and it is scaled by multiplying it with $(\delta_{P\max}/\beta_{\max})$ to make $C'_{l\delta P}$ comparable with $C_{l\beta}$, as shown in equation 8.5. This vector is almost

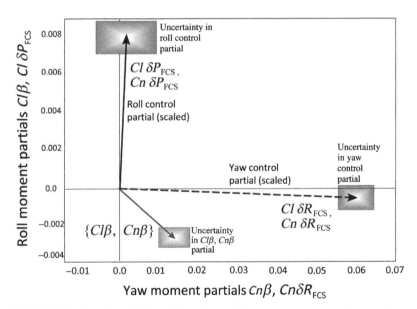

FIGURE 8.7   Partials of Roll and Yaw Control Moments versus Partials due to Beta

vertical indicating that roll control demand δP_{FCS} produces mostly a roll torque with very little coupling in yaw. For good controllability the magnitude of $C_{l\beta}$ should obviously be less than the magnitude of the scaled $C'_{l\delta P}$, and the magnitude of $C_{n\beta}$ should be less than the magnitude of the scaled $C'_{n\delta R}$, as shown in the equations 8.5 and 8.6 below.

$$
|C_{l\beta}| < \left(\frac{\delta_{P\max}}{\beta_{\max}}\right)|C_{l\delta P}| \quad \text{or} \quad |C_{l\beta}| < \left(\frac{1}{U_{P\max}\,\beta_{\max}}\right)|C_{l\delta P}| \quad \text{and}
$$

$$
\{C'_{l\delta P} \quad C'_{n\delta P}\} = \left\{\left(\frac{1}{U_{P\max}\,\beta_{\max}}\right)C_{l\delta P} \quad \left(\frac{1}{U_{P\max}\,\beta_{\max}}\right)C_{n\delta P}\right\}
$$

(8.5)

$$
|C_{n\beta}| < \left(\frac{\delta_{R\max}}{\beta_{\max}}\right)|C_{n\delta R}| \quad \text{or} \quad |C_{n\beta}| < \left(\frac{1}{U_{R\max}\beta_{\max}}\right)|C_{n\delta R}| \quad \text{and}
$$

$$
\{C'_{l\delta R} \quad C'_{n\delta R}\} = \left\{\left(\frac{1}{U_{R\max}\,\beta_{\max}}\right)C_{l\delta R} \quad \left(\frac{1}{U_{R\max}\,\beta_{\max}}\right)C_{n\delta R}\right\}
$$

(8.6)

In Figure 8.7 the magnitudes of the two scaled control partials are greater than the moment partials due to beta in both roll and yaw directions. The control partials are almost orthogonal to each other, decoupled, and each affecting their corresponding direction. They are the product of an efficiently designed mixing-logic matrix. Since we are dealing with partials we do not have negative directions as we did in the maximum/minimum control demand diagrams. Notice that the yaw control partial $C_{n\delta R_{\text{FCS}}}$ is an order of magnitude bigger than the roll control partial $C_{l\delta P_{\text{FCS}}}$, because the roll moment of inertia in this vehicle is much smaller than the yaw inertia, plus the disturbance partial due to beta is bigger in yaw and requires greater controllability in yaw. The disturbance partials are smaller than the control partials in both directions, as they should be, and this is an indicator that the vehicle has good controllability against the expected lateral disturbances.

The square rectangles at the tips of the vector partials represent the effects of the uncertainties in the aero derivatives. The uncertainties in $C_{n\beta}$ and $C_{l\beta}$ are represented by a rectangle at the tip of the disturbance partial vector. The uncertainties in the control surface derivatives are initially defined for each individual surface separately as a function of surface deflection. They must be converted from individual surfaces to uncertainties per FCS axis demands: $\delta C_{l\delta P_{\text{FCS}}}$, $\delta C_{l\delta R_{\text{FCS}}}$, $\delta C_{n\delta R_{\text{FCS}}}$, $\delta C_{m\delta Q_{\text{FCS}}}$, etc. which requires involvement of the mixing logic.

The rectangle at the tip of the yaw control vector partial $(C_{n\delta R_{\text{FCS}}}, C_{l\delta R_{\text{FCS}}})$ represents the possible variations of this vector due to the combined uncertainties of the surfaces involved in that direction. Similarly, the rectangle at the tip of the roll control vector partial $(C_{n\delta P_{\text{FCS}}}, C_{l\delta P_{\text{FCS}}})$ represents the possible variations of this vector due to the combined surface uncertainties in that direction. Equations 8.10 and 8.11 describe how the uncertainties are combined from individual aerosurface panels to uncertainties in vehicle axes demands using the mixing-logic matrix.

8.2.4   Vector Diagram Partials of Acceleration per Acceleration Demand

The fourth category of vector diagrams are partials of accelerations achieved in two control directions per acceleration demands along those directions. They are calculated at a specific flight condition. They are used for evaluating the vehicle open-loop maneuverability in certain directions by the magnitude and direction of its response along the demanded directions. In a good design the accelerations per acceleration demand vectors must be pointing toward the demanded directions and their magnitudes should be close to unity. This is not a strict requirement but only a guideline because when the control loop is closed the FCS takes care of the cross-coupling imperfections. In an ideal situation the open-loop plant is diagonalized by the mixing-logic matrix and the transfer path between the acceleration demands and vehicle accelerations becomes the identity matrix. It means that the vehicle accelerations become equal to the accelerations demanded by the FCS (unit vectors pointing toward the demanded directions), and they are orthogonal to each other, meaning that they are perfectly decoupled from each other. This ideal situation, however, is practically impossible to achieve at all times. The acceleration partials are useful in evaluating the efficiency of the effector mixing-logic matrix (K_{mix}) in achieving the accelerations demanded by the FCS and in providing a certain amount of decoupling between the control axes. Perfect diagonalization should not be expected, but a certain amount of diagonal dominance in the controlled directions should at least be achieved. It makes it easier to design a closed-loop control system.

Referring to Figure 8.3, the moments and forces on the vehicle generated by a flight control input is the vector $M_\delta = C_M K_{\mathrm{mix}} \delta_{\mathrm{FCS}}$, where δ_{FCS} is the control acceleration demands vector in (rad/sec^2) for rotational, or in (ft/sec^2) for translational. The size of vector δ_{FCS} varies from three (for three angular accelerations) to six (also including translational acceleration demands). The matrix K_{mix} converts the flight control demands to effector deflections or throttle commands. The matrix C_M converts the deflections to a vector of three moments and up to three forces. The rotational and translational accelerations of the vehicle are obtained by multiplying out the matrices in equation 8.7. In the 6×6 mass matrix below, the 3×3 matrix in the upper left corner is the vehicle inertia dyadic, and the vehicle mass is in the lower right corner along x, y, and z. The product is a matrix of 6×6 maximum dimension and its elements consist of the acceleration partials.

$$
\begin{bmatrix} \dot{p} \\ \dot{q} \\ \dot{r} \\ \ddot{x} \\ \ddot{y} \\ \ddot{z} \end{bmatrix}_{\mathrm{control}} = \begin{bmatrix} I_{XX} & -I_{XY} & -I_{XZ} & 0 & 0 & 0 \\ -I_{XY} & I_{YY} & -I_{YZ} & 0 & 0 & 0 \\ -I_{XZ} & -I_{YZ} & I_{ZZ} & 0 & 0 & 0 \\ 0 & 0 & 0 & M_V & 0 & 0 \\ 0 & 0 & 0 & 0 & M_V & 0 \\ 0 & 0 & 0 & 0 & 0 & M_V \end{bmatrix}^{-1} C_M K_{\mathrm{mix}} \begin{pmatrix} \delta_P \\ \delta_Q \\ \delta_R \\ \delta_X \\ \delta_Y \\ \delta_Z \end{pmatrix}_{\mathrm{FCS}} \qquad (8.7)
$$

If we assume that the vehicle has sufficient effectors to span all control directions (three rotations and all three translations) and if the mixing-logic matrix is properly designed to provide dynamic decoupling between the axes, as described in Chapter 5, the multiplication of matrices in equation 8.7 becomes the 6×6 identity matrix. In general, however, not all translational directions are directly controllable by the FCS and some of the rows in equation 8.7 are ignored because they are uncontrollable or weakly controllable, plus the mixing matrix is intended to affect only the controllable directions. In addition, the selection of the mixing-logic matrix is sometimes influenced by other performance factors, such as the LCDP, and it is not always ideal from the acceleration partials point of view.

For example, Figure 8.8 shows the partials of accelerations per acceleration demands in roll and yaw directions. The solid vertical vector pointing toward roll is the vehicle acceleration response per roll acceleration demand $\{\dot{P}/\delta P_{FCS}, \dot{R}/\delta P_{FCS}\}$. The dashed horizontal vector pointing toward yaw is the vehicle acceleration response per yaw acceleration demand $\{\dot{P}/\delta R_{FCS}, \dot{R}/\delta R_{FCS}\}$. The axis units are in (rad/sec^2) per (rad/sec^2). Both angular acceleration partial vectors are pointing toward their intended directions. They are almost orthogonal to each other (close to being decoupled) and their magnitudes are close to unity as a result of the mixing-logic matrix. It means that the vehicle will almost perfectly achieve the accelerations demanded, open-loop with minimal cross-coupling. This is a good property for control design.

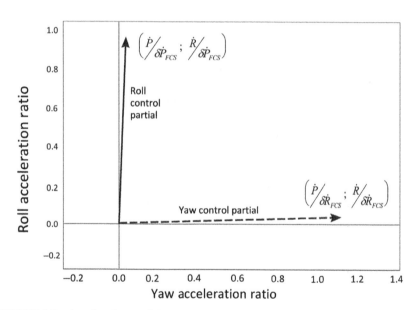

FIGURE 8.8   Acceleration Partials Vector Diagram in Roll and Yaw, Showing Small Coupling Between Axes

8.3   CONVERTING THE AERO UNCERTAINTIES FROM INDIVIDUAL SURFACES TO VEHICLE AXES

In vector diagram analysis the control moments and force vectors for the combined effector system are calculated in the vehicle axes. The uncertainties, however, are generated by the aero group for each individual control surface separately. When many aerosurfaces are used for multiple-axes control, none of these surfaces are specifically dedicated to control a single axis. An acceleration demand is decoded by the mixing-logic matrix and is converted into multiple surface deflections. Since the vector diagrams are presented in vehicle axes the effects of the individual surface uncertainties must also be combined to reflect uncertainties in vehicle axes. The uncertainties from each individual surface must, therefore, be transformed to moment, force, or acceleration uncertainties in body axes and summed up for all surfaces along the vehicle axes according to each surface's participation in the control directions. Since the mixing-logic matrix defines the structure of the effectors combination, the selection of the (K_{mix}) matrix will also combine the aero uncertainties and determine the size of the uncertainty rectangles at the tips of the control vectors.

We will now calculate the uncertainty rectangles in the control partial vector diagrams and also in the maximum control moment/force vector diagrams from the uncertainties in the individual aerosurface derivatives provided by the aero group. Let us consider a flight vehicle that is controlled by (n) aerosurfaces which are trimmed at a certain deflections vector: $\delta_T = (\delta_1 \cdots \delta_n)$. The moment and force variations due to the individual aerosurface aerodynamic uncertainties are obtained from matrix equation 8.8, where each element of the matrix is an uncertainty in the moment/force surface derivative.

$$
\begin{bmatrix} \delta L \\ \delta M \\ \delta N \\ \delta F_X \\ \delta F_Y \\ \delta F_Z \end{bmatrix} = \overline{Q} S_r \begin{bmatrix} b\,\delta C_{l\delta 1} & b\,\delta C_{l\delta 2} & b\,\delta C_{l\delta 3} & b\,\delta C_{l\delta n} \\ \overline{c}\,\delta C_{m\delta 1} & \overline{c}\,\delta C_{m\delta 2} & \overline{c}\,\delta C_{m\delta 3} & \overline{c}\,\delta C_{m\delta n} \\ b\,\delta C_{n\delta 1} & b\,\delta C_{n\delta 2} & b\,\delta C_{n\delta 3} & b\,\delta C_{n\delta n} \\ \delta C_{A\delta 1} & \delta C_{A\delta 2} & \delta C_{A\delta 3} & \delta C_{A\delta n} \\ \delta C_{Y\delta 1} & \delta C_{Y\delta 2} & \delta C_{Y\delta 3} & \delta C_{Y\delta n} \\ \delta C_{Z\delta 1} & \delta C_{Z\delta 2} & \delta C_{Z\delta 3} & \delta C_{Z\delta n} \end{bmatrix} \begin{pmatrix} \delta_1 \\ \delta_2 \\ \vdots \\ \delta_n \end{pmatrix}_T \tag{8.8}
$$

The FCS output demands are related to the individual aerosurface deflections by the mixing-logic matrix as shown in equation 8.9, where $\left(\delta P \quad \delta Q \quad \delta R \right)_{FCS}$ are the FCS rotational acceleration demands and $\left(\delta X \quad \delta Y \quad \delta Z \right)_{FCS}$ are the translational acceleration FCS demands.

$$
\begin{pmatrix} \delta_1 \\ \delta_2 \\ \vdots \\ \delta_n \end{pmatrix} = K_{mix} \begin{pmatrix} \delta_P \\ \delta_Q \\ \delta_R \\ \delta_X \\ \delta_Y \\ \delta_Z \end{pmatrix}_{FCS} \tag{8.9}
$$

The number of columns in matrix K_{mix} varies between 3 and 6, as already discussed. By substituting equation 8.9 to equation 8.8 we can obtain the relationship in equation 8.10 which calculates the magnitudes of the moment and force uncertainties in body axes as a function of flight control demands. Note that the uncertainties are defined always positive and, therefore, the absolute value of (K_{mix}) is used in equation 8.10 to cover for the worst uncertainty combination.

$$
\begin{bmatrix} \delta L \\ \delta M \\ \delta N \\ \delta F_X \\ \delta F_Y \\ \delta F_Z \end{bmatrix} = \bar{Q} S_r \begin{bmatrix} b\,\delta C_{l\delta 1} & b\,\delta C_{l\delta 2} & b\,\delta C_{l\delta 3} & b\,\delta C_{l\delta 4} \\ \bar{c}\,\delta C_{m\delta 1} & \bar{c}\,\delta C_{m\delta 2} & \bar{c}\,\delta C_{m\delta 3} & \bar{c}\,\delta C_{m\delta 4} \\ b\,\delta C_{n\delta 1} & b\,\delta C_{n\delta 2} & b\,\delta C_{n\delta 3} & b\,\delta C_{n\delta 4} \\ \delta C_{A\delta 1} & \delta C_{A\delta 2} & \delta C_{A\delta 3} & \delta C_{A\delta 4} \\ \delta C_{Y\delta 1} & \delta C_{Y\delta 2} & \delta C_{Y\delta 3} & \delta C_{Y\delta 4} \\ \delta C_{Z\delta 1} & \delta C_{Z\delta 2} & \delta C_{Z\delta 3} & \delta C_{Z\delta 4} \end{bmatrix} abs\,[K_{mix}] \begin{pmatrix} \delta P \\ \delta Q \\ \delta R \\ \delta X \\ \delta Y \\ \delta Z \end{pmatrix}_{FCS} \quad (8.10)
$$

After multiplying the matrices in equation 8.10, and let us assume, for example, that together with the three rotations only one translational direction along x is directly controllable. The uncertainties matrix equation for the 4-DOF becomes as shown in equation 8.11, and the actual moment and force uncertainties are obtained from the LHS of the equation as a function of the FCS acceleration demand.

$$
\begin{bmatrix} \delta L \\ \delta M \\ \delta N \\ \delta F_X \\ \delta F_Y \\ \delta F_Z \end{bmatrix} = \begin{bmatrix} \delta L_P & \delta L_Q & \delta L_R & \delta L_X \\ \delta M_P & \delta M_Q & \delta M_R & \delta M_X \\ \delta N_P & \delta N_Q & \delta N_R & \delta N_X \\ \delta X_P & \delta X_Q & \delta X_R & \delta X_X \\ \delta Y_P & \delta Y_Q & \delta Y_R & \delta Y_X \\ \delta Z_P & \delta Z_Q & \delta Z_R & \delta Z_X \end{bmatrix} \begin{pmatrix} \delta P \\ \delta Q \\ \delta R \\ \delta X \end{pmatrix}_{FCS} \quad (8.11)
$$

8.3.1 Uncertainties in the Control Partials

The uncertainties in the moment and force derivatives with respect to the flight control demands are obtained directly from the matrix elements of equation 8.11. For example, the uncertainty in the pitch moment derivative due to pitch control demand (virtual elevon) is $\delta C_{m\delta Q} = \delta M_Q/\bar{Q}S\bar{c}$. The uncertainty in rolling moment derivative due to roll control demand (virtual aileron) is $\delta C_{l\delta P} = \delta L_P/\bar{Q}Sb$ and the uncertainty in rolling moment derivative due to yaw demand is $\delta C_{l\delta R} = \delta L_R/\bar{Q}Sb$. These uncertainties are used in the partial moment per FCS demand or the partial force per FCS demand vector diagrams. They define the size of the rectangles at the tips of the control vectors. The rectangles are also scaled proportionally to the vector magnitudes, as described in Figure 8.6 and in equations 8.3–8.6.

8.3.2 Uncertainties due to Peak Control Demands

Having calculated the uncertainties in the control vector partials we can now extend the method to calculate the moment and force uncertainties when the flight control demands are maximized either in positive or negative directions. Equation 7.20 calculates the maximum acceleration that can be demanded by the FCS along a controlled axis before at least one of the effectors saturates, for example, the maximum pitch control demand is:

$$\delta_{Q\,\text{Max}_{FCS}} = \frac{1}{U_{Q\,\text{Max}}} \tag{7.20}$$

To calculate, therefore, the magnitude of the maximum moment (or force) uncertainty in a certain direction we must multiply the corresponding row in equation 8.11 with the maximum FCS demand applied in that direction. For example, the uncertainties in the pitch moment and in the x and z forces (nondimensional) as a result of the uncertainties in the control surface coefficients, when the pitch control demand is at its maximum positive position, are obtained from equation 8.12a. The uncertainties in the same three directions when the forward acceleration demand is at its maximum positive position are calculated by equation 8.12b. Notice how they depend on the max demand.

$$\begin{pmatrix} \delta C_M \\ \delta C_X \\ \delta C_Z \end{pmatrix}_{Q\,\text{Max}} = \begin{bmatrix} \delta M_Q / Q\,\bar{c}S_{\text{ref}} \\ \delta X_Q / Q\,S_{\text{ref}} \\ \delta Z_Q / Q\,S_{\text{ref}} \end{bmatrix} 1/U_{Q\,\text{Max}}; \qquad \begin{pmatrix} \delta C_M \\ \delta C_X \\ \delta C_Z \end{pmatrix}_{X\,\text{Max}} = \begin{bmatrix} \delta M_X / Q\,\bar{c}S_{\text{ref}} \\ \delta X_X / Q\,S_{\text{ref}} \\ \delta Z_X / Q\,S_{\text{ref}} \end{bmatrix} 1/U_{X\,\text{Max}}$$

$$(8.12\ a\ \&\ b)$$

Similarly in yaw, equation 8.13a calculates the uncertainty magnitudes in roll and yaw control moments and also in side force when the yaw control demand is maximized in the positive direction. It is caused by uncertainties in the aerosurface coefficients. Similarly, the uncertainties in the same coefficients are calculated when the roll control demand is maximized in the positive direction, using equation 8.13b.

$$\begin{pmatrix} \delta C_L \\ \delta C_N \\ \delta C_Y \end{pmatrix}_{R\,\text{Max}} = \begin{bmatrix} \delta L_R / Q\,bS_{\text{ref}} \\ \delta N_R / Q\,bS_{\text{ref}} \\ \delta Y_R / Q\,S_{\text{ref}} \end{bmatrix} 1/U_{R\,\text{Max}}; \qquad \begin{pmatrix} \delta C_L \\ \delta C_N \\ \delta C_Y \end{pmatrix}_{P\,\text{Max}} = \begin{bmatrix} \delta L_P / Q\,bS_{\text{ref}} \\ \delta N_P / Q\,bS_{\text{ref}} \\ \delta Y_P / Q\,S_{\text{ref}} \end{bmatrix} 1/U_{P\,\text{Max}}$$

$$(8.13\ a\ \&\ b)$$

Equations 8.12 and 8.13 are used to calculate the uncertainties in the maximum control moment and force vector diagrams. They define the size of the uncertainty rectangles at the tips of the control vectors generated from a maximum control demand. Notice that the trim positions of the aerosurfaces are not necessarily centered and the max deflection in the positive direction from trim is not necessarily equal to the peak deflection in the opposite direction. For example, we proved in equation 7.21b that the peak FCS demand in the negative pitch direction is different from the max positive pitch FCS demand, and it is defined by equation 8.14, where $U_{Q\text{Min}}$ is obtained from equation 7.18b.

$$\delta_{-Q\,\text{Max}_{\text{FCS}}} = \frac{1}{U_{Q\,\text{Min}}} \tag{8.14}$$

This means that the uncertainty rectangles around the tips of the control vectors generated from a peak negative control demand will be different from the rectangles at the tips of the control vectors generated from a max positive control demand, because the size of the two controls are different, and the effectors authority may be different in the reverse direction. The uncertainties in the pitch moment and in the x and z forces as a result of the uncertainties in the control surface coefficients, when the pitch control demand is at its peak negative position are obtained from equation 8.15a, and the uncertainties in the same three directions when the forward acceleration demand is at its peak negative position are obtained from equation 8.15b.

$$\begin{pmatrix} \delta C_M \\ \delta C_X \\ \delta C_Z \end{pmatrix}_{Q\,\text{Min}} = \begin{bmatrix} \delta M_Q / Q\,\bar{c}S_{\text{ref}} \\ \delta X_Q / Q\,S_{\text{ref}} \\ \delta Z_Q / Q\,S_{\text{ref}} \end{bmatrix} {}^{1}\!/_{U_{Q\,\text{Min}}} ; \quad \begin{pmatrix} \delta C_M \\ \delta C_X \\ \delta C_Z \end{pmatrix}_{X\,\text{Min}} = \begin{bmatrix} \delta M_X / Q\,\bar{c}S_{\text{ref}} \\ \delta X_X / Q\,S_{\text{ref}} \\ \delta Z_X / Q\,S_{\text{ref}} \end{bmatrix} {}^{1}\!/_{U_{X\,\text{Min}}}$$

$$\text{(8.15 a \& b)}$$

Similarly, the uncertainties in the lateral coefficients when the roll and yaw acceleration control demands are maximized in the negative directions are obtained from equations 8.16.

$$\begin{pmatrix} \delta C_L \\ \delta C_N \\ \delta C_Y \end{pmatrix}_{R\,\text{Min}} = \begin{bmatrix} \delta L_R / Q\,bS_{\text{ref}} \\ \delta N_R / Q\,bS_{\text{ref}} \\ \delta Y_R / Q\,S_{\text{ref}} \end{bmatrix} {}^{1}\!/_{U_{R\,\text{Min}}} ; \quad \begin{pmatrix} \delta C_L \\ \delta C_N \\ \delta C_Y \end{pmatrix}_{P\,\text{Min}} = \begin{bmatrix} \delta L_P / Q\,bS_{\text{ref}} \\ \delta N_P / Q\,bS_{\text{ref}} \\ \delta Y_P / Q\,S_{\text{ref}} \end{bmatrix} {}^{1}\!/_{U_{P\,\text{Min}}}$$

$$\text{(8.16 a \& b)}$$

8.3.3 Acceleration Uncertainties

The equations described above can also be used for calculating the uncertainty rectangles at the tips of the maximum acceleration vector diagrams. In this case the moment and force uncertainties must be converted to acceleration uncertainties by multiplying them with the inverse of the inertias and mass matrix below.

$$\begin{bmatrix} I_{XX} & -I_{XY} & -I_{XZ} & 0 & 0 & 0 \\ -I_{XY} & I_{YY} & -I_{YZ} & 0 & 0 & 0 \\ -I_{XZ} & -I_{YZ} & I_{ZZ} & 0 & 0 & 0 \\ 0 & 0 & 0 & M_V & 0 & 0 \\ 0 & 0 & 0 & 0 & M_V & 0 \\ 0 & 0 & 0 & 0 & 0 & M_V \end{bmatrix}$$

9

FLIGHT CONTROL DESIGN

There is a huge amount of work that is available on flight control system (FCS) design, and the purpose of this book is not to dwell on it but to highlight only some of the most commonly used methods and to present some design examples. Specifically, we will focus on two of the most commonly used methods, the H_2 or linear quadratic regulator (LQR), and the H-infinity (H_∞) which are frequently used in the design of multivariable systems. The advantage of the multivariable design methods over the classical single-loop methods is because the classical methods have difficulty in systems with a large number of possible connections between multiple sensors and actuators. It is desirable, therefore, to have formal methods that automatically synthesize optimal feedback controllers based on specified criteria. The secret of optimal control is to construct a cost function that captures the design requirements and will direct the optimization algorithm to produce control laws that satisfy the requirements. The control design requirements can be formulated as follows. For a given plant model $P(s)$ shown in Figure 9.1, with external inputs u_1 and outputs y_1, and with inputs and outputs u_2 and y_2 available to a feedback controller, find a linear control law K so that the closed-loop system is stable, and also the transfer function norm (either H_2 or H_∞) between u_1 and y_1 known as the "lower linear fractional transformation" (LFT) is minimized and satisfies the design sensitivity requirements. The control system K, for example, can be designed to reduce the vehicle's accelerometer response (y_1) to wind turbulence excitation (u_1).

The H_2 and the H_∞ control design methods are based on a "synthesis model" (SM) which is a state-space system that includes the basic system dynamics, that is, the

Performance Evaluation and Design of Flight Vehicle Control Systems, First Edition. Eric T. Falangas.
© 2016 by Eric T. Falangas. Published 2016 by John Wiley & Sons, Inc.

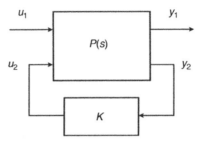

FIGURE 9.1   Control System K is Expected to Stabilize the Plant $P(s)$ and to Minimize Sensitivity Between u_1 and y_1

dynamics that we want to control. It is mainly rigid-body dynamics but it may also include some of the low frequency structural modes in order to be actively attenuated by the control system. We must make sure, of course, that the actuator bandwidth is high enough to respond at those frequencies. The SM is not just plant dynamics but it includes also the designer's performance requirements expressed indirectly in terms of matrices that penalize the activity of the controls and states.

9.1   LQR STATE-FEEDBACK CONTROL

Consider the following state-space system in equation 9.1 which represents the basic plant dynamics. F is the state feedback matrix to be designed, x is the state vector, u is the control input, and z is the system output.

$$
\begin{aligned}
\dot{x} &= A x + B u \\
z &= C x \\
u &= F x
\end{aligned}
\tag{9.1}
$$

In Figure 9.2, $P(s)$ represents the SM that includes our optimization requirements. It consists of the plant dynamics plus some additional inputs and outputs that are used in the optimization process. It is excited by an external vector input u_1 and has an

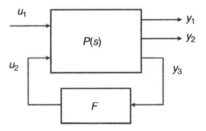

FIGURE 9.2   Synthesis Model for LQR Design With State Feedback

external output vector y_1 to be minimized for good sensitivity. Minimizing, however, only the vector y_1 would lead to a very big control usage and a large system bandwidth so we also include a second output in the optimization criteria y_2 that penalizes the controls and prevents the controls of the closed-loop system from becoming too big. The third output vector y_3 in Figure 9.2 represents the measurement vector which in this case is equal to the state vector x, and it is used in the feedback loop. The matrix F closes the state-feedback loop, stabilizes the system, and minimizes the H_2-norm of the transfer function between the vector u_1 and the combined output vector $[y_1, y_2]$ by minimizing the time-domain cost function J, in equation 9.2, where the matrix R is positive definite, and Q is positive semidefinite.

$$J = \int_0^\infty \left[z^T Q z + u^T R u \right] dt \tag{9.2}$$

Equation 9.3 shows the SM in state-space form which is presented in the H_2 optimization algorithm. The output y_1 penalizes the original system outputs z by means of matrix $Q^{1/2}$. The output y_2 penalizes the controls u_2 by means of matrix $R^{1/2}$.

$$
\begin{pmatrix} \dot{x} \\ y_1 \\ y_2 \\ y_3 \end{pmatrix} =
\left[
\begin{array}{c|cc}
 & x & u_1 \quad u_2 \\
\hline
A & I & B \\
Q^{1/2}C & 0 & 0 \\
0 & 0 & R^{1/2} \\
I & 0 & 0
\end{array}
\right] \tag{9.3}
$$

The selection of matrices Q and R may sometimes require a few iterations for accomplishing a satisfactory trade-off between sensitivity to external disturbances and control system bandwidth. A good compromise must be eventually reached that avoids the magnitude of the control signals from exceeding the capability of the actuators. Otherwise, the performance expectations from the system should be revised or the actuators be improved. The solution of the LQR optimization problem using state feedback is presented in equation 9.4, where X is obtained by solving the Riccati equation using the following Hamiltonian matrix.

$$
\begin{aligned}
F &= -R^{-1}B^T X \\
X &= \mathbf{Ric} \begin{bmatrix} A & -BR^{-1}B^T \\ -C^T QC & -A^T \end{bmatrix}
\end{aligned} \tag{9.4}
$$

The following MATLAB® function solves this Riccati equation problem and calculates the state-feedback gain F, where X is the solution of the Riccati equation. Q and R are the weight matrices on the outputs z and on the controls u_2. E provides the closed-loop eigenvalues which are the roots of matrix (A-BF).

$$[F, X, E] = LQR(A, B, Q, R)$$

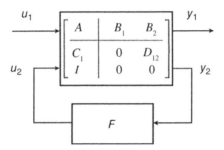

FIGURE 9.3   Synthesis Model for the *H*-Infinity Full State-Feedback Formulation

9.2   *H*-INFINITY STATE-FEEDBACK CONTROL

The *H*-infinity full state-feedback control problem has a state-feedback formulation and solution that is similar to the LQR problem, but with some differences. The problem is to find a state-feedback matrix *F* that stabilizes the plant in Figure 9.3 and achieves a bound on the infinity norm gain of $G_{11}(s)$, shown in equation 9.5. That is, to bound the maximum gain over the entire frequency spectrum.

$$\|G_{11}\|_\infty < \gamma \qquad (9.5)$$

where $G_{11}(s)$ is the closed-loop transfer function system between u_1 and y_1, and γ is an upper bound requirement on the gain. However, the following conditions must be satisfied:

- The system must be stabilizable from both inputs, u_1 and u_2
- The system must be detectable from y_1
- The matrix $D_{12}^T C_1 = 0$
- The matrix $D_{12}^T D_{12} = I$

In this formulation the matrix C_1 plays the role of Q, and D_{12} plays the role of R in the LQR formulation. The criteria output y_1 takes care of both $[y_1,\ y_2]$ vector requirements in the LQR formulation which justifies the condition why the output y_1 must have two orthogonal parts, as shown in equation 9.6.

$$D_{12}^T C_1 = \begin{bmatrix} 0 & R^{1/2} \end{bmatrix} \begin{bmatrix} Q^{1/2} \\ 0 \end{bmatrix} = 0 \qquad (9.6)$$

A stable solution exists for $\|G_{11}\|_\infty < \gamma$ by solving the Riccati equation in 9.7, assuming that the Hamiltonian in the Riccati formulation has no imaginary eigenvalues. Otherwise we must lower our expectations and increase γ, until the Hamiltonian condition is satisfied.

$$X = \mathbf{Ric} \begin{bmatrix} A & \frac{1}{\gamma^2}B_1 B_1^T - B_2 B_2^T \\ -C_1^T C_1 & -A^T \end{bmatrix}$$

$$F = -B_2^T X$$

(9.7)

9.3 *H*-INFINITY CONTROL USING FULL-ORDER OUTPUT FEEDBACK

The following *H*-infinity solution is used when a full state measurement is not available. The controller in this case $K(s)$ is the same order as the SM $P(s)$ because it contains an estimator of the state vector x. The state-space realization of $P(s)$ in this case is shown in Figure 9.4. It is expected to satisfy a given upper bound γ on the infinity norm of the transfer function between the disturbance u_1 and the output y_1 vectors. The vector y_1 consists of a combination of variables made up of output criteria and controls.

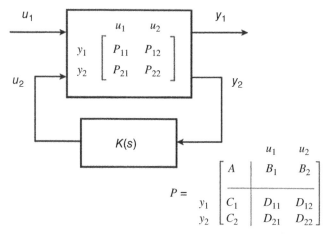

FIGURE 9.4 Synthesis Model for the Output Feedback *H*-Infinity Problem Formulation

The solution to the standard formulation requires the following assumptions on $P(s)$. However, some of those assumptions can be relaxed because it is possible to transform more general formulations to this standard form by proper scaling and loop-shifting operations prior to the optimization process.

1. The system must be stabilizable from both inputs u_1 and u_2
2. The system must be detectable from both outputs y_1 and y_2

3. The matrix $D_{12}^T D_{12} = I$, that is, number of y_1 outputs $\geq$ number of actuators in u_2
4. The matrix $D_{21} D_{21}^T = I$, that is, number of u_1 inputs $\geq$ number of sensors in y_2
5. The matrices $D_{11} = 0$ and $D_{22} = 0$

The solution to this H-infinity optimization problem is as follows:

$$X = \text{Ric} \begin{bmatrix} A - B_2 D_{12}^T C_1 & \gamma^{-2} B_1 B_1^T - B_2 B_2^T \\ -\tilde{C}_1^T \tilde{C}_1 & -\left(A - B_2 D_{12}^T C_1\right)^T \end{bmatrix}$$

$$Y = \text{Ric} \begin{bmatrix} \left(A - B_1 D_{21}^T C_2\right)^T & \gamma^{-2} C_1^T C_1 - C_2^T C_2 \\ -\tilde{B}_1 \tilde{B}_1^T & -\left(A - B_1 D_{21}^T C_2\right) \end{bmatrix}$$

$$\tilde{B}_1 = B_1 \left(I - D_{21}^T D_{21}\right)$$

$$\tilde{C}_1 = \left(I - D_{12} D_{12}^T\right) C_1$$

$$F = - \left(B_2^T X + D_{12}^T C_1\right)$$

$$H = - \left(Y C_2^T + B_1 D_{21}^T\right)$$

$$Z = (I - \gamma^{-2} YX)^{-1}$$

(9.8)

For a solution to exist the following constraints must be satisfied:

1. The matrices X and Y must be positive semidefinite.
2. The maximum eigenvalue of XY or spectral radius $\rho\left(XY\right) < \gamma^2$.
3. The Hamiltonians for X and for Y in the Riccati domain must not have purely imaginary eigenvalues.

The above conditions must be satisfied in order for the controller $K(s)$ to satisfy the requirement of bounding the infinity norm gain on the closed-loop transfer function $G_{11}(s)$ between u_1 and y_1, where γ is an upper bound requirement on the gain.

$$\|G_{11}\|_\infty < \gamma \tag{9.9}$$

Otherwise, we must increase γ until all of the conditions are satisfied. When γ is increased and approaches infinity the solution to this problem becomes the same as the linear quadratic gaussian (LQG) problem, which is not as robust as H-infinity and, therefore, not presented here because it is a subset to this problem. As γ decreases, the Hamiltonian eigenvalues shift closer to the imaginary axis until one of the constraints is violated. The solution to this problem is iterative and it is repeated until a minimum γ is reached. The transformation of the original to the standard SM formulation through scaling and loop shifting must be repeated for each new value of γ.

Figure 9.5 shows the state-space structure of the H-infinity controller and its interconnections with $P(s)$. The controller contains an estimate $\hat{x}$ of the plant's state vector x. The output estimate is $\hat{y}$. The actuator command is $u_2 = F\hat{x}$.

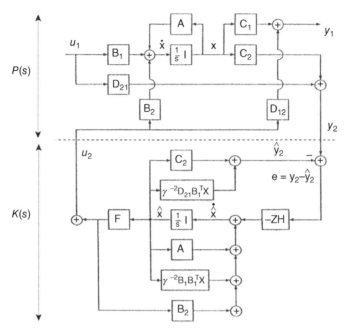

FIGURE 9.5   Structure of the *H*-infinity Controller and Its Interconnections With the Synthesis Model

9.4   CONTROL DESIGN EXAMPLES

We will now present three design examples using flight vehicles of different design requirements and use them to synthesize control laws and to demonstrate the control methods described in this section. The design, analysis, and simulations are performed in MATLAB® using flight vehicle models developed in Flixan®. The first example analyzes a reentry vehicle in different phases along its trajectory. The second example is a rocket plane that is controlled by aerosurfaces and a throttling engine. The third example demonstrates a redesign of the Space Shuttle ascent FCS using *H*-infinity for improving sensitivity to wind-gust disturbances. The work files for these examples are available in Flixan.com/Examples.

9.5   CONTROL DESIGN FOR A REENTRY VEHICLE

The vehicle in this example is a reentry glider that uses six aerosurfaces and six Reaction Control System (RCS) jets for control. Both, aerosurfaces and RCS jets are used to control this vehicle in different modes of operation, starting with early reentry, midphase, and approach/landing phases. The purpose of this example is to demonstrate how to design simple control laws for a vehicle that is controlled with a combination of aerosurfaces and RCS throttling jets. Since our emphasis here is

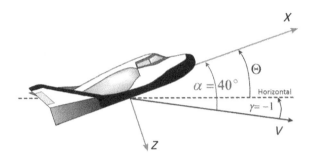

FIGURE 9.6   Vehicle Orientation Angles During Early Reentry

in control design the jets are assumed to be analog at this point and not "on–off" thrusters. Their thrust can be varied continuously between zero and a maximum thrust of 100 (lb). Bang–bang control is beyond our current scope, however, continuous control laws can be modified for bang–bang thrusters. All three operating phases will be studied because they are driven by different requirements and we will design control laws and analyze system performance at selected flight conditions during each phase. The analysis files for this example are located in Flixan.com/Examples/Reentry Glider/.

1. During early reentry phase the dynamic pressure is low. The vehicle is trimmed at a high angle of attack (approximately 40°) by using the jets and properly positioning the body flap. In this orientation the drag coefficient is high which helps to slow the vehicle down by exposing its shielded front and bottom surface sections to the airflow. The flight path angle γ is shallow, approximately $-1°$ to protect the vehicle from coming down too soon through the thicker atmospheric layers and overheating. The pitch control system which is initially controlling alpha at 40°, is gradually reducing it to smaller positive angles, and later it switches to γ-control. In the lateral directions the FCS controls roll attitude about the velocity vector V_0 and performs roll maneuvers at constant alpha. When the dynamic pressure exceeds approximately 5 pounds per square feet, the vehicle rolls about the velocity vector and banks approximately 70° which prevents it from bouncing back off into space. Further down it performs several bank reversal maneuvers while the angle of attack slowly decreases. The bank reversals at high Mach numbers are used for managing the excess energy.

2. The midphase begins when the vehicle is still controlling the angle of attack at smaller positive values, and the control system switches to flight path angle control, initially at a steady shallow value of $\gamma = -1°$. Although the dynamic pressure is sufficient to control the vehicle using aerosurfaces alone, the RCS is also partially active in order to augment controllability and to prevent the possibility of roll reversals. As it approaches landing the flight path angle comes down steeper to $-20°$ in order to maintain high speed which is needed for the final flare maneuver prior to landing.

3. During the approach and landing phase the RCS jets are turned off and control is performed only with the six aerosurfaces. The vehicle altitude and speed are

controlled independently with the proper combination of the aerosurfaces. In lateral the heading direction and its alignment with the runway is controlled by rolling about the velocity vector V_0.

9.5.1 Early Reentry Phase

During the early reentry phase of the descent trajectory the dynamic pressure is very low and the vehicle uses mostly the reaction control jets in combination with the aerosurfaces to control its attitude and mainly the angle of attack at 40°. The vehicle begins its descent trajectory from orbit by firing the orbital maneuvering thrusters to slow it down and placing it to a shallow reentry flight path angle $\gamma = -1°$. Then the RCS is used to rotate the vehicle at $\alpha = 40°$ while the dynamic pressure is still low. This optimal angle is selected for heat protection. The RCS continues to control the angle of attack, initially at 40° and gradually the aerosurfaces are used to reduce it to smaller alphas. As the dynamic pressure increases and the aerosurfaces become more effective, the RCS is gradually phased out and used in parallel with the surfaces as a backup. During this early phase the vehicle uses only four of its aerosurfaces, the two flaps and two rudders. The body flap is used for trimming and the speed brake is reserved for near landing. The combined RCS and aerosurface FCS maintains the required alpha during this shallow flight path angle to protect the vehicle and prevent it from burning as a result of atmospheric friction.

Although alpha is constant, the bank angle ϕ, however, is changing, and as soon as the dynamic pressure exceeds 5 (psf) the vehicle rolls about the velocity vector at approximately 70° to one side that prevents it from bouncing back off into space. It continues performing several bank reversal maneuvers as it slowly decreases the angle of attack. The roll reversals are also used for managing the excess energy since the speed brake is not active during hypersonic speeds. During this phase we are not interested to control the normal and axial accelerations. In pitch we must accurately control the angle of attack and indirectly the flight path angle because this controls heating. In the lateral direction we should be able to perform roll maneuvers about the velocity vector V_0 because this minimizes sideslip and lateral loading.

9.5.1.1 Control Design and Analysis during Early Reentry
The control system consists of roll, pitch, and yaw loops by combining RCS and aerosurfaces. It controls the angle of attack in pitch along a prescheduled command and also enables it to perform roll maneuvers about the velocity vector. There are several control approaches that can be used for combining the aerosurfaces and RCS jets together. One approach is to let the aerosurface control system operate in the inner loop and use the RCS as an outer loop, acting as a backup system when the errors exceed a certain limit. In this example, however, we will use the two separate control systems operating in parallel. One system is using the four aerosurfaces and the other uses the six RCS jets. In the flight condition chosen to analyze, the dynamic pressure is sufficient so that each system is capable of controlling the vehicle independently from the other.

9.5.1.2 Pitch and Lateral Design Models
The LQR method will be used in this example to design the state-feedback control laws. This method requires a state-space

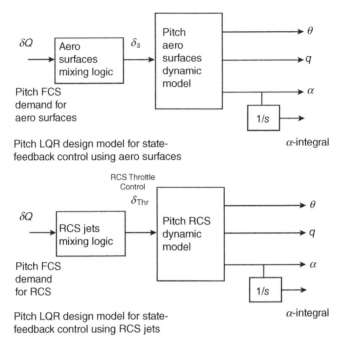

Pitch LQR design model for state-feedback control using aero surfaces

Pitch LQR design model for state-feedback control using RCS jets

FIGURE 9.7   Two Design Models for Pitch LQR Control: (a) Using Aerosurfaces; (b) Using RCS Jets

system that consists of the vehicle rigid-body dynamics. In fact two design systems will be created, one for pitch and one for lateral LQR designs. The first step, however, is to design the effector combination matrices, that is, a mixing matrix for the four aerosurfaces KSmix and a mixing matrix for the six RCS jets KJmix. They define how the effectors must combine in order to achieve the demanded control accelerations in roll, pitch, and yaw. Then we must generate the pitch and lateral rigid-body vehicle models that will be used in the LQR design. The design models must include the corresponding columns of the mixing matrices, so the input to the pitch model is pitch acceleration demand, and the two inputs to the lateral model are roll and yaw acceleration demands from the FCS. The variations in velocity and altitude states (δV and δh) are not included in the pitch model because during this phase we are not interested to control these variables which affect mainly the phugoid mode. The integral of the angle of attack is included in the pitch state vector of the design model in order to maintain better alpha-tracking. So the pitch state vector in the design model consists of: (θ, q, α, and α-integral).

Figure 9.7 shows the two LQR design models for the pitch axis. The top one is for aerosurfaces control and the bottom one is for RCS throttle control. They are used to design two (1 × 4) state-feedback matrices KSq and KJq for surface and jet control, respectively. The state vector and the control δQ must be penalized in the LQR optimization algorithm by the proper selection of the Q and R matrices. In this case we attach a heavier weight on the Q-elements that correspond to the α

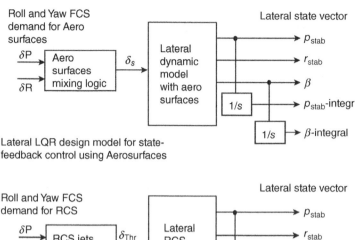

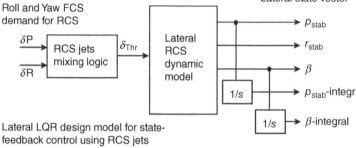

FIGURE 9.8 Two Design Models for Lateral LQR Control: (a) Using Aerosurfaces; and (b) Using RCS Jets

and α-integral states and a smaller penalty on θ because we intend to achieve good α-tracking and we do not care so much about θ deflections and transients. The R is not matrix but scalar in this case because we only have one control δQ. We must adjust R in order to achieve a good compromise between control bandwidth and α-tracking performance.

In the lateral direction the vehicle is expected to perform roll maneuvers about the velocity vector V_0, which is at $\alpha = 40°$. The roll and yaw rates in the lateral design model must, therefore, be in stability axes, that is, roll rate is measured about the velocity vector, and yaw rate is perpendicular to pitch and roll. Commanding the vehicle to rotate about the velocity vector V_0 minimizes the transient in the sideslip angle β, so it makes sense to derive the LQR gains based on the stability axis model and command roll directly in the stability axis. A roll command about the velocity vector corresponds to a simultaneous roll and yaw command about the body axis, which reduces the beta transient. The state vector of the lateral design model is also augmented to include two additional states, the integral of beta and the integral of p_{stab}, which is the roll rate about V_0. This achieves further minimization of the lateral loads and better tracking of the roll commands. So the state vector of the augmented design model consists of: (p_{stab}, r_{stab}, β, p_{stab}-integral, and β-integral). The inputs are roll and yaw acceleration demands.

Figure 9.8 shows the two LQR design models for the lateral directions. The top one is for aerosurfaces control and the other for RCS jet throttle control. They are used to design two 2 × 5 state-feedback matrices KSpr and KJpr for surface and jet control, respectively. The control vectors in both cases consist of two elements (δP, δR) for roll and yaw acceleration demands, respectively. The state vector and the control (δP, δR) must be penalized in the LQR optimization by the proper selection of the Q and R matrices. In this case we attach a heavier weight on the Q-elements that correspond to the β, β-integral, p_{stab}, and p_{stab}-integral states because we intend to achieve good tracking of the roll commands and to minimize the lateral loads due to sideslip. We must also adjust the values of the two elements in matrix R that penalize the roll and yaw controls in order to achieve a good compromise between control bandwidth and performance.

9.5.1.3 LQR Design

The following MATLAB® script loads the four LQR design models, shown in Figures 9.7 and 9.8, and the simulation model. It also loads the two mixing-logic matrices KSmix and KJmix for the aerosurfaces and RCS jets, respectively. It transforms the rates of the lateral design systems from body to stability axes. Then it performs the four LQR designs, that is, pitch and lateral using the aerosurfaces, and also using the RCS jets. It generates four state-feedback matrices. Two pitch (KJq, and KSq), and two lateral (KJpr and KSpr), for using jets and aerosurfaces separately.

```
d2r=pi/180; r2d=180/pi;
[Apj, Bpj, Cpj, Dpj] = pdes_rcs;        % Load Pitch RCS Jet Design Model
[Aps, Bps, Cps, Dps] = pdes_surf;       % Load Pitch aero-surf Design Model
[Alj, Blj, Clj, Dlj] = ldes_rcs;        % Load Lateral RCS Jet Design Model
[Als, Bls, Cls, Dls] = ldes_surf;       % Load Lateral aero-surf Design Model
[Ave, Bve, Cve, Dve] = vehi_sim;        % Simulation Model 6-dof
load KSmix.mat -ascii;                  % Load Surfaces Mix Logic (4 x 3)
load KJmix.mat -ascii;                  % Load RCS Jets Mix Logic (6 x 3)
alfa0=41.5; V0=24124.0; Thet0=40.3;
   ge=32.174;                           % Additional Vehicle Parameters
calfa=cos(alfa0*d2r);
   salfa=sin(alfa0*d2r);                % for Body to Stability Transform
% Convert Lateral State Vector from Body to Stability Axes, Outputs=States
Al2= Clj*Alj*inv(Clj); Bl2= Clj*Blj;    % for the Jets
Cl2= Clj*inv(Clj);       Dl2 = Dlj;
% Lateral LQR Design Using Only the RCS Jets
[Al5,Bl5,Cl5,Dl5]= linmod('LRdes5x');   % 5-state model {p,r,bet,pint,betint}
R=[1,1]*20; R=diag(R);                   % CS LQR Weights R=[1,1]*20
Q=[1 1 0.1 0.5 0.005]*3; Q=diag(Q);      % CS LQR Weights Q=[1 1 0.1 0.5 0.005]*3
[KJpr,s,e]=lqr(Al5,Bl5,Q,R);             % Perform LQR design on Jets
% Convert Lateral State Vector from Body to Stability Axes, Outputs=States
Al2= Cls*Als*inv(Cls); Bl2= Cls*Bls;    % Transform from Body to Stability
Cl2= Cls*inv(Cls);       Dl2= Dls;      % for the Surfaces
% Lateral LQR Design Using Only the Aero-Surfaces
[Al5,Bl5,Cl5,Dl5]= linmod('LRdes5x');   % 5-state model {p,r,bet,pint,betint}
R=[1,1]*20; R=diag(R);                   % CS LQR Weights R=[1,10]*2
Q=[1 1 0.1 0.2 0.002]*2; Q=diag(Q);      % CS LQR Weights Q=[1 1 1 0.04 0.04]*0.4
[KSpr,s,e]=lqr(Al5,Bl5,Q,R);             % Perform LQR design on Jets
% Pitch LQR Design Using Only the RCS Jets
Apd=Apj; Bpd=Bpj; Cpd=Cpj; Dpd=Dpj;
```

```
[Ap4,Bp4,Cp4,Dp4]= linmod('Pdes4x');    % 4-state des model {theta,q,alfa,alf_int}
Ap5= Cp4*Ap4*inv(Cp4); Bp5= Cp4*Bp4;    % Convert to Output=State={theta,q,alfa,
                                           alf_int}
Cp5 = Cp4*inv(Cp4);      Dp5= Dp4;
R=4; Q=[0.05 0.5   1 1]; Q=diag(Q);      % RCS LQR Weights {theta,q,alfa,alf_int)
[KJq,s,e]=lqr(Ap5,Bp5,Q,R);              % Perform LQR design on Surf
% Pitch LQR Design Using Only the 4 Aero-Surfaces
Apd=Aps; Bpd=Bps; Cpd=Cps; Dpd=Dps;
[Ap4,Bp4,Cp4,Dp4]= linmod('Pdes4x');    % 4-state des model {theta,q,alfa,alf_int}
Ap5= Cp4*Ap4*inv(Cp4); Bp5= Cp4*Bp4;    % Convert to Output=State={theta,q,alfa,
                                           alf_int}
Cp5= Cp4*inv(Cp4);       Dp5= Dp4;
R=2; Q=[0.05 0.5   1 1]; Q=diag(Q);      % AS LQR Weights {theta,q,alfa,alf_int}
[KSq,s,e]=lqr(Ap5,Bp5,Q,R);              % Perform LQR design on Surf
```

Figures 9.9 and 9.10 show the closed-loop simulation models for pitch and lateral axes, respectively. The closed-loop simulation systems are used for evaluating the performance of the control system design and for adjusting the LQR optimization. The command in the pitch system is angle of attack α_{comd} since the vehicle is intended to maintain a constant $\alpha = 40°$ during this phase. The command in the lateral direction is roll angle ϕ_{comd} which is a rotation about the velocity vector V_0 and is used for performing the bank maneuvers that manage the excess energy and control the heading direction. Both systems consist of two state-feedback control loops that were designed independently and are operating in parallel, a loop that uses the aerosurfaces and another parallel loop that controls the vehicle by varying the jet thrusts. The surfaces state-feedback matrices stabilize the state vectors by generating the proper acceleration demands in pitch, roll, and yaw. The mixing-logic matrices in the simulation models are not assimilated with the vehicle system as it was in the design models of Figures 9.7 and 9.8, but they are included as separate blocks in order to be able to measure and to evaluate the control responses δ. The surfaces mixing-logic matrix KSmix converts the control system demands to surface deflections that close the surfaces loop to control alpha and the roll angle. Similarly for the RCS

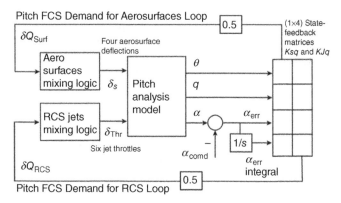

FIGURE 9.9 Closed-Loop Model for the Pitch Control System Shows the Aerosurface and the RCS State-Feedback Loops Operating in Parallel

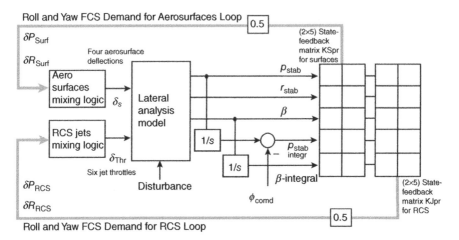

FIGURE 9.10   Closed-Loop Model for the Lateral Control System Shows the Aerosurface and the RCS State-Feedback Loops Operating in Parallel

loop. The RCS state-feedback matrices stabilize the state vectors by generating the proper acceleration pitch, roll, and yaw demands. The RCS mixing-logic matrix KJmix converts the RCS control demands to jet throttle commands that close the RCS loop to control alpha and the roll angle. Both control loops were designed to achieve approximately the same bandwidth so they are equally effective. Since they are operating in parallel their control gains are halved, so each loop is contributing half of the amount that it should be contributing if the other loop was absent.

The LQR design process may be repeated a few times after adjusting some of the elements in the Q and R matrices until a satisfactory compromise between performance and control usage is achieved. The elements that adjust the penalties on the states and controls are modified based on the system responses to the commands. For example, if one of the state variables to be controlled produces large errors, then we increase the magnitude of the corresponding element in the Q matrix and repeat the LQR process until the tracking performance is acceptable. If the surface deflections, on the other hand, are too big, or the throttle commands on the jets exceed ± 1, which means that they are exceeding their throttling capabilities, then the penalty on the corresponding controls in the R matrix must be increased.

9.5.1.4 Simulations   The following two simulations demonstrate the vehicle's response to guidance commands, a $1°$ command in α and a $30°$ command in ϕ. In the first simulation case shown in Figure 9.11 the vehicle is commanded to change the angle of attack $1°$, typically from its trim position at $40°$. The α-command is accomplished by a combination of surface deflection plus thrusting of the RCS jets. Notice that both flaps and rudders are deflecting upward (negative) from trim to create a positive vehicle pitch. The front and rear vertical thrusters are also assisting in the pitch-up maneuver by throttling differentially to create a positive moment. An increase in alpha causes negative z-acceleration.

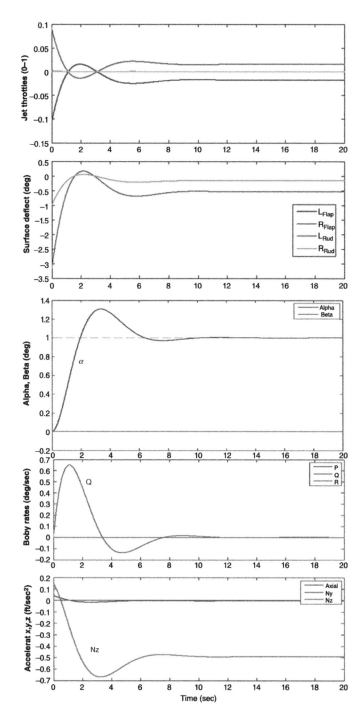

FIGURE 9.11   (a) Aerosurfaces and RCS Jets Are Both Active During the Pitch Maneuver; (b) Vehicle Response to $\alpha = 1°$ Command

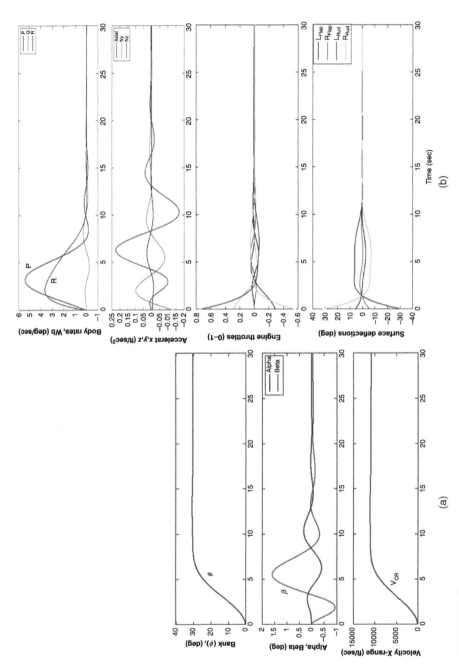

FIGURE 9.12 (a) The 30° Bank Maneuver Causes Small Transients in the Aerodynamic Angles and a Steady Cross-Range Velocity; (b) Aero-Surfaces and RCS Jets Are Both Active During the Maneuver; They Operate Differentially to Provide the Necessary Roll and Yaw Torques

Phi-Command Simulation   The second simulation in Figure 9.12 shows how the vehicle responds when it is commanded to roll 30° about the velocity vector V_0, (ϕ-stability). It performs the roll maneuver in 7 sec, using a combination of aerosurfaces and RCS. The roll and yaw body rates are both responding to the ϕ-stability command. Both the RCS and the aerosurface control systems are equally contributing toward the maneuver, the RCS by differential throttling, and the surfaces by differential deflections (mainly the two rudders). The steady bank angle also causes a steady cross-range velocity V_{CR}.

9.5.2 Midphase

During midphase the dynamic pressure is sufficient to control the vehicle by using the aerosurfaces alone without any assistance from the RCS. The RCS, however, is still used for backup and also during the LCDP transitioning periods where the aileron and rudder are not very reliable. So in the lateral axes, both the aerosurface and RCS jets are active in parallel for roll/yaw control in a similar manner as the previous phase. In the pitch axis, however, only the aerosurfaces are used for control, that is, the two flaps and the two rudders. The RCS is for backup. The body flap is only for trimming and it is scheduled open loop. Excess energy is dissipated by roll maneuvering about the velocity vector, as before. The control surfaces are commanded open-loop to their prescheduled trim positions and the deflection commands from flight control are added to the trim positions. Initially the angle of attack is controlled at 40°, as in the previous phase, and it is gradually reduced to 10°. The control mode changes to γ-control, and the descent flight path angle is gradually reduced from $\gamma = -1.2°$ at first, and then toward the end of the midphase it comes down steeper to $-18°$ in order to maintain high speed as it approaches Mach 2. This also causes the dynamic pressure to increase.

9.5.2.1 *Control Design*   The lateral design is very similar to the one described in the previous phase and it will not be repeated. The flight parameters, however, are different and the control gains and bandwidth are also different. In the pitch axis the control system is different than before and also the design model. We are no longer interested in controlling α, but now we are expected to track a certain γ-command which is calculated from the guidance system that controls the descent rate. The flight path angle is also controlled by pitching but the command is no longer on α but on γ. Changing γ implies a change in the vehicle longitudinal direction which is produced by the aerodynamic normal force created by changing the angle of attack. This obviously takes longer to control than changing α and it requires a thicker atmosphere.

The LQR design model for the pitch control is shown in Figure 9.13. It does not include the jets but only the four aerosurfaces. The design algorithm attempts to optimize the flight path angle which is constructed by combining pitch attitude and angle of attack ($\gamma = \theta - \alpha$). The state vector consists of: (γ, γ-integral, q, and α). The augmented design plant includes also the 4×3 control surfaces mixing matrix "KSmix" which transforms the pitch demand to four surface deflections. The surface

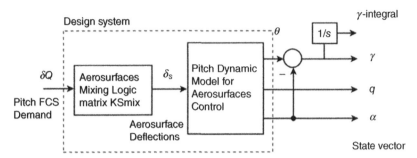

FIGURE 9.13 Augmented Design Model for LQR; It Includes the Surface Mixing Matrix and γ-Integral is Included in the State Vector

mixing matrix is designed for this flight condition and it is different from the matrix used during early reentry. Only the second column of matrix KSmix is used in this model because it consists only of pitch states. The other two columns are used in the lateral design model. The following MATLAB® file loads the systems and matrices and performs the LQR design. Figure 9.14 is a closed-loop simulation model used

```
% File Init.m for Initialization and Perform Control Design
d2r=pi/180; r2d=180/pi;
[Ave, Bve, Cve, Dve] = vehicle_sim;          % Simulation Model 6-dof
[Aps, Bps, Cps, Dps] = pdes_as;              % Load Pitch aero-surf Design Model
[Avls,Bvls,Cvls,Dvls]= ldes_as;             % Load Lateral aero-surf Design Model
[Avlj,Bvlj,Cvlj,Dvlj]= ldes_rcs;            % Load Lateral RCS Jet Design Model
load KSmix.mat -ascii;                       % Load Surfaces Mix Logic (4 x 3)
load KJmix.mat -ascii;                       % Load RCS Jets Mix Logic (4 x 3)

alfa0=14.5; V0=2649.4; Thet0=13.3; ge=32.174;   % Additional Parameters for Sim
calfa=cos(alfa0*d2r); salfa=sin(alfa0*d2r);     % for Body to Stability Transform

% Pitch LQR Design Using the 4 Aero-Surfaces, States: {gamm_int,gamma,q,alfa}
[Ap4,Bp4,Cp4,Dp4]= linmod('Pdes4x');         % 4-state des model (gami,gama,q,alfa)
Ap5= Cp4*Ap4*inv(Cp4); Bp5= Cp4*Bp4;         % Convert to Output=State={gami,gama,q,alfa
Cp5= Cp4*inv(Cp4);     Dp5= Dp4;
R=4; Q=[2 5 2 2]*5; Q=diag(Q);               % LQR Weights {gami,gama,q,alfa}
[Kq,s,e]=lqr(Ap5,Bp5,Q,R)                    % Perform LQR design on Surf
save Kq.mat Kq -ascii

% Lateral LQR Design Using Only the RCS Jets
Alj= Cvlj*Avlj*inv(Cvlj); Blj= Cvlj*Bvlj;    % Tranform Vector from Body to Stability
Clj= Cvlj*inv(Cvlj);      Dlj= Dvlj;         % for the RCS Jets
[Al5,Bl5,Cl5,Dl5]= linmod('LRdes5x');        % 5-state model {p,r,bet,pint,betint)
R=[1,1]*4; R=diag(R);                        % RCS LQR Weights R=[1,10]*2
Q=0.2 0.2 1 0.01]*1; Q=diag(Q);              % RCS LQR Weights Q=[1 1 1 0.04 0.04]*0.4
[KJpr,s,e]=lqr(Al5,Bl5,Q,R)                  % Perform LQR design on Jets
save KJpr.mat KJpr -ascii

% Lateral LQR Design Using Only the Aero-Surfaces
Als= Cvls*Avls*inv(Cvls); Bls= Cvls*Bvls;    % Tranform Vector from Body to Stability
Cls= Cvls*inv(Cvls);      Dls= Dvls;         % for the Surfaces
[Al5,Bl5,Cl5,Dl5]= linmod('LSdes5x');        % 5-state model {p,r,bet,pint,betint)
R=[1,1]*2; R=diag(R);                        % AS LQR Weights R=[1,10]*2
Q=[2 0.2 0.2 1 0.01]*0.8; Q=diag(Q);         % AS LQR Weights Q=[1 1 1 0.04 0.04]*0.4
[KSpr,s,e]=lqr(Al5,Bl5,Q,R)                  % Perform LQR design on Surfaces
save KSpr.mat KSpr -ascii
```

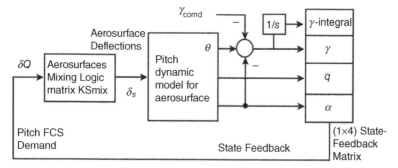

FIGURE 9.14   Pitch Axis Closed-Loop State-Feedback System Using Only Aerosurfaces

for preliminary evaluation of the pitch LQR design and for adjusting the LQR weight matrices. The mixing logic matrix is included as a separate block, and the 1×4 state-feedback matrix Kq that was created by the LQR algorithm is used to close the state-feedback loop.

9.5.2.2   Coupled, Pitch, and Lateral Axes Simulation

We will now create a more realistic simulation model that includes both pitch and lateral axes combined and also the actuator dynamics. The vehicle rates used in the previous lateral design models were calculated relative to the velocity vector, and hence the lateral control system gains were intended to receive roll and yaw rates in stability axes. The vehicle rates, however, are typically measured in the body axes and in our next simulation, shown in Figure 9.15, the rates in the vehicle model are defined in body axes. A transformation, therefore, is included that converts the roll and yaw rates from body to stability axes which are needed by the lateral FCS. The pitch FCS controls the flight path angle γ, and the lateral FCS controls roll about the velocity V_0. Notice that the (α, β) variables in the state feedback were replaced with normal and lateral accelerations (N_z, N_y), respectively. This is not always possible and usually alpha and beta estimators are necessary, but in this particular case the accelerometer signals are almost proportional to the aerodynamic angles which are not directly measured.

The FCS calculates the roll, pitch, and yaw acceleration demands which are inputs to the mixing matrices. There are two sets of control demands, a 3-axes set for the aerosurfaces, and a roll/yaw set for the RCS jets. The mixing-logic matrices convert them to effector commands. The matrix KSmix converts the roll, pitch, and yaw demands to deflections for the four aerosurfaces (two flaps and two rudders), and the matrix KJmix converts the roll and yaw demands to throttle commands (0 to ± 1) for the six RCS jets. The vehicle model also includes a gust input that is shaped by a low-pass filter to produce a smooth gust velocity impulse of 20 (ft/sec). The gust direction is selected to excite both pitch and lateral axes. The FCS and the body to stability axis transformation are shown in detail in Figures 9.16–9.18.

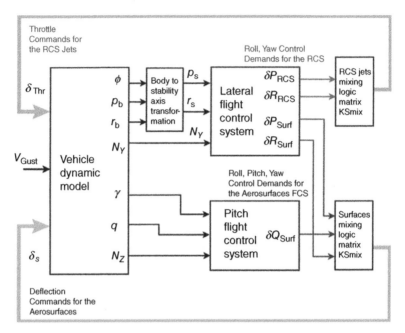

FIGURE 9.15   Simulation Model With a Body to Stability Rates Transformation and a Combination of Aerosurfaces Plus Throttling Jets

Simulation Results: The simulation model of Figure 9.15 is setup to receive three inputs: a gamma command, a roll command about the velocity vector, and a wind-gust disturbance. We will use this model to generate two simulation runs: (a) a step command in roll in combination with a wind-gust disturbance, and (b) a step command in gamma without gust.

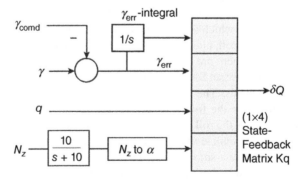

FIGURE 9.16   Pitch State-Feedback Control System Using Only Aerosurfaces; It Has an N_z filter and an N_z to Alpha Conversion Gain

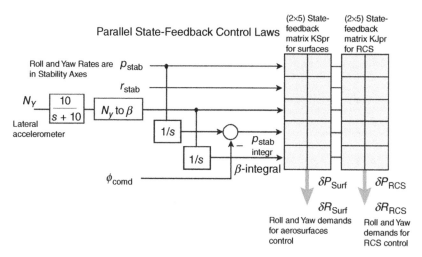

FIGURE 9.17   Lateral State-Feedback Control System Using Both Aerosurfaces and RCS Jets; It Has an N_y Filter and an N_y to Beta Conversion Gain

Phi-Command with Gust: A 30° roll step command is applied at $t = 0$ sec, and a wind velocity impulse of 20 (ft/sec) is applied later at $t = 20$ sec. The vehicle performs the roll maneuver about the velocity vector V_0 with a very small transient in beta, as expected, see Figure 9.19. Both body rates (p and r) respond together in order to rotate the vehicle about the velocity vector. The gust direction is toward the vehicle with a $-z$ velocity component causing a temporary increase in altitude, $-z$ acceleration, and a transient in velocity. Mainly the two flaps, but also the two rudders, deflect differentially to produce the necessary torque and rotation about V_0. The surfaces also deflect to compensate against the pitch transient caused by the gust. The jets also respond to both command and gust. Initially they throttle differentially to generate the roll and yaw torques required to rotate about V_0. The FCS gains were adjusted in the LQR design to prevent the jets from violating their maximum throttle limits (which are ± 1). A throttle value of ± 1 corresponds to a thrust ± 100 (lb) which is defined in the vehicle data.

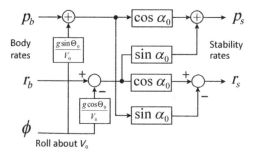

FIGURE 9.18   Body to Stability Rates Transformation, Because Stability Rates Are Used for State Feedback

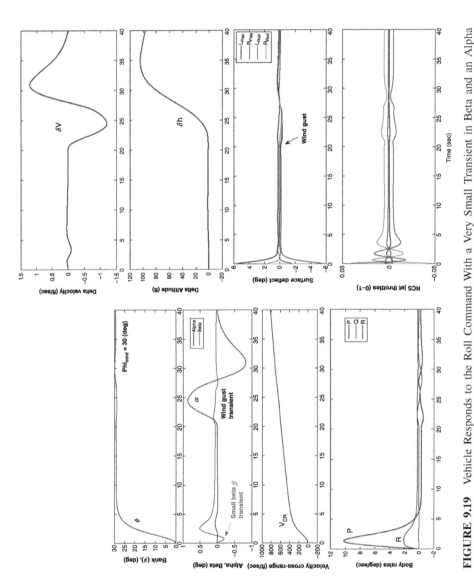

FIGURE 9.19 Vehicle Responds to the Roll Command With a Very Small Transient in Beta and an Alpha Response to the Gust; Roll and Yaw Body Rates Respond Together to the Roll Command, The Gust Causes a Transient in Velocity and an Increase in Altitude; Both, Aerosurfaces and Jets Respond to the Command and Gust With Differential Action.

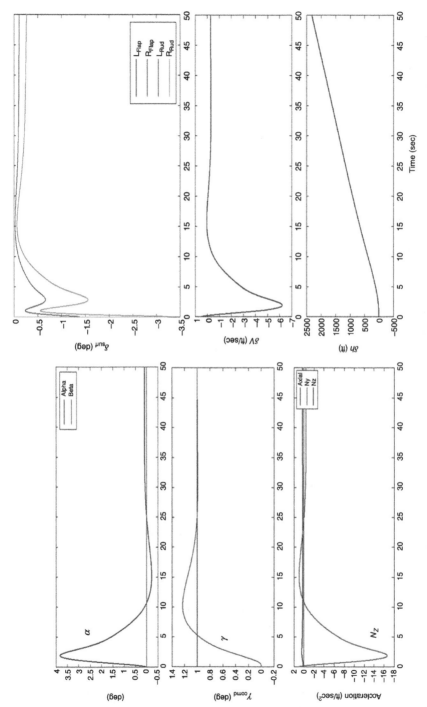

FIGURE 9.20 Vehicle Pitches Up to Increase Alpha and Lift in Order to Gain Altitude and Gamma; Aerosurfaces Deflect Negative (Up) to Cause a Positive Alpha and an Increase in Altitude

Flight Path Change Command: In the second case, shown in Figure 9.20, the vehicle is commanded to perform one degree change in flight direction γ. Both flaps and elevons deflect momentarily in the negative direction causing the vehicle to pitch-up and to generate a positive alpha. The positive alpha generates the normal force required for the flight path angle to start climbing in the positive direction and for the altitude to increase.

9.5.3   Approach and Landing Phase

The approach and landing phase begins below Mach 1 where the dynamic pressure climbs to about 200 psf. The vehicle maintains a positive angle of attack but the flight path γ drops to a much steeper angle of $-22°$ that is needed for maintaining high speed in order to be able to perform a successful flare before landing. The vehicle also executes the heading alignment maneuver where it rolls on one side and circles to align its direction with the runway and levels off prior to landing. During this final approach and landing phase the FCS does not use the RCS jets but all six control surfaces are active. It is no longer dissipating extra energy by roll maneuvering but it controls velocity by modulating drag using the speed brake which is partially deployed, in combination with the body flap. It regulates altitude also by deflecting the flaps in combination with pitching which affect the normal acceleration N_z. The altitude and velocity commands are coordinated by the guidance system which calculates the proper commands for achieving the required terminal position and velocity conditions, while it prevents the angle of attack from stalling or reaching unacceptably high dynamic pressures. In the lateral direction the vehicle uses the aileron (differential flap) for maneuvering and for maintaining its alignment with the runway against crosswinds. We will show how to design simple control laws for the approach and landing phase and demonstrate the system performance with a landing simulation.

9.5.3.1   *Vehicle Design Model*   The condition that will be analyzed in this phase is when the vehicle descends at a very steep flight path angle $\gamma = -22°$, $\alpha = 6°$, at an altitude of 5500 (ft), and velocity of 450 (ft/sec). This is a few seconds before it performs the landing flare. The dynamic model in this phase has six control inputs because it uses all six aerosurfaces. It also has a wind-gust disturbance input that excites both pitch and lateral axes. Figure 9.21 shows the design models used for longitudinal and lateral LQR control. The longitudinal design model is different from the previous two phases because it must now include variations in altitude and velocity variables (δh, δV) in the state vector since we are attempting to control these two variables which are also measurable. Alpha integral is not included because we are not directly controlling alpha in this case, and also for the benefit of better tracking altitude and speed. Alpha is also included in the design states, with a relatively low weight in the Q matrix, however, in order to keep it stable. The lateral design model is not very different from before. It does not use RCS but only the aerosurfaces, see Figure 9.22. It does not include β-integral in the state vector for the benefit of better

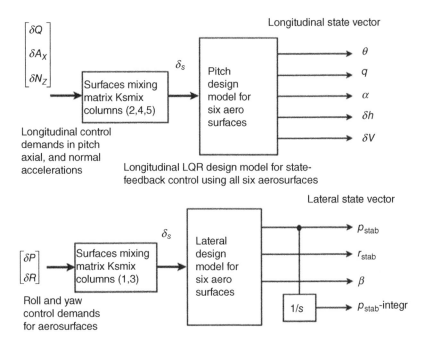

FIGURE 9.21   Design Models for Longitudinal and Lateral Control Which Include the Corresponding Columns of the Surfaces Mixing Matrix

regulating the heading direction which is a lot more important during landing than minimizing the lateral loads.

9.5.3.2 Surface Mixing Matrix and Control Design   The FCS in this flight condition controls not only the three rotations, but also speed and the rate of descent. A 6×5 control surface mixing-logic matrix KSmix is designed, that in addition to the three rotations includes two additional columns for translational control along x and z. It converts the five FCS demands (roll, pitch, yaw, plus axial and normal accelerations) to six aerosurface commands. It means that the aerosurfaces, as a system, must have the authority to control the vehicle in five directions. 5-DOF control is not necessary, however, because pitch and N_z controls (which correspond to KSmix columns 2 and 5) are combined together to control altitude. Roll and yaw are controlled by columns 1 and 3, and speed is controlled by column 4. The vehicle rate of descent and speed are controlled by the proper calculation of altitude and velocity change commands. In the absence of forward thrust the control commands which are coordinated by the guidance system must provide acceptable alpha, descent rate, and speed commands in order to prevent the vehicle from stalling or developing excessive dynamic pressures.

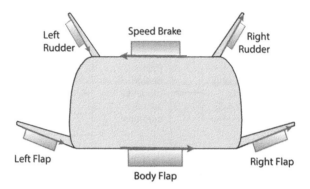

FIGURE 9.22 All Six Aerosurfaces are Used During the Approach and Landing Phase

The mixing-logic matrix KSmix, in Figure 9.23, allocates control to the aerosurfaces based on direction demands. The longitudinal model has three control input demands, which are pitch, axial, and N_z acceleration demands that correspond to KSmix matrix columns (2, 4, and 5) and they are included in the design model as shown. The axial acceleration demand δA_X affects the descent speed by a combination of speed brake and body-flap deflections. The pitch and normal acceleration demands (δQ and δN_Z) are not used independently to control these variables but they are combined together to regulate altitude. This provides better performance than using only pitch for altitude control, because if affects altitude not only by pitching but it is assisted with some additional lift from the surfaces. Figure 9.22 shows the vehicle from behind and the directions of the six aerosurface hinge vectors during the approach and landing phase.

Figure 9.23 shows the mixing-logic matrix with its inputs and outputs. It shows how the aerosurfaces combine together in order to achieve accelerations along the five demand directions. The roll axis is controlled mainly by the differential deflection of the two flaps. Positive pitch is obtained mainly by the negative deflection of the two rudders (up) and with some contribution from the body flap. Yaw is obtained

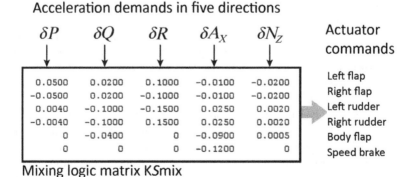

FIGURE 9.23 Aerosurfaces Combination Matrix for the Five Directions

```
% LQR Design & Param Initialization file init.m
d2r=pi/180;  r2d=180/pi;
[Aps, Bps, Cps, Dps] = pitch_des;       % Load Pitch aero-surf Design Model
[Als, Bls, Cls, Dls] = later_des;       % Load Lateral aero-surf Design Model
[As,  Bs,  Cs,  Ds]  = vehicle_sim;     % Simulation Model 6-dof
load KSmix.mat -ascii; Kmix=KSmix;      % Load Surfaces Mix Logic (6 x 5)

alfa0=6.0; V0=465.37; Thet0=-16.0; ge=32.174;   % Additional Vehicle Parameters
calfa=cos(alfa0*d2r); salfa=sin(alfa0*d2r);     % for Body to Stability Transform

% Lateral LQR Design Using States: (ps, rs, beta, phi-stab)
% Convert Lateral State Vector from Body to Stability Axes, Outputs=States
[A14,B14,C14,D14]= linmod('Ldes4x');    % 5-state model (p,r,beta,pint)
A15= C14*A14*inv(C14);  B15= C14*B14;   % Stability axis System
C15= C14*inv(C14);      D15= D14;
R=[1,1]*1; R=diag(R);                   % LQR Weights R=[1,1]*2
Q=[10 2 1.5 20]; Q=diag(Q);             % LQR Weights Q=[10 2 0.5 20]
[Kpr,s,e]=lqr(A15,B15,Q,R);             % Perform LQR design on Jets
save Kpr_T1860.mat Kpr -ascii           % Lateral State-Feedback Gain

% Pitch LQR Design Using the 6 Aero-Surfaces, States: (theta,q,alfa,dH,dV)
[Ap4,Bp4,Cp4,Dp4]= linmod('Pdes6x');    % Include Kmix in design model
R=[3,1,1]; R=diag(R);                   % LQR Control Weights (Pitch,Nz,Ax)
Q=[0.0001 0.1 0.01 14 4]; Q=diag(Q);    % LQR State Weights (theta,q,alfa,dH,dV)
[Kq,s,e]=lqr(Ap4,Bp4,Q,R);             % Perform LQR design on Surf
save Kq_T1860.mat Kq -ascii             % Longitudinal State-Feedback Gain
```

FIGURE 9.24 MATLAB® Script for LQR Design During the Approach and Landing Phase

by the differential deflection of the two rudders because their rotations are defined in opposite directions. Forward acceleration, or less drag in this case, is obtained by negative deflections of the speed brake (down) and body flap (up), since their hinge directions are opposite. Normal acceleration (toward the earth) is obtained mainly by deflecting the two flaps negative (up). Other surfaces are also contributing by counteracting the induced pitching moment. The script file in Figure. 9.24 is used to load the design and simulation models and matrices in MATLAB® and to design the state-feedback flight control matrices Kq and Kpr for the longitudinal and the lateral directions using the LQR method.

9.5.3.3 Simulation Model

The performance of the LQR designs is evaluated first by using the pitch and lateral closed-loop models shown in Figure 9.25. The Q and R weight matrices in the LQR process are adjusted for a good trade-off between performance and control usage. The columns of the surface mixing matrix K_{Smix} are split between the two systems. The longitudinal system receives the pitch, axial, and normal acceleration demands from the longitudinal state-feedback matrix K_q, and the lateral system receives the roll and yaw acceleration demands from the lateral state-feedback matrix K_{pr}.

In the simulations the aerodynamic angles α and β, which are assumed to be measurable in the LQR designs, they are replaced with filtered normal and lateral accelerometer measurement (N_z, N_y), as already described. Prior to landing the vehicle aligns the heading direction with the runway and controls its rate of descent and speed. A pitch and lateral coupled simulation model is used to create the descent

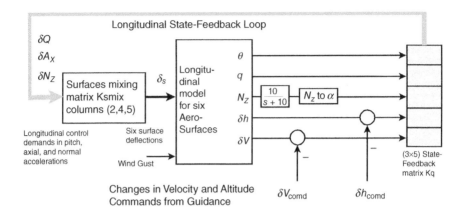

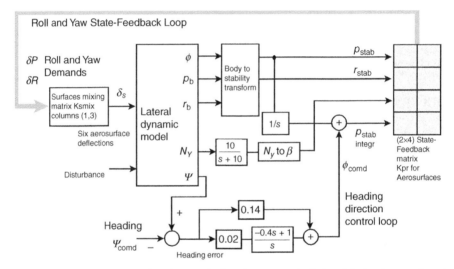

FIGURE 9.25   Longitudinal and Lateral Closed-Loop Analysis Models for Approach and Landing

FIGURE 9.26   Approach and Landing of an Experimental Vehicle

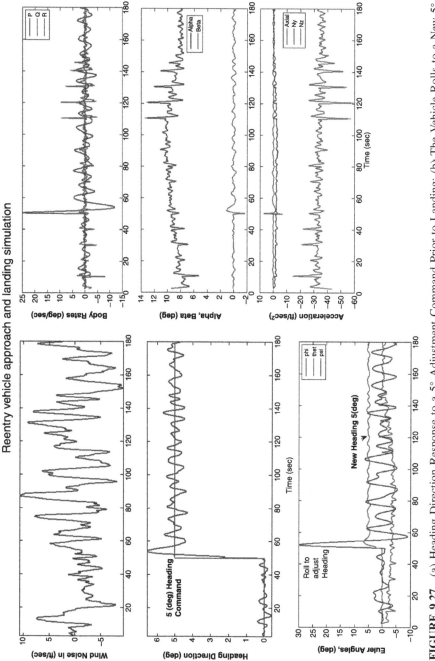

FIGURE 9.27 (a) Heading Direction Response to a 5° Adjustment Command Prior to Landing; (b) The Vehicle Rolls to a New 5° Heading and Returns to Level φ. It Continues to Oscillate in Roll Because of the Crosswind Disturbance. The Angles of Attack and Sideslip, and the Accelerations are Also Affected by the Wind Disturbance

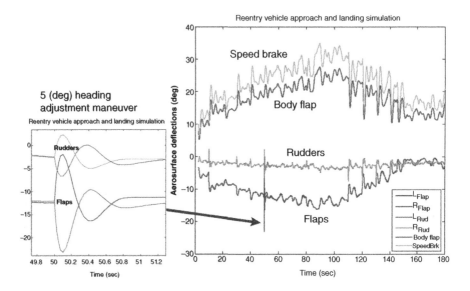

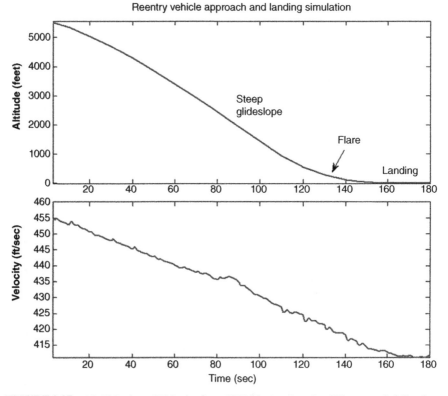

FIGURE 9.27   (c) Altitude and Velocity from 5500 (feet) to Landing Where $\gamma = 0$, following a Steep Glideslope and a Pitch-Up Flare; Includes Deflections of the Six Control Surfaces Relative to Trim Positions

and landing trajectory of the reentry vehicle starting from an altitude of 5500 (ft), $\alpha = 6°$, and a speed of 460 (ft/sec). The lateral control system receives a 5° heading correction command (ψ_{cmd}) from guidance to align its heading with the runway. The heading error becomes a roll command that temporarily rolls the vehicle to change its heading direction.

A noise generator creates a random wind-gust disturbance with amplitudes varying ±10 (ft/sec). The lateral system responds to the 5° change in heading command by banking to the right for a short period and returning to level when the new heading is achieved. The disturbance due to the crosswinds causes ±5° oscillations in roll which is acceptable. The noise excitation also introduces a negligible ±0.5° disturbance in the vehicle heading all the way to landing, see Figure 9.27.

The longitudinal control system receives coordinated altitude and speed change commands (δh_{cmd}, δV_{cmd}) from guidance which control its rate of descent and speed. Figure 9.27c shows the altitude and velocity versus time responses. The altitude follows a steep gamma glide slope, the pitch-up flare, and landing where γ is brought to zero just prior to touchdown. The effect of the wind disturbance is visible in the velocity plot.

The top part of Figure 9.27(c) shows the deflections of the control surfaces. These are increments relative to their trim positions. Notice how the speed brake and body-flap deflections increase prior to the pitch-up flare to slow down the rate of descent. The flaps and rudders control altitude and the angle of attack as needed for landing. Notice also the differential deflections of the two flaps and the two rudders in order to perform the roll/yaw maneuver at $t = 50$ sec for the heading alignment.

9.6   ROCKET PLANE WITH A THROTTLING ENGINE

In this example we will design and analyze a small rocket-plane vehicle that is controlled by five aerosurfaces and a fixed engine similar to the vehicle shown in Figure 9.28. The control surfaces are: two flaps which are almost horizontal used

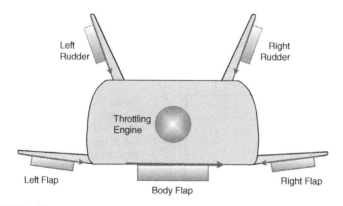

FIGURE 9.28   Flight Vehicle With Five Control Surfaces and a Throttling Engine

mostly for roll control, two V-shaped rudders that are used for pitch and yaw control, and a body flap for trimming and assisting in pitch control. The engine has a nominal thrust of 2000 (lb). It does not gimbal but it is used mainly for speed control and it can vary its thrust between 0 and 4000 (lb) by throttling. We will analyze a flight condition at Mach 0.85, altitude 40,000 (ft) and dynamic pressure of 150 (psf), and design a longitudinal control system for altitude and velocity control. The longitudinal control system consists of two control loops: an altitude control loop that is accomplished by controlling the normal acceleration by means of pitch control, and a velocity control loop by varying the engine thrust (throttling).

A state-feedback pitch control system will be designed by using the LQR method. The speed is controlled by an independently designed throttle control Proportional Integral Derivative (PID) system that adjusts the engine thrust, or rather the throttle in this case since the thrust and maximum throttle are included in the dynamic model. The throttle control varies as a function of velocity error, acceleration, and velocity integral and the PID gains are calculated to achieve the bandwidth requirement of the throttle system. The throttle control input to the dynamic model must vary between 0 and ± 1. Zero throttle command corresponds to nominal engine thrust of 2000 (lb). A throttle command of $+1$ corresponds to a maximum thrust of 4000 (lb), and a throttle command of -1 corresponds to zero thrust. The phugoid states (δh and δV) are not included in the state-feedback controller but the controller requires an angle of attack estimator since alpha is not directly measurable in this case. We will design control laws for this vehicle, perform stability analysis, and evaluate its ability to track input commands and also its response to gusts. The lateral axis will not be examined because it is similar to our previous example. The data files for this example can be downloaded from Flixan.com/Examples/Rocket Plane/.

9.6.1 Design Model

We must first calculate a mixing-logic matrix K_{mix} for this longitudinal axis example that will combine the control surface deflections and the engine throttle command in order to satisfy the two acceleration FCS demands: δA_X along the vehicle x-axis and δQ for pitch control. The K_{mix} matrix is shown in Figure 9.29. It converts the

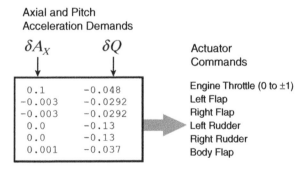

FIGURE 9.29 Effector Combination Matrix Kmix That Converts the Control Acceleration Demands to Effector Commands

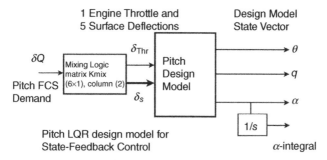

FIGURE 9.30   LQR Design Model for the Pitch State-Feedback Control Loop

two longitudinal demands to a combination of engine throttling plus control surface deflections that provide the demanded accelerations open loop. It premultiplies the plant model and minimizes the dynamic coupling between the two longitudinal directions because it attempts to diagonalize the forward open-loop system. Axial acceleration is accomplished mainly by throttling up the engine in combination with small flap deflections in order to counteract the pitching moment induced due to throttling. Similarly, pitching is produced by deflecting the flaps and rudders upward but this reduces drag and requires throttling down in order to maintain the same acceleration.

The function of the longitudinal control system is to independently control altitude and speed using the available controls. The design approach in this case is to control altitude by pitching and thus inducing variations in normal acceleration and not by regulating the altitude state directly as in our previous example. Figure 9.30 shows the LQR design model for the pitch control loop. It does not include the phugoid states (δh and δV), but we introduce instead an alpha integral state that will help us regulate the angle of attack more precisely. The design model consists of the pitch vehicle plant that has six inputs: an engine throttle control and five surface deflections which are all contributing to pitch control. The plant is premultiplied by the second column of the effector mixing matrix K_{mix} which corresponds to the pitch demand δQ.

9.6.2   LQR Control Design

The following m-file loads the pitch analysis and design systems, the mixing-logic matrix, the actuator model, and calculates the control gains using the LQR method. The state-feedback vector closes the loop from the four states: pitch attitude, pitch rate, alpha, and alpha-integral. R is a scalar in this case and penalizes the pitch control δQ. The diagonal matrix Q penalizes mainly alpha and alpha-integral.

```
% Initialization File
d2r=pi/180; r2d=180/pi;
[Ap,Bp,Cp,Dp]= vehi_pitch_rb;          % Pitch Rigid Body Model
[Ad,Bd,Cd,Dd]= vehi_pdes_4x;           % LQR Design Model
```

```
load KPITMIX.mat KPITMIX -ascii          % Mixing Logic
zet=0.7; omg=40;                         % Actuator Model
Aact=[0, 1; -omg^2, -2*zet*omg]; Bact=[0; omg^2];
Cact=[1, 0]; Dact=0;
% LQR Design
Q=[1.0e-6, 0.02, 0.5, 1.5]; Q=diag(Q); R=0.03;
Kpg= LQR(Ad,Bd,Q,R)                      % LQR State-Feedback
```

9.6.3   Simulation of the Longitudinal Control System

The longitudinal closed-loop control system is shown in Figure 9.31. The vehicle model used in the simulation includes the phugoid variables (δh and δV). The control system consists of two loops:

1. Velocity PID control loop that controls the vehicle ground speed by issuing a δA_X demand causing variations in the engine thrust. Thrust variations are implemented by a throttle command (δ_{Thr}) that varies between -1 and $+1$, as a function of the velocity error, and it is generated by the mixing matrix.
2. Altitude control loop that uses the state-feedback gain to control the vehicle altitude by issuing a pitch demand δQ and commanding the aerosurfaces (δ_s) to change the angle of attack and the normal acceleration.

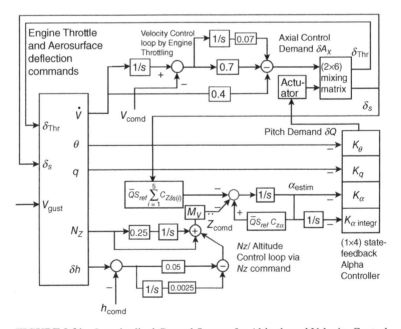

FIGURE 9.31   Longitudinal Control System for Altitude and Velocity Control

9.6.3.1 State Estimator   The LQR state-feedback controller inputs come from the four system states: pitch attitude, rate, alpha, and alpha-integral. Two of the states, however, alpha and alpha-integral are not directly measurable and an estimator is included in order to estimate the angle of attack α and α-integral, from the normal acceleration and from the predicted aerodynamic forces. This is described in detail in Section 3.12, and it basically solves the normal force equation and converges to the actual angle of attack when the vehicle aerodynamic coefficients and mass are reasonably known. Its inputs are normal acceleration $\ddot{z}$ and deflections δ_s from the five control surfaces. The estimated α and α-integral replace the actual state variables which are expected for the state feedback.

9.6.3.2 Delta-Altitude Command   The altitude control system includes a PID that receives an altitude increment command from guidance and issues a normal acceleration command to the alpha estimator which is translated to an alpha command. When the altitude error is zero, the α-estimate converges to α. Otherwise, the α-estimate includes a bias that causes the altitude to change in the direction of the h-command.

9.6.3.3 Simulation Results   Figures 9.32a–9.32c show the system's response to simultaneous altitude and velocity change commands. A wind-gust velocity impulse

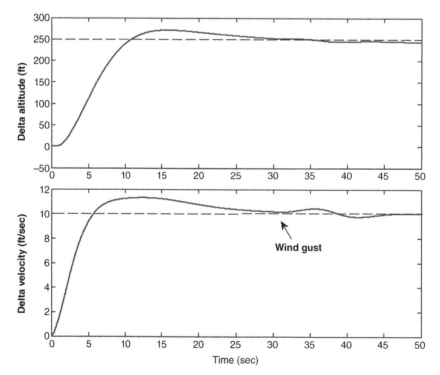

FIGURE 9.32   (a) Altitude and Velocity Response to Simultaneous Altitude and Velocity Step Commands and Also to a Wind Gust Occurring at 30 sec

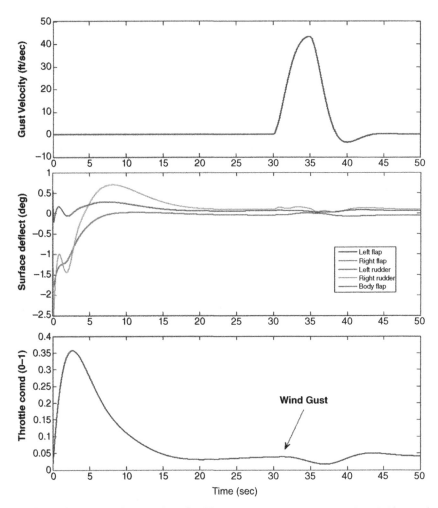

FIGURE 9.32   (b) Surfaces and Engine Throttle Response to Commands and Also to the Wind Gust

is also applied to the simulation model that is shaped by a low-pass filter and it excites the vehicle dynamics. The commands are: a 250 (ft) step increase in altitude, and a 10 (ft/sec) increase in ground speed. The gust is 40 (ft/sec) and it occurs at 30 sec.

To perform this maneuver the vehicle temporarily increases its thrust 35% and deflects its surfaces in the negative (up) direction that cause an increase in alpha, which accelerates the vehicle upward (negative N_z) and increases altitude. Both flaps and rudders are contributing to the pitch maneuver. Eventually the altitude and velocity approach the commanded values and the surface controls decay back to the original trim deflections, while the thrust decreases to a value 5% higher than its original nominal thrust. The flight path angle γ starts from zero at level flight, increases to 2.5 (deg) and it comes back to zero (horizontal flight) as the aircraft reaches a steady altitude.

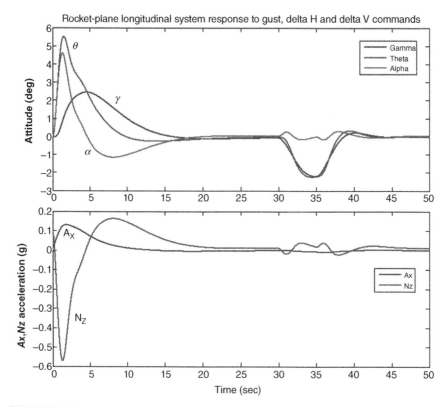

FIGURE 9.32   (c) Attitude and Acceleration Response to Commands and to the Wind Gust

The gust occurs at 30 sec and it is in the negative z direction creating a lift. The vehicle responds with a pitch-down in theta and gamma while trying to maintain a level altitude. It responds by deflecting the surfaces down and throttling down to avoid gaining altitude. The transient in the normal acceleration is relatively small. Notice that gamma is relative to the air mass and not the ground.

9.6.4   Stability Analysis

The system stability in the classical sense is measured by opening one loop at a time and calculating the frequency response across the two open ends of the open-loop system while the other control loops are closed. For example, the altitude system's stability is measured by opening the loop at the connection between the state-feedback output and the actuator input, which is the pitch demand δQ in Figure 9.31, and calculating the frequency response between the two opened points. The phase and gain at each frequency are then plotted on a Nichols chart and the phase and gain margins are measured between the locus and the critical point (0 dB, $-180°$), as shown in Figure 9.33. In this example the altitude control system has approximately 55° of

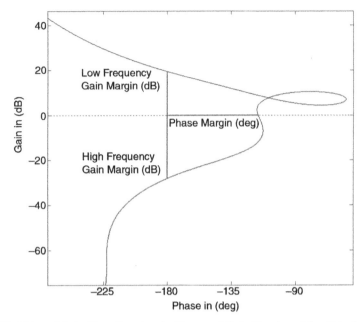

FIGURE 9.33 Nichols Chart Showing the Phase and Gain Margins of the Altitude Control Loop With the Velocity Loop Closed

phase margin, 20 dB of low frequency gain margin, and 27 dB of high frequency gain margin, which is very good.

9.7 SHUTTLE ASCENT CONTROL SYSTEM REDESIGN USING *H*-INFINITY

The Space Shuttle FCS was designed in the mid-1970s using classical methods, before the robust control design methods where invented. It was designed almost entirely on stability criteria for control, using phase and gain margins, without taking directly into consideration the system's sensitivity to wind gusts. When performing sensitivity analysis, the results show that at high dynamic pressures the lateral system is very vulnerable to wind-gust disturbances because there is an aerodynamic resonance at 0.9 (rad/sec), see Figure 9.34. This sensitivity is further aggravated by the fact that the disturbance itself is cyclic and it occurs on average at approximately the same frequency as the resonance, causing roll/yaw oscillatory excitations by interacting with the big vertical stabilizer, as the Shuttle ascends at a steep angle through the atmospheric layers that have different airspeeds.

One might think; how would we redesign the Shuttle ascent control system using modern and robust control design methods? At high dynamic pressures the emphasis of the control system, second to stability, is not in tracking the commands from

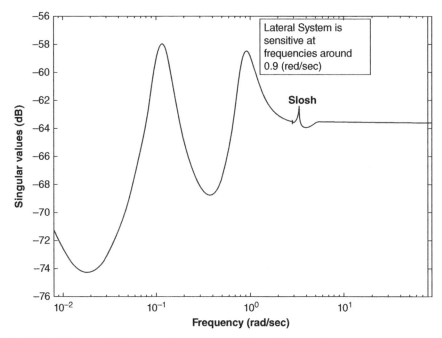

FIGURE 9.34 Sensitivity Function Between Gust Velocity Disturbance to Alpha/Beta Response. There is an Undesirable Resonance at 0.9 (rad/sec) That Coincides With the Wind Excitation Frequency

guidance, but in minimizing the normal and lateral loading on the vehicle caused by the winds. The command following capability is somewhat compromised at Max-Q. The *H*-infinity method is by definition very suitable in accomplishing this performance versus sensitivity trade-off, because, first of all, it guarantees stability, and second it can be made to minimize the magnitude of the transfer function between the gust disturbance input and the vehicle response criteria. The criteria variables to be optimized in this flight condition are the normal and lateral loads which are proportional to the angles of attack and sideslip. This criteria minimization is equivalent to the classical load-relief function that was described in the previous Shuttle example in Section 5.9. The analysis and data files for this example can be downloaded from Flixan.com/Examples/Shuttle Ascent/H-Infinity Design.

9.7.1 Pitch Axis *H*-Infinity Design

As we mentioned earlier, at Max-Q the command performance is sacrificed in order to gain sensitivity reduction against lateral wind disturbances. The load-relief system causes the vehicle's response to commands to slowdown and increases the steady-state error, because instead of responding to attitude commands the vehicle turns toward the airflow, as soon as the control bandwidth allows, in order to reduce normal

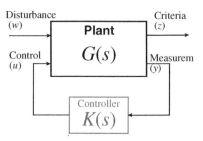

FIGURE 9.35 Synthesis Model $G(s)$ Used in H-infinity Design. The Control System $K(s)$ Minimizes the Infinity-Norm of the Transfer-Function Between w and z

and lateral wind loading. The inherent static stability also makes it easier for the vehicle to turn toward the flow.

In this example we will use the H-infinity algorithm that was described in Section 9.3. The design is based on a SM of the vehicle plant, typically consisting only of the rigid-body dynamics. The SM, depicted by $G(s)$ in Figure 9.35, has two sets of input vectors and two sets of output vectors. The input vectors are disturbances w and controls u. The output vectors are the criteria to be optimized (z) and measurements (y), which are processed by the control system $K(s)$ to generate the controls u. The criteria to be minimized (z) are not necessarily measureable variables but they can be defined by equations. The H-infinity algorithm synthesizes the dynamic controller $K(s)$ that closes the control loop between the measurements and the controls. The controller minimizes the infinity norm of the transfer function between the wind disturbance w and the criteria z. In this case, however, we know that the disturbance is likely to occur at around 1 (rad/sec). It is possible, therefore, to introduce the disturbance as a second-order resonance, shown in Figure 9.36 and represented by a filter in Figure 9.37, that can be incorporated in the SM. This filter resonance would influence the H-infinity algorithm and provide additional control at that frequency and as a result it would further attenuate the disturbance, as we shall see.

Before we begin the control design we must first design the TVC matrix that will convert the roll, pitch, and yaw FCS demands to five engine deflections. We must also create the pitch and lateral systems that will be used in the H-infinity design. The pitch H-infinity SM is a state-space system to be optimized by the H-infinity algorithm, see Figure 9.38. The pitch column of the TVC matrix is included in the SM, and the inputs to the SM are: pitch control acceleration demand (δQ) in (rad/sec^2), and the wind-gust velocity disturbance (w) in (ft/sec). They are both scalars. The design plant outputs are: pitch attitude θ, pitch rate q, and angle of attack α. The augmented pitch axis H-infinity SM also includes the filter resonance of Figure 9.37 that is tuned to 1 (rad/sec). The resonance is excited by the angle of attack α which is the most important variable to be minimized. The filter introduces two additional states x_1 and x_2 to the SM that become part of the measurement vector y and also to the criteria output vector z that will be optimized by the design algorithm. The x_1 and x_2 filter states in the criteria are functions of α and they are expected to provide additional attenuation at the 1 (rad/sec) resonance.

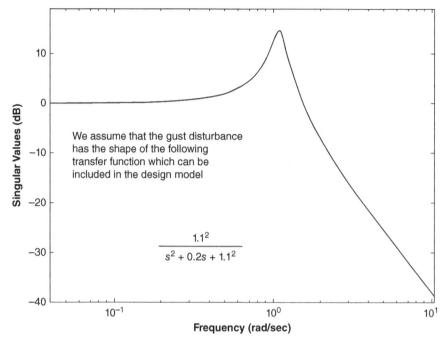

FIGURE 9.36   The Disturbance is Captured by a Second-Order Resonance Tuned at Around 1 (rad/sec)

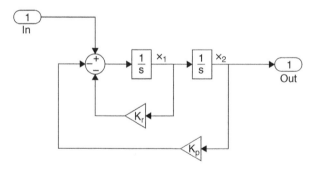

FIGURE 9.37   Compensator Resonance Tuned at 1 (rad/sec) to be Included in the Synthesis Model

The criteria variables include the control u that will adjust the system bandwidth. The criteria penalties are included in the diagonal matrix Q. Notice that the pitch attitude θ is not penalized much because, as we already explained, the attitude tracking is compromised at high dynamic pressures. Penalizing the control prevents it from becoming too big when the control loop is closed. Increasing the penalty weight on the control u in matrix Q decreases the control bandwidth. Increasing the penalties

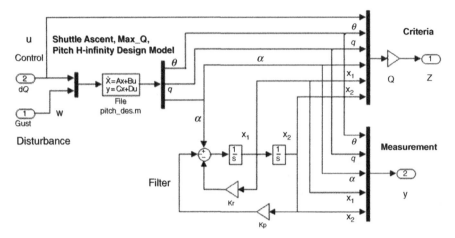

FIGURE 9.38   Pitch Synthesis Model Includes the Design Plant and the Disturbance Filter

```
[Ad, Bd, Cd, Dd]= pitch_des;              % Pitch design plant
[An, Bn, Cn, Dn]= pitch_anal;             % Simple analysis plant
[Ap, Bp, Cp, Dp]= plant;                  % Plant w actuat,sensors, TWD
load TVC.mat -ascii                       % TVC Matrix


%... Beta Estimator parameters
Mass=93215; Sref= 2690; Qbar=745.4; Cza=-0.0574;
Thr=[470000, 470000, 470000, 0.245e7, 0.245e7]';

%... Hinfinity Design ...
%        [DQtvc, thet, q,    alfa, x1,   x2]
Q=diag([0.06, 0.005, 0.2, 0.03, 0.03, 0.03]);
om=1.0; zt=0.14; Kp=om^2; Kr=2*zt*om;
[Ad, Bd, Cd, Dd]= linmod('Design'); sys= SS(Ad,Bd,Cd,Dd);
sysd=MKTITO(sys,5,1);
[K,CL,GAM,INFO] = hinfsyn(sysd);
Kp3=INFO.KFI(:,1:5)
%Kp3 = [0.3042  3.4539  0.5105  0.0711 -0.2670];


[Ao, Bo, Co, Do]= linmod('Anal4');
[As, Bs, Cs, Ds]= linmod('Sensit4');
w=logspace(-3, 2, 10000);                 % Define Frequ Range
syso= SS(Ao,Bo,Co,Do);                    % Create SS System
syss= SS(As,Bs,Cs,Ds);                    % Create SS System
figure(1); Nichols(syso,w)                % Plot Nichol's Chart
figure(2); Bode(syso,w)                   % Plot Bode
figure(3); Sigma(syss,w)                  % Plot Sensitivity
```

FIGURE 9.39   MATLAB® Script That Performs LQR Design and Analysis in Pitch

that correspond to the states: (α, x_1, and x_2), improves α−sensitivity to disturbances but it increases the control bandwidth. The measurements vector consists of all five states: three vehicle and two filter states. Notice that the α-state is not measurable. It must be estimated using an alpha estimator from the N_z measurement and the gimbal deflections δy.

The MATLAB® file in Figure 9.39 loads the systems and the TVC matrix and performs the control design and analysis. It uses the *H*-infinity method to design a 1 × 5 state-feedback gain from the five-states vector $(\theta, q, \alpha, x_1, x_2)$ which are also measurements, although α is not directly measureable and it must be estimated. The state-feedback matrix minimizes the norm of the transfer function and hence the sensitivity between the gust disturbance and the criteria defined. Having designed the state-feedback controller the next step is to analyze the α sensitivity to the gust disturbance in the frequency domain by calculating the singular-value (SV) response between the gust input and the angle of attack using the closed-loop model in Figure 9.40. Notice that the actual angle of attack α is used for output in the sensitivity calculation. In the state-feedback, however, the angle of attack is estimated using an alpha-estimator block, shown in detail in Figure 9.41. The alpha-estimator solves equation 9.10 for α as was described in Section 3.12. The filtered pitch rate derivative compensates the accelerometer measurement because it is not located at the vehicle CG and it also measures rotational accelerations.

$$M\,\ddot{z} = \bar{Q}\,S_{\text{ref}}\,C_{Z\alpha}\alpha + \sum T_i\,\delta_{yi} \qquad (9.10)$$

The input to the disturbance filter (which calculates the states x_1 and x_2) is α-estimate and not real α. Figure 9.42 is a SV frequency response plot that shows the alpha sensitivity of the closed-loop system to wind-gust disturbances. The sensitivity function has a big notch at approximately 1 (rad/sec) to provide additional attenuation at the expected disturbance frequency.

The block diagram in Figure 9.43 is similar to 9.40 but it has the loop opened at the control system output. It is used for calculating the frequency response and analyzing the pitch axis stability. The Nichols plot in Figure 9.44 shows the phase and gain margins in the pitch control loop. The liquid oxygen slosh resonance is phase stabilized. The liquid hydrogen resonance is very small.

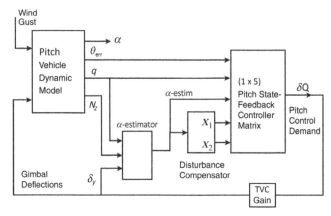

FIGURE 9.40   Closed-Loop Model Used for Calculating the Sensitivity. Alpha is Replaced with α-Estimate

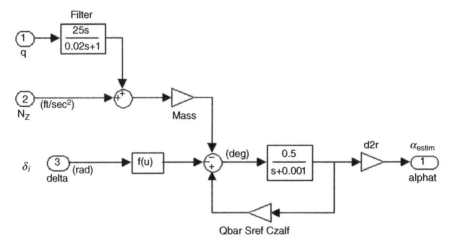

FIGURE 9.41 The Angle of Attack Estimator is Using Measurements From the Normal-Accelerometer, Gimbal Deflections, and Pitch Rate

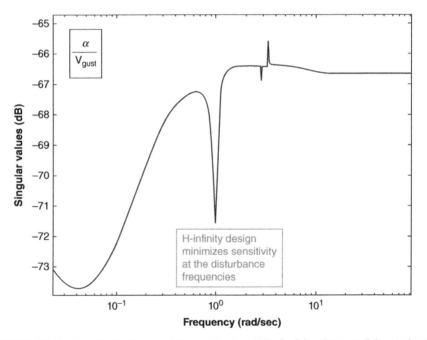

FIGURE 9.42 Sensitivity Function Between the Gust Velocity Disturbance and the Angle of Attack. It Shows How the Disturbance Filter Provides Additional Attenuation at the Expected Disturbance Frequency Which is at 1 (rad/sec)

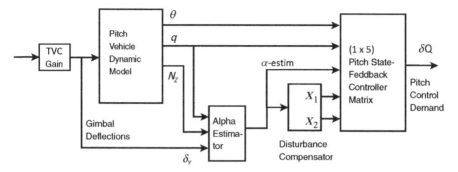

FIGURE 9.43   Model Used for Pitch Axis Stability Analysis

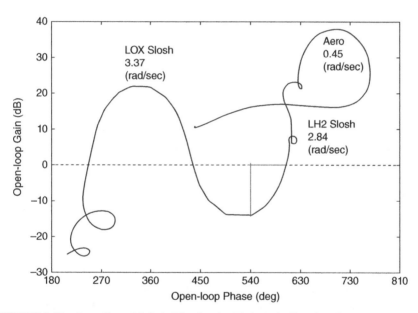

FIGURE 9.44   Open-Loop Nichols Plot for the Pitch Axis Showing the Phase and Gain Margins

9.7.2   Lateral Axes *H*-infinity Design

A very similar approach is used for redesigning the Shuttle lateral FCS, and this is where the original design had poor sideslip sensitivity to wind-gust disturbances. The *H*-infinity control system must provide a load-relief feedback signal from the sideslip angle β which is estimated from the lateral accelerometer (N_y), similar to the pitch axis.

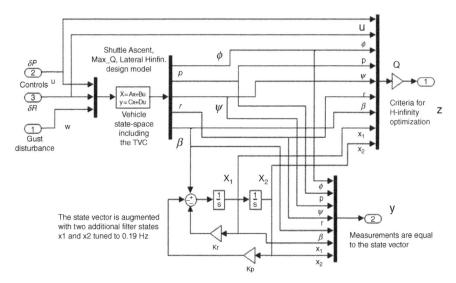

FIGURE 9.45   Lateral *H*-infinity Synthesis Model That Includes the Disturbance Compensator Filter

Figure 9.45 shows the *H*-infinity SM for the lateral axes. The lateral vehicle state-space design model contains only the rigid-body dynamics (no slosh, TWD, or flex). It includes, however, the TVC matrix at the input, that is, only columns 1 and 3 that correspond to roll and yaw demands. So the control inputs *u* are roll and yaw demands (δP and δR) in (rad/sec^2), and the disturbance w is wind-gust velocity in (ft/sec). The 1 (rad/sec) disturbance resonance is also included in the SM output, and it is excited by the sideslip β, which is the key variable to be minimized by the *H*-infinity design since β is an indicator of lateral loading. The second order resonance introduces two additional states x_1 and x_2 that become part of the measurement *y* and criteria *z* output vectors. The criteria consist of lateral states including the compensator states x_1 and x_2 which amplify beta at the resonance frequency in order to penalize it further in the optimization process. The criteria variables also include the controls $u = (\delta P, \delta R)$ which are penalized in order to prevent high bandwidths and avoiding too much control usage. The criteria penalties are introduced in the diagonal matrix *Q*. Notice that the attitudes (ϕ, ψ) are not penalized much because command following is compromised at high dynamic pressures. Increasing elements of the *Q* matrix that correspond to the controls *u* decreases the control bandwidth. Increasing on the other hand the magnitude of the *Q* elements that correspond to the states: (β, x_1, and x_2) reduces sideslip beta, but it also increases the magnitude of the controls and the closed-loop system bandwidth. The measurements vector consists of the seven states: five vehicle and two filter states. Notice that in the flight control system implementation, shown in Figure 9.46, the β-state is not directly measurable and it must be estimated, mainly from N_y and the yaw gimbal deflections δz.

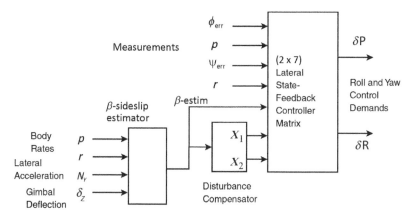

FIGURE 9.46   Lateral Flight Control System Includes the Disturbance Attenuator Filter and β-estimator

The MATLAB® file in Figure 9.47 loads the systems and the TVC matrix and performs the control design and sensitivity analysis in MATLAB®. It uses the *H*-infinity method to design a 2 × 7 state-feedback gain matrix that closes the loop from the 7-states vector (ϕ, p, ψ, r, β, x_1, x_2) that appear also in the measurements. The *H*-infinity derived state-feedback matrix minimizes the sensitivity function between the gust disturbance and the criteria defined.

```
r2d=180/pi; d2r=pi/180;
om=1.2; zt=0.14; Kp=om^2; Kr=2*zt*om; %frq; % Disturb Filter Characteristics
[Ad, Bd, Cd, Dd]= later_des;          % Shuttle Ascent, Max_Q, Lateral Hinf Design Model
[An, Bn, Cn, Dn]= later_anal;         % Shuttle Ascent, Max_Q, Lateral Analysis with Slosh & TVC
[Ap, Bp, Cp, Dp]= plant;              % Shuttle Plant Model at Max-Q (Flex Vehicle, Actuators, Sens
load TVC.mat -ascii                   % Shuttle Stage-1 TVC Matrix at Max-Q
load orig.mat                         % from Classical Design

%... Alpha/Beta Estimator parameters
Mass=93215; Sref= 2690; Qbar=745.4; Cyb=-0.0353; Cza=-0.0574;
Thr=[470000, 470000, 470000, 0.245e7, 0.245e7]';

%... Hinfinity Design ...
%       [DPtvc,DPtvc, phi,  p,   psi,   r,   beta, x1,   x2]
Q=diag([0.09, 0.12, 0.001, 0.1, 0.001, 0.1, 0.24, 0.13, 0.07]);
[Ad, Bd, Cd, Dd]= linmod('design');
sys = SS(Ad,Bd,Cd,Dd);
sysd=MKTITO(sys,7,2);
[K,CL,GAM,INFO] = hinfsyn(sysd,7,2);
K13=INFO.KFI(:,1:7);
Kp3 = [0.3042  3.4539  0.5105  0.0711 -0.2670];

% [Ao, Bo, Co, Do]= linmod('Anal2');
% [As, Bs, Cs, Ds]= linmod('Sensitiv2');
[Ao, Bo, Co, Do]= linmod('Anal5');
[As, Bs, Cs, Ds]= linmod('Sensitiv4');
%w=logspace(-3, 2.5, 10000);          % Define Frequ Range
syso= SS(Ao,Bo,Co,Do);               % Create SS System
syss2= SS(As,Bs,Cs,Ds);              % Create SS System
figure(1); Nichols(syso,w)           % Plot Nichol's Chart
figure(2); Bode(syso,w)              % Plot Bode
sig2=Sigma(syss2,w);                 % SV Bode
figure(3); loglog(w,sig1,'r',w,sig2,'b')   % Plot SV Bode
```

FIGURE 9.47   MATLAB® Script That Performs LQR Design and Analysis in Roll and Yaw

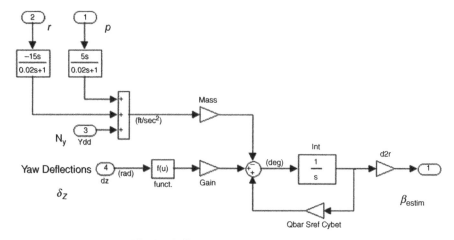

FIGURE 9.48   Angle of Sideslip Estimator

A β-estimator is used to estimate β which is required for the state feedback, shown in Figure 9.48. It uses measurements from the lateral accelerometer, gimbal deflections, and body rates, and it solves equation 9.11 for β. The rate derivatives compensate the accelerometer measurement because it is not located at the vehicle CG and it is also measuring rotations. Notice that the inputs to the disturbance filter (which calculates the states x_1 and x_2) are from β-estimate and not the actual β.

$$M\ddot{y} = \bar{Q} S_{\text{ref}}\, C_{Y\beta}\beta + \sum T_i\, \delta_{zi} \tag{9.11}$$

After designing the state-feedback controller the next step is to analyze the β-sensitivity to the lateral wind-gust disturbance. This is performed in the frequency domain by calculating the SV frequency response between the gust input and the angle of sideslip using the closed-loop model in Figure 9.49. Notice that in the sensitivity calculation the actual angle of sideslip β and not the estimate is used for output.

Figure 9.50 shows the SV sensitivity response of the combined pitch and lateral systems. The sensitivity function is calculated by applying a skewed wind disturbance that excites both pitch and lateral directions and measuring both α and β together. The sensitivity plot compares the original Shuttle ascent design versus the H-infinity design. It demonstrates that the H-infinity design improves sensitivity, both at low frequencies against wind shear, and also in the range between 0.5 and 1.5 (rad/sec) which is where the cyclic wind-gust disturbances occur.

Figure 9.51 shows the open-loop Bode and Nichols plots for classical frequency response analysis showing the phase and gain margins of the yaw loop. Notice that the compensator has introduced an additional resonance in the open-loop frequency response, at 1 (rad/sec). It is the proper phasing of this resonance that achieves the additional sensitivity attenuation at the disturbance frequency. The LOX mode is strong but it is properly phase stable which contributes toward attenuating the mode in closed loop.

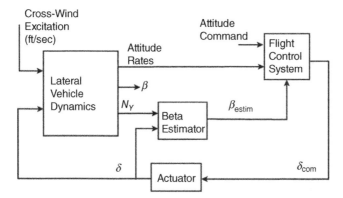

FIGURE 9.49   Closed-Loop Lateral Control System

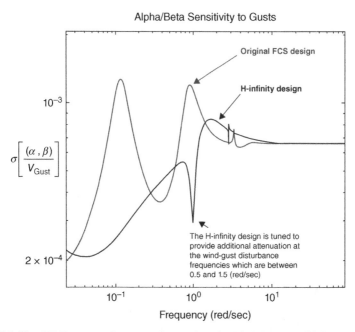

FIGURE 9.50   SV Frequency Response Comparison in Wind-Gust Sensitivity Between the Original Shuttle Design and the *H*-infinity Design Showing an Overall Improvement in the *H*-Infinity Design.

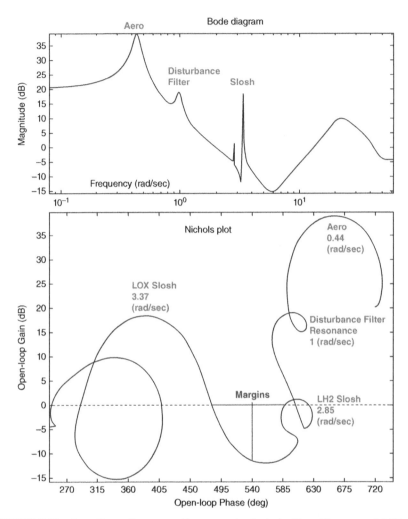

FIGURE 9.51   Open-Loop Frequency Response Analysis in Yaw, Shows an Additional Resonance at 1 (rad/sec) Introduced by the Disturbance Compensator

9.7.3   Sensitivity Comparison Using Simulations

Figures 9.53 and 9.54 compare sensitivity to wind disturbances between the original and the H-infinity designs using simulations. Both systems are excited with the same cyclic disturbance shown in Figure 9.52 consisting of a combination of frequencies between 0.5 and 1.5 (rad/sec). The simulation results show that the H-infinity design significantly improves the system's sensitivity to cyclic gusts, mainly in the lateral direction. The results also show that in addition to improving the lateral loading, the roll and yaw attitude errors and rates, and gimbal deflections are significantly reduced by the H-infinity control design. The improvement, however, in the pitch direction is not as substantial because the original design was not as sensitive to noise as the lateral was.

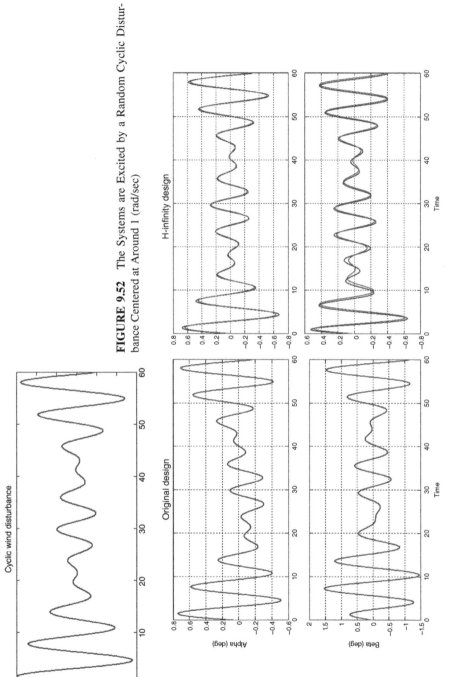

FIGURE 9.52 The Systems are Excited by a Random Cyclic Disturbance Centered at Around 1 (rad/sec)

FIGURE 9.53 Angle of Attack and Sideslip Response to the Cyclic Wind Disturbance. The *H*-infinity Responses Also Show the Estimated Alpha and Beta

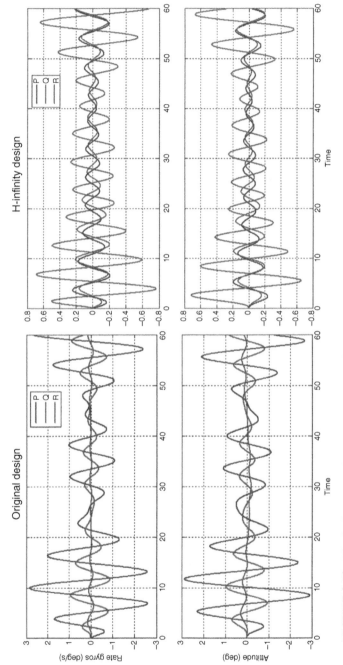

FIGURE 9.54 The Body Rates and Attitude Errors are Also reduced by the *H*-infinity Design, and also the Gimbal Deflections

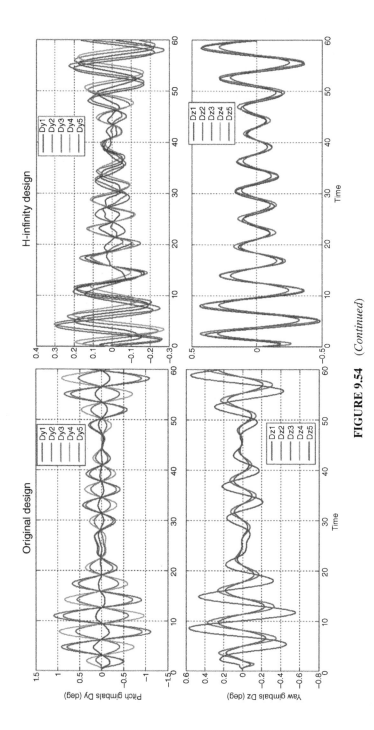

FIGURE 9.54 (*Continued*)

9.8   CREATING UNCERTAINTY MODELS

Uncertainty models are used for analyzing the robustness of a control system to uncertainties. Robustness is the ability of the control system to tolerate uncertainties, either internal or external. The ability of a system to handle external disturbances is measured by the effect of the disturbances on some sensitive outputs, for example, an optical sensor, or a structural load sensor. Well-known methods are used (using SVs) for analyzing a system's sensitivity to external disturbances. The question is how do we analyze a system's robustness to internal parameter variations, and how much parameter variations can a system tolerate before it becomes unstable, or stops functioning properly? Parameter uncertainties can be seen as imprecise knowledge of the plant model parameters, such as the mass properties, moments of inertia, aerodynamic coefficients, vehicle altitude, dynamic pressure, center of gravity. The uncertainties in a model are defined in terms of variations in the actual plant parameters, above or below their specified values. These uncertainties are called *"Structured,"* in contrast with the *"Unstructured"* uncertainties which are described in the frequency domain in terms of maximum amplitude error in the transfer function model.

In this section we will present a method that is used for creating state-space models that can be used to analyze the robustness of systems that have bounded and known uncertainties in their parameters. Each parameter variation is "pulled out" of the uncertain plant model and it is placed inside a diagonal block Δ that contains only the uncertainties. The remaining plant is assumed to be known (best guess). The Δ block is attached as shown in Figure 9.55 to the known plant $M(s)$ by means of n input/output "wires," where n is the number of plant uncertainties. In essence, if $M(s)$ is the plant model representing a flight vehicle we are creating n additional inputs and outputs to the plant that connect with the uncertainties block Δ, which is a block diagonal matrix $\Delta = \text{diag}(\delta_1, \delta_2, \delta_3, ... \delta_n)$. The individual elements δ_i of the Δ block may be scalars or matrices and each element represents a real uncertainty in the plant. Its magnitude represents the maximum possible variation of the corresponding parameter p_i above or below its nominal value. The parameters may be aerodynamic coefficient variations from the specified values, moment of inertia variations, thrust

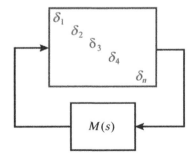

FIGURE 9.55   Uncertainties are Extracted From the Plant $M(s)$ and Placed in a Diagonal Δ Block

variations, etc. Note that $M(s)$ in addition to the plant model, also includes the control system in closed-loop form and $M(s)$ is stable. In essence the internal uncertainties Δ block is "pulled out" of the closed-loop plant $M(s)$ and is connected to it by the additional inputs and outputs.

The nominally stable closed-loop system $M(s)$ in Figure 9.55 configuration is defined to be robust to a set of parameter variations $\delta_{(i)}$ which are included in the Δ block if it remains stable regardless of all possible variations of those parameters as long as the magnitude of each variation from nominal does not exceed the maximum defined uncertainty $\delta_{(i)}$. The structured singular value SSV or () (μ) is the perfect tool for analyzing this type of robustness problems. It is performed in the frequency domain similar to the sensitivity analysis. To make the analysis easier, the plant $M(s)$ inputs and outputs are scaled so that the individual elements of the diagonal uncertainty block Δ can now vary between $+1$ and -1. This is simple scaling. The gains of the parameter variations $\delta_{(i)}$ are absorbed in $M(s)$ and the magnitudes of the new uncertainties are now bounded to be less than 1. The value of $1/\mu(M)$ represents the magnitude of the smallest perturbation that will destabilize the normalized closed-loop system $M(s)$. According to the small gain theorem, the closed-loop system is robust as long as $\mu(M)$ across the normalized block Δ is less than one at all frequencies. But the question is how do we extract the uncertainties out of the model?

The μ-methods use the SSV method and apply criteria for analyzing three types of closed-loop system situations: (a) "nominal performance", that is sensitivity to external disturbances alone without any parameter variations, (b) "robustness" to parameter uncertainties, and (c) "robust performance" which is simultaneously satisfying robustness to uncertainties while maintaining an acceptable sensitivity to external disturbances. In this section we will describe a method used for extracting the uncertainties out of the plant, and it is called the Internal Feedback Loop (IFL) method. The augmented state-space model (without the Δ block) is then used to analyze robustness using μ-analysis methods. We will also present an algorithm that creates uncertainty models for robustness analysis by using an existing vehicle modeling program. This algorithm is implemented in the Flixan program.

9.8.1   The Internal Feedback Loop Structure

The IFL concept allows internal parameter variations in a plant to be treated like external disturbances in the system. This representation allows us to use μ-tools for robustness analysis, or to apply H_∞ plus other robust methods to design control systems that can tolerate internal parameter variations. To utilize the IFL concept the state-space system must be written in the following form:

$$\begin{bmatrix} \dot{x} \\ y \end{bmatrix} = \left\{ \begin{bmatrix} A & B \\ C & D \end{bmatrix} + \begin{bmatrix} \Delta A & \Delta B \\ \Delta C & \Delta D \end{bmatrix} \right\} \begin{bmatrix} x \\ u \end{bmatrix} \qquad (9.12)$$

Suppose that there are (l) independently perturbed parameters $p_1, p_2, \ldots p_l$ with bounded parameter variations δp_i, where $|\delta p_i| \leq 1$. The perturbation matrix

$\Delta P = [\Delta A, \Delta B; \Delta C, \Delta D]$ can be decomposed with respect to each parameter variation as follows:

$$\Delta_i = -\sum_{i=1}^{L} \delta p_i \begin{pmatrix} \alpha_x^{(i)} \\ \alpha_y^{(i)} \end{pmatrix} \begin{pmatrix} \beta_x^{(i)} & \beta_u^{(i)} \end{pmatrix} \tag{9.13}$$

The perturbation matrix ΔP is assumed to have a rank-1 dependency with respect to each parameter (p_i). Where for each parameter p_i,

$\alpha_x^{(i)}$ and $\alpha_y^{(i)}$ are column vectors
$\beta_x^{(i)}$, and $\beta_u^{(i)}$ are row vectors

The plant uncertainty ΔP due to all perturbations can be written in the following form:

$$\Delta P = -\begin{pmatrix} M_x \\ M_y \end{pmatrix} \Delta \begin{pmatrix} N_x & N_u \end{pmatrix} = -M\Delta N \tag{9.14}$$

where M_x and M_y are stacks of column vectors, and N_x and N_u are stacks of row vectors as shown below:

$$M_x = \begin{bmatrix} \alpha_x^{(1)} & \alpha_x^{(2)} & \cdots & \alpha_x^{(L)} \end{bmatrix}; \quad M_y = \begin{bmatrix} \alpha_y^{(1)} & \alpha_y^{(2)} & \cdots & \alpha_y^{(L)} \end{bmatrix}$$
$$N_x = \begin{bmatrix} \beta_x^{(1)} \\ \vdots \\ \beta_x^{(L)} \end{bmatrix}; \quad N_u = \begin{bmatrix} \beta_u^{(1)} \\ \vdots \\ \beta_u^{(L)} \end{bmatrix} \tag{9.15}$$

and where $\Delta = \text{diag} [\delta p_1, \delta p_2, \delta p_3, \ldots \delta p_l]$ is the diagonal block of Figure 9.55 containing the uncertainties. Notice, that in order to simplify the implementation, the columns of matrices M_x and M_y and the rows of matrices N_x and N_u are scaled, so that the elements of the diagonal block Δ have unity upper bound. Now let us introduce two new variables (z_p and w_p) and rewrite the equations in the following system form in order to express it as a block diagram.

$$z_p = N_x x + N_u u \quad \text{and} \quad w_p = -\Delta z_p \tag{9.16}$$

The perturbed state-space system can be expressed by the following augmented representation which is the same as the original system in the upper left side, with some additional input and output vectors, an input and an output for each parameter uncertainty.

$$\begin{pmatrix} \dot{x} \\ y \\ z_p \end{pmatrix} = \begin{bmatrix} A & B & M_x \\ C & D & M_y \\ N_x & N_u & 0 \end{bmatrix} \begin{pmatrix} x \\ u \\ w_p \end{pmatrix} \tag{9.17}$$

If we further separate the plant inputs u into disturbances w and controls u_c. That is, $u = [w, u_c]$, and if we also separate the plant outputs y into performance criteria z and control measurements y_m, the above system is augmented as shown in equation 9.18.

$$
\begin{pmatrix} \dot{x} \\ z \\ y_m \\ z_p \end{pmatrix} = \begin{bmatrix} A & B_1 & B_2 & M_x \\ C_1 & D_{11} & D_{12} & M_w \\ C_2 & D_{21} & D_{22} & M_{ym} \\ N_x & N_w & N_{uc} & 0 \end{bmatrix} \begin{pmatrix} x \\ w \\ u_c \\ w_p \end{pmatrix}
\tag{9.18}
$$

The above formulation is useful for μ-synthesis or robustness/performance analysis using μ-methods. It is shown in block diagram form in Figure 9.56. The uncertainties block Δ is normalized to unity by scaling the columns in matrices M_x, M_w, and M_{ym} and the rows in matrices N_x, N_w, and N_{uc} after dividing with the square root of the corresponding SV. The normalized parameter variations block Δ is connected to the plant by means of the inputs w_p and the outputs z_p. When the controller feedback loop between y_m and u_c is closed, the controller $K(s)$ which is designed based on the nominal plant $P(s)$, is also expected to keep the plant stable despite all possible variations inside Δ block. This property is defined as robust stability. In addition to robust stability the controller must also satisfy performance requirements between the disturbances w and the criteria z not only for the nominal plant (which is called nominal performance), but also for the perturbed plant that has the uncertainty loop closed via the Δ block. This property is known as robust performance. It means, that the system must be stable and it must satisfy the expected performance criteria between w and z, despite all possible internal plant variations captured in Δ, which are normalized and their individual magnitudes δ_{pi} cannot exceed 1.

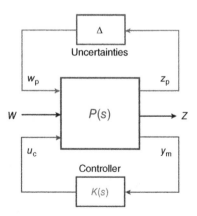

FIGURE 9.56   Robustness Analysis Block Showing the Uncertainties IFL Loop, the Control Feedback Loop, the Disturbances (w), and Performance Outputs (z)

This system can also be represented in matrix transfer function form as follows:

$$
\begin{pmatrix} z_p \\ z \\ y_m \end{pmatrix} = \begin{bmatrix} G_{11} & G_{12} & G_{13} \\ G_{21} & G_{22} & G_{23} \\ G_{31} & G_{32} & G_{33} \end{bmatrix} \begin{pmatrix} w_p \\ w \\ u_c \end{pmatrix}
$$

where

$$
w_p = -\Delta z_p \quad \text{and} \quad u_c = -K(s) y_m \tag{9.19}
$$

After closing the loop with a stabilizing controller $K(s)$ the closed-loop system is represented with the following transfer function matrix:

$$
\begin{pmatrix} z_p \\ z \end{pmatrix} = \begin{bmatrix} T_{11}(s) & T_{12}(s) \\ T_{21}(s) & T_{22}(s) \end{bmatrix} \begin{pmatrix} w_p \\ w \end{pmatrix} \quad \text{and} \quad w_p = -\Delta z_p
$$

where

$$
T_{11}(s) = G_{11} - G_{13} K \left(I + G_{33} K \right)^{-1} G_{31}
$$

$$
T_{12}(s) = G_{12} - G_{13} K \left(I + G_{33} K \right)^{-1} G_{32}
$$

$$
T_{21}(s) = G_{21} - G_{23} K \left(I + G_{33} K \right)^{-1} G_{31}
$$

$$
T_{22}(s) = G_{22} - G_{23} K \left(I + G_{33} K \right)^{-1} G_{32}
$$

The above transfer functions are used to evaluate system robustness and performance of the closed-loop system.

Robust Stability: Stability robustness with respect to parameter uncertainty is determined by the transfer function $T_{11}(s)$. Smaller $\|T_{11}\|_\infty$ allows larger parameter uncertainty for closed-loop stability. The closed-loop system is considered to be robustly stable with respect to the parameter perturbations block Δ, where $\|\Delta\| \leq 1$, when the $\mu\{T_{11}(\omega)\} < 1$ at all frequencies (ω).

Nominal Performance: Nominal performance is used to calculate the system's sensitivity to excitations and it is obtained from the transfer function $T_{22}(s)$. This transfer function must be scaled by multiplying its inputs with the max magnitude of the excitations and by dividing its outputs with the max allowable error. The system meets nominal performance when the scaled $\|T_{22}(\omega)\|_\infty < 1$ at all frequencies (ω). For example, maximum wind-gust velocity disturbance must not exceed the maximum allowable angle of attack dispersion.

Robust Performance: Robust performance is achieved when the system meets the performance and robustness requirements together. This happens when the following condition is satisfied at all frequencies.

$$
\mu \begin{bmatrix} T_{11}(s) & T_{12}(s) \\ T_{21}(s) & T_{22}(s) \end{bmatrix} < 1
$$

9.8.2 Implementation of the IFL Model

In order to extract the parameter uncertainties block, augment the plant model by including the additional inputs and outputs as already described, and to to apply the IFL method, the analyst obviously needs a modeling program, such as Flixan®, that implements the vehicle equations of motion described in Chapter 3, processes the data, and generates the vehicle state-space models. An additional algorithm must be implemented that calls the vehicle modeling program multiple times, beginning by processing the nominal plant, and repeating the data processing for each parameter variation, in order to create the uncertainty state-space model, by using the following sequence:

1. The modeling program is used initially to process the nominal set of vehicle data and to create the "known" plant state-space model $[A, B; C, D]$.
2. One (and only one) of the vehicle data parameters must be modified at a time, either increased or decreased from its nominal value by an amount that is equal to the maximum expected variation (δp_1) and the vehicle data is reprocessed by the vehicle modeling program to create a new state-space system $[A_1, B_1, C_1, D_1]$ that corresponds to parameter #1 variation. The matrix difference between the nominal and the perturbed state-space models is calculated:
$$\begin{bmatrix} \Delta A_1 & \Delta B_1 \\ \Delta C_1 & \Delta D_1 \end{bmatrix} = \begin{bmatrix} A_1 & B_1 \\ C_1 & D_1 \end{bmatrix} - \begin{bmatrix} A & B \\ C & D \end{bmatrix}$$
3. This matrix is decomposed using Singular Value Decomposition (SVD) to calculate the column vectors $\alpha_x^{(1)}$ and $\alpha_y^{(1)}$ and the row vectors $\beta_x^{(1)}$, and $\beta_u^{(1)}$, as shown in equation 9.13.
4. Restore the previous parameter to its original value and modify another parameter #2 in the vehicle input data by an amount (δp_2) that represents the maximum variation of this parameter, as in step 2. Repeat steps 2 and 3 and calculate the vectors $\alpha_x^{(2)}$, $\alpha_y^{(2)}$, $\beta_x^{(2)}$, and $\beta_u^{(2)}$.
5. Select another parameter to perturb and repeat steps (2 and 3) until there is no more uncertain parameters to vary. Stack the row and column vectors as shown in equation 9.15 to create matrices consisting of stacks of column vectors: M_x and M_y and stacks of row vectors: N_x and N_u.
6. These matrices are then used to create the additional inputs and outputs in the state-space model as shown in equations 9.17 and 9.18. The columns of matrices M_x and M_y and the rows of matrices N_x and N_u must also be scaled according to the magnitude of the uncertainties δp_i so that the interconnections correspond to a unity normalized Δ-block.

Robustness is then analyzed after closing the control loop as shown in Figure 9.56 and calculating the μ-frequency response of the plant across its interconnections with Δ, that is, between w_p and z_p, with the control loop $K(s)$ closed. More robustness analysis examples and workfiles are presented and can be downloaded from the website: http://flixan.com/Examples.

10

VEHICLE DESIGN EXAMPLES

In this chapter we present three conceptual vehicle design examples that will be used to illustrate the material presented in the previous chapters. The examples are fictitious and they do not represent any actual flight vehicles. The first example analyzes a reentry glider returning from space where we will study the various descent phases separately and design control laws for each phase. The second example is a launch vehicle that is controlled by multiple types of engines and aerosurfaces. In the third example we will design and analyze the momentum management control system of a large space station controlled by an array of control moment gyros (CMGs). We will study these examples in detail and analyze vehicle performance, stability, controllability, design control systems, and create simulations in MATLAB®. They include analysis details which are often not included in textbooks, technical papers, and typical presentations. The analysis files of the examples presented, plus additional examples can be downloaded from the website Flixan.com/Examples/.

10.1 LIFTING-BODY SPACE-PLANE REENTRY DESIGN EXAMPLE

In this example we will study a lifting-body aircraft that is used as a transportation vehicle from space. It is capable of returning from space by gliding and landing autonomously controlled by its aerosurfaces. Figure 10.1 shows the vehicle shape and its seven aerosurfaces. It has two elevons, a rudder, and four body flaps (two upper and two lower). The hinge vectors indicate the directions of positive aerosurface

Performance Evaluation and Design of Flight Vehicle Control Systems, First Edition. Eric T. Falangas.
© 2016 by Eric T. Falangas. Published 2016 by John Wiley & Sons, Inc.

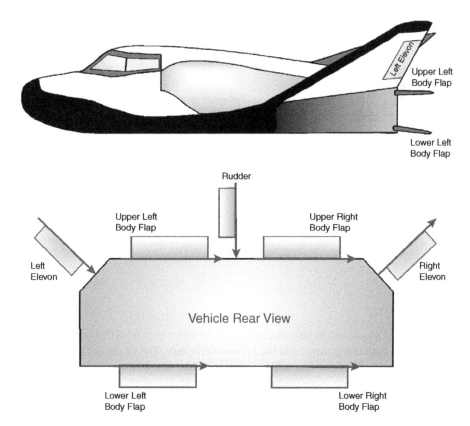

FIGURE 10.1   Seven Control Surfaces and Hinge Directions Are Shown From the Back of the Vehicle

rotations. Having multiple aerosurfaces makes it easier for the vehicle to trim and to be controlled without a need for RCS. However, RCS is also available but it is only used for maneuvering and controlling attitude at low dynamic pressures and also as a back-up system during descent. The primary function of the elevons and rudder is to provide roll, pitch, and yaw control. The four body flaps are mainly for trimming and for speed-brake control. However, they are also used to provide some flight control assistance to the elevons.

The purpose of this example is to demonstrate a complete flight control design of a typical reentry vehicle from deorbit to landing, beginning with a preliminary performance and controllability analysis, control law synthesis at selected Mach points, and performing linear dynamic analysis and simulation. It teaches the student how to create dynamic models for flight control design and linear analysis, how to design simple control laws in MATLAB®, and how to generate dynamic models for analyzing robustness to uncertain parameters. The analysis concludes by creating a 6-degrees of freedom (DOF) nonlinear reentry simulation from deorbit to landing using MATLAB/Simulink®.

10.1.1   Control Modes and Trajectory Description

The reentry trajectory begins when the dynamic pressure is sufficient for the vehicle to trim and to be controlled using the seven aerosurfaces alone without any assistance from the RCS jets. The descent trajectory is separated into four different phases having different control requirements and different control modes of operation. The analysis is, therefore, separated into four sections that describe and analyze in detail the four control modes, which are as follows:

1. The hypersonic phase where the Mach number varies between 28 and 20, and the flight-path angle γ is at a very shallow dive of $-1°$ to avoid overheating due to aerodynamic friction. The angle of attack is controlled at $30°$ that provides better heat protection due to shielding. In the lateral directions the control system is able to perform roll maneuvers and to control the heading direction. The vehicle rolls about the velocity vector V_0 to reduce lateral loading due to sideslip.
2. The normal acceleration N_z-control, during which phase the vehicle is tracking an almost steady N_z acceleration command from guidance and it gradually transitions to flight-path angle γ-control mode.
3. The flight-path angle γ-control, during which phase the flight control system (FCS) tracks a flight-path γ_{cmd} angle which is commanded by the closed-loop guidance. It also performs a heading alignment maneuver prior to approach and landing phase in order to align its direction with the runway.
4. The approach and landing phase, where the longitudinal guidance controls both altitude and speed. The speed brake is partially deployed and velocity is controlled by modulating drag. In lateral, the heading guidance controls the flight direction against cross-winds by controlling the roll angle.

We begin the trim and controllability analysis with a preliminary reentry trajectory that is generated from a point-mass simulation. The trajectory is separated in four segments that correspond to the four control phases described. We will examine each phase separately by trimming the effectors and analyzing static performance along each trajectory segment. We will use contour plots and vector diagrams to analyze performance and maneuverability, generate dynamic models at selected flight conditions, perform flight control designs, and analyze stability and robustness to uncertainties at selected trajectory points. Separate control analysis and detail documentation will be presented for each control mode, including simulations. We will finally verify the control design by creating a 6-DOF nonlinear simulation for the entire reentry flight from deorbit to landing in Simulink® using the control laws derived from the linear analysis.

10.1.2   Early Hypersonic Phase Using Alpha Control

After deorbiting and during the early reentry phase the vehicle uses the RCS jets to maintain a $30°$ constant angle of attack which optimizes aeroheating. Atmospheric

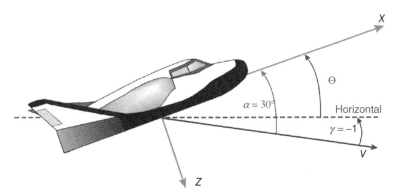

FIGURE 10.2   The Flight-Path Angle γ is $-1°$ and the Angle of Attack α is $30°$ During Early Reentry

reentry begins at an altitude of 250,000 (ft) and at Mach 28. The vehicle maintains a mostly negative shallow flight-path angle γ of approximately $-1°$, see Fig. 10.2, and it rolls about its velocity vector V_0 in order to avoid from bouncing back off into space. As the dynamic pressure increases the aerosurfaces are used to trim and control the angle of attack at $30°$. Figure 10.3 shows some of the trajectory parameters in the hypersonic region between Mach (28 and 19). The angle α is gradually reduced and the control system eventually switches to normal acceleration N_z-control.

10.1.2.1 Control Surface Trimming   Before analyzing the vehicle performance we must use the trimming algorithm described in Chapter 6 to trim the positions of the aerosurfaces in order to balance the vehicle moments along this hypersonic trajectory. Only the moments are trimmed during this phase because the speed brake is not active and no translational trimming is required. During actual flight, the precalculated effector trim positions are commanded open-loop (scheduled) as a function of the flight condition or time. The deflection commands produced by the FCS are superimposed on the prescheduled commands, as we will show later in the 6-DOF nonlinear simulation in Section 10.1.6. Figure 10.4 shows the trim deflections for the seven aerosurfaces as a function of trajectory time during this high alpha hypersonic region.

Notice that the two elevons trim at $4.5°$, the two upper body flaps trim at $-5°$, and the lower left and lower right body flaps both trim at $33°$ in this high alpha hypersonic condition. They come down to smaller values, however, later when the angle of attack is reduced.

10.1.2.2 Performance Parameters Along the Trajectory   Having obtained the trim positions of the aerosurfaces, our next goal is to check the static performance and stability parameters along the trajectory that were described in Chapter 7. Before evaluating the vehicle performance, however, the analyst must select a mixing-logic matrix that defines how the seven surfaces are combed together to control the three

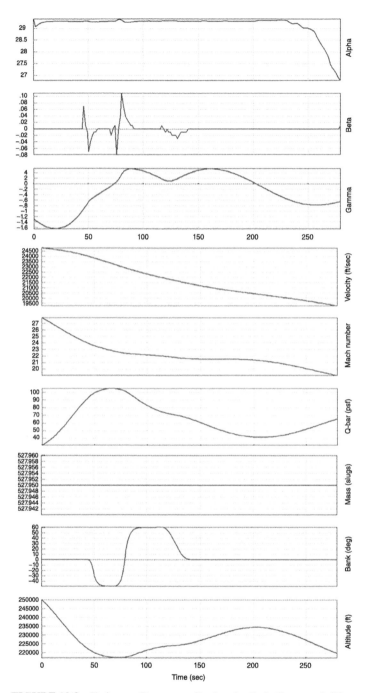

FIGURE 10.3   Trajectory Parameters During the Early Hypersonic Phase

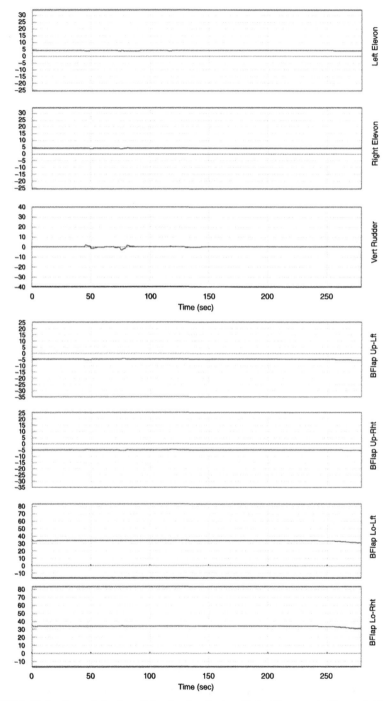

FIGURE 10.4   Control Surface Trim Angles During Early Hypersonic Phase, 2 Elevons, Rudder, and 4 Body-Flaps

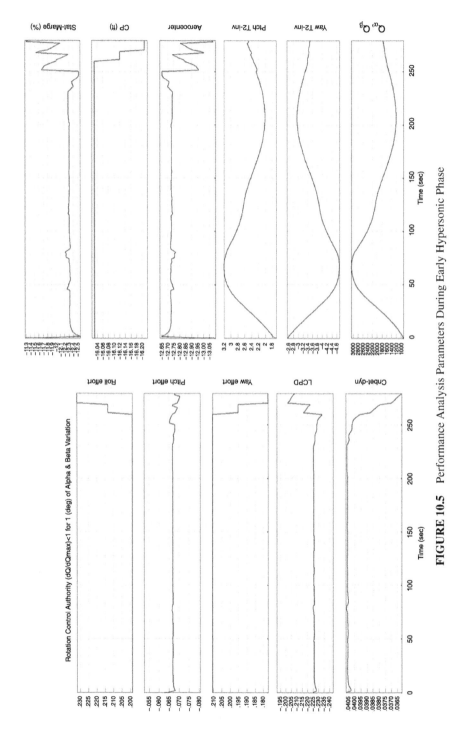

FIGURE 10.5 Performance Analysis Parameters During Early Hypersonic Phase

311

axes. The matrix K_{mix} determines the effector allocation along roll, pitch, and yaw, and the control effectiveness strongly depends on this matrix, as already described in Section 7.2. It is typically scheduled as a function of Mach and α similar to the control gains, but in this case we will use a constant K_{mix}. We must also define the maximum expected magnitude of the wind disturbance in terms of α_{max} and β_{max} angles, which in this case are both set to $\pm 1°$. The dispersion in $\pm \alpha_{max}$ may also be due to a command from guidance and the vehicle should have the control authority to achieve it.

The performance analysis results in Figure 10.5 show that in the longitudinal direction the vehicle is unstable with a peak time to double amplitude 0.31 sec. In the lateral direction it is statically stable with a Dutch-roll resonance reaching 4.6 (rad/sec). The peak (Q-α, Q-β) loading with $\pm 1°$ dispersion in α_{max} and β_{max} angles is 3000 (psf-deg) which is acceptable. The control effort against α_{max} and β_{max} dispersions is less than 0.5 in all three axes which allows sufficient control authority for gust disturbances and other functions. The $C_{n\beta}$-dynamic is positive which means that the vehicle is directionally stable, but the (LCDP) ratio is negative and small in magnitude. It means that without RCS the roll control will be reversed and slow, which may be acceptable since the roll maneuvers are slow during this phase.

10.1.2.3 Controllability Analysis Using Vector Diagrams

Vector diagrams are two-dimensional plots used for analyzing the vehicle controllability in two directions at a fixed flight condition. We are selecting an arbitrary flight condition that corresponds to $t = 150$ sec, Mach 27, in the middle of the early hypersonic phase. In Figure 10.6 we compare the control moment capability of the aerosurfaces in roll and yaw against the moments generated by the wind-shear disturbance that is defined in terms of $\pm \beta_{max}$. It shows the roll and yaw moments (nondimensional) produced when the roll and yaw FCS demands are maximized (before saturating the aerosurfaces). The two opposite and closer to horizontal vectors corresponds to maximizing the $\pm$yaw FCS demands ($\delta R_{\pm FCS\_Max}$), and the other two vectors which are closer to vertical correspond to maximum $\pm$roll FCS demands ($\delta P_{\pm FCS\_Max}$). The directions of the control moment vectors are pointing mostly toward their intended directions but with a significant amount of coupling in each other's direction.

The two small vertical vectors show the disturbance moments generated due to variations in the angle of sideslip $\pm \beta_{max}$. The disturbance is mainly in roll, $+\beta_{max}$ generates a negative rolling moment because the vehicle has significant dihedral effect. The rectangles at the tips of the arrows show the uncertainty in C_l and in C_n in this flight condition. This vector diagram is used to determine if the vehicle has sufficient control authority to counteract the disturbance moments by examining the control vectors against the disturbance vectors and if the controls are powerful in their intended directions to counteract the aerodynamic moments due to $\pm \beta_{max}$ dispersions along the roll and yaw axes.

The vector partials in Figure 10.7 shows the variation in roll and yaw moment per roll and yaw acceleration demands in (rad/sec^2). The horizontal vector pointing

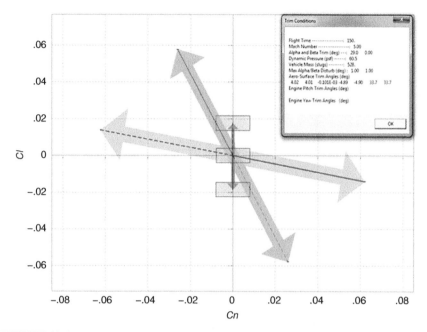

FIGURE 10.6   Roll and Yaw Max Moments due to Max Control Demands Against Max Disturbance Moments due to $\pm\beta_{max}$

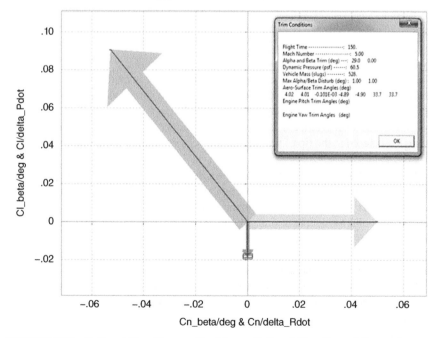

FIGURE 10.7   Roll and Yaw Moment Partials per Roll and Yaw Control Demands versus $\{C_{n\beta}, C_{l\beta}\}$ Partials

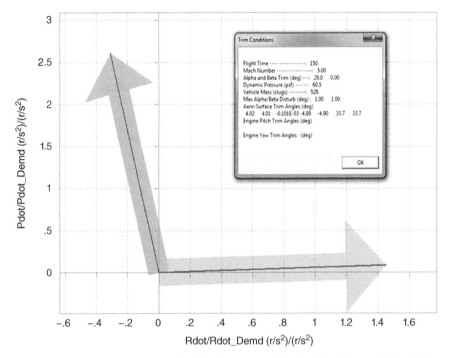

FIGURE 10.8   Vector Diagram of Roll and Yaw Acceleration Partials per Roll and Yaw Acceleration Demands

toward the right is the moment partials per yaw demand $\{C_{n\delta R_{FCS}}, C_{l\delta R_{FCS}}\}$ which is entirely in the yaw direction. The upward pointing vector is moment partials per roll demand $\{C_{n\delta P_{FCS}}, C_{l\delta P_{FCS}}\}$ which is mostly in the roll direction but it also couples into yaw. The vectors pointing downward are the $\{C_{n\beta}, C_{l\beta}\}$ partials. Notice that $C_{l\beta}$ is negative due to the dihedral and it is much bigger in magnitude than $C_{n\beta}$. The rectangle centered at the tip of the $\{C_{n\beta}, C_{l\beta}\}$ vector is due to the uncertainties in the partials.

   The vector diagram in Figure 10.8 shows the partials of accelerations per acceleration demands in roll and yaw. The upward pointing vector shows the roll/yaw acceleration per roll acceleration demand $\{\dot{P}/\delta P_{FCS}, \dot{R}/\delta P_{FCS}\}$, and the horizontal vector is the roll/yaw accelerations per yaw demand $\{\dot{P}/\delta R_{FCS}, \dot{R}/\delta R_{FCS}\}$. The axes units are in (rad/sec^2) per (rad/sec^2). In an ideal situation they should be unit vectors pointing in their corresponding direction, up and right. This would achieve perfect open-loop control. This situation, however, is rarely achievable open-loop, plus it would be unreliable. It is not necessary to diagonalize the plant because the control feedback compensates for imperfections. They are, however, close to being orthogonal and they are pointing in the proper directions and this is sufficient for good flight control design.

10.1.2.4   Dynamic Modeling, Control Design, and Stability Analysis   We will now create a dynamic model of the reentry vehicle at a fixed flight condition, in the hypersonic region at $t = 12$ (sec), using Flixan and the equations described in Chapter 3. We will demonstrate how to design pitch and lateral control laws in MATLAB®, analyze control system stability in the frequency domain, and simulate its performance when tracking α and ϕ commands. The same process can be repeated to analyze other flight conditions along the trajectory. The control laws will eventually be used in a 6-DOF simulation and interpolated between design points.

LQR Control Design: Two separate dynamic models are created for control design and linear analysis. The output rates in the first model are defined in stability axes (roll rate is about the velocity vector), and it is used for control design. The output rates in the second dynamic model are body rates, and it is used for linear analysis and simulations. A mixing-logic matrix "K_{mixM27}" is also created to convert the (roll, pitch, and yaw) flight control demands to seven aerosurface deflection commands. The participation of the four body flaps is de-emphasized in the calculation of K_{mixM27} because we would prefer the elevons and rudder to be more active in flight control having greater control authority and bandwidth. The pitch and lateral design models are combined with the mixing logic and the Linear Quadratic Regulator (LQR) method is used to calculate state-feedback gains. The following file loads the simulation and design systems and the surface mixing matrix into MATLAB®, and performs the pitch and lateral LQR designs.

```
d2r=pi/180; r2d=180/pi;
[Aps, Bps, Cps, Dps] = pitch_des;          % Load Pitch aero-surf Design Model
[Als, Bls, Cls, Dls] = later_des;          % Load Lateral aero-surf Design Model
[Ave, Bve, Cve, Dve] = vehicle_sim;        % Simulation Model 6-dof
load KmixM27.mat -ascii; Kmix=KmixM27;     % Load Surfaces Mix Logic (7 x 3)

alfa0=29.274; V0=24675.1; Thet0=27.69; ge=32.174;% Additional Vehicle Parameters
calfa=cos(alfa0*d2r); salfa=sin(alfa0*d2r);      % for Body to Stability Transform

% Convert Lateral State Vector from Body to Stability Axes, Outputs=States
[A14,B14,C14,D14]= linmod('Ldes5x');       % 5-state model (ps,rs,bet,pint,betint)
A15= C14*A14*inv(C14);  B15= C14*B14;      % Stability axis System
C15= C14*inv(C14);      D15= D14;

% Lateral LQR Design Using Only the RCS Jets  % States: [ps,rs,bet,psint,betint]
R=[1,1]*5; R=diag(R);                       % Cntrl LQR Weights R=[1,1]*5
Q=[1 0.4 0.5 0.2 0.005]*3; Q=diag(Q);       % State LQR Weights Q=[1 0.4 0.5 0.2 0.005]
[Kpr,s,e]=lqr(A15,B15,Q,R)                  % Lateral LQR design
save Kpr_M27_0.mat Kpr -ascii

% Pitch LQR Design Using the 7 Aero-Surfaces  % States: (gami,gama,q,alfa,alfint)
[Ap4,Bp4,Cp4,Dp4]= linmod('Pdes4xb');       % 4-state des model (gami,gama,q,alfa,alfint)
R=4; Q=[0.001 0.1 20.0 100]; Q=diag(Q);     % LQR Weights (gama,q,alfa,alf_int)
[Kq,s,e]=lqr(Ap4,Bp4,Q,R)                   % Perform LQR design on Surf
save Kq_M27_0.mat Kq -ascii
```

The pitch design model consists of states $\{\theta, q, \text{and } \alpha\}$. It is augmented to include also α-integral in the state vector. The phugoid states (δh and δV) are not included in the design model. The pitch controller is a (1×4) state-feedback gain matrix of (θ, q, α, α-integral) states. It regulates the angle of attack, which is initially at 30° and gradually it is reduced to smaller values, as required to control heating. The control deflections are mainly in the two elevons but the four body flaps are also participating by smaller amounts. This is adjusted by the Q and R matrices in the LQR algorithm. The lateral design model consists of states $\{p_s, r_s, \text{and } \beta\}$, where the rates are relative to the velocity vector. It is augmented to include also the integral of the roll rate p_s-integral and β-integral in the state vector. The stability axis model is preferred over the body axis model in the lateral LQR design because the vehicle is commanded to roll about the velocity vector. The lateral LQR state-feedback controller is a (2×5) gain matrix of $(p_s, r_s, \beta, p_s\text{-integral}, \beta\text{-integral})$ states, and performs roll maneuvers by rolling the vehicle about the velocity vector V_0. Rotating about V_0 minimizes the beta transients and the lateral loads during turns.

Linear Simulation Model: The LQR design during Early Reentry is evaluated by using a linear simulation model for the Mach 27 case, shown in Figure 10.9, and commanded to maneuver in roll. The simulation is used for analyzing the system's response to commands and wind disturbances. The output rates in this model are body rates since the rate gyro measurements are in the body axes. A body to stability axis transformation block is, therefore, included to convert the (p and r) body rates to stability rates (p_{stab} and r_{stab}) because the LQR controller expects roll and yaw rates relative to the velocity vector V_0. During a roll maneuver the surface deflections are mainly in rudder and differential elevons and the transient in beta is small which minimizes the lateral load, as expected.

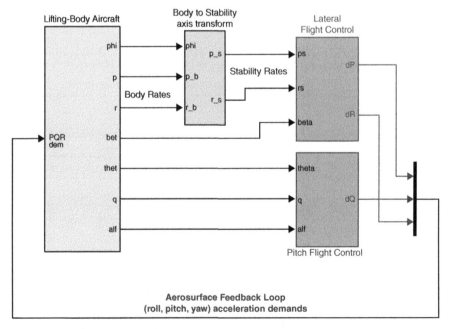

FIGURE 10.9   Simple Simulation Model for Early Reentry Phase

Stability Analysis: The system in Figure 10.9 can be reconfigured to analyze the stability margins of the system in the frequency domain by opening one loop and closing the other two. The Nichols plots in Figure 10.10 show the phase and gain margins for the Mach #27 case in the pitch and roll directions.

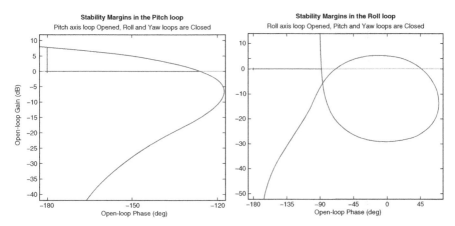

FIGURE 10.10   Nichols Plots Showing the Phase and Gain Margins in Pitch and Roll Axes During Early Reentry Phase

10.1.3   Normal Acceleration Control Mode

As the α-command begins to come down from 30°, at $t = 270$ sec, the FCS begins to transition from the alpha control mode to the N_z-control mode by using feedback from the normal accelerometer. Initially it attempts to maintain a steady N_z of -33.5 (ft/sec^2) and the command is gradually reduced to provide a comfortable 1g acceleration for the crew. The angle of attack is also gradually reduced and the descent rate is determined by the flight-path angle γ that begins to come down steeper. Figure 10.11 shows some of the variables during this phase of the trajectory between Mach 19 and Mach 5.

10.1.3.1   *Control Surface Trimming*   We must also trim the vehicle along this N_z-control segment of the trajectory and determine the new trim positions of the aerosurfaces. Only the moments are trimmed during this phase (no translational trimming is required). Figure 10.12 shows the trim deflections for the seven aerosurfaces during N_z-control phase as a function of the trajectory time. Notice that the trim angles of the lower left and lower right body flaps have come down to smaller angles in comparison with the previous positions in the alpha control region. They now trim at approximately 5° instead of 33° earlier. This is because the angle of attack has come down from 30°. The elevon trim angles did not change much.

10.1.3.2   *Hinge Moments Along the Trajectory*   Equation 7.49 is used to calculate the moments at the hinges of the control surfaces, as function of surface deflection, alpha, Mach number, dynamic pressure, and hinge moment coefficients. The hinge moments are used in sizing the actuator torques, and they are shown in Figure 10.13 plotted against the trajectory time.

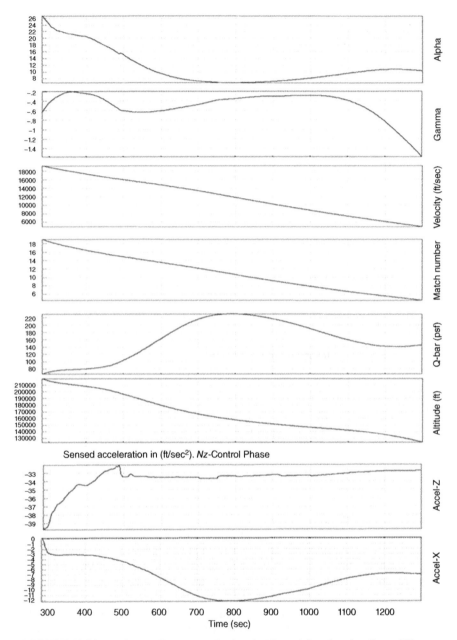

FIGURE 10.11   Trajectory Parameters During the Normal Acceleration Control Phase

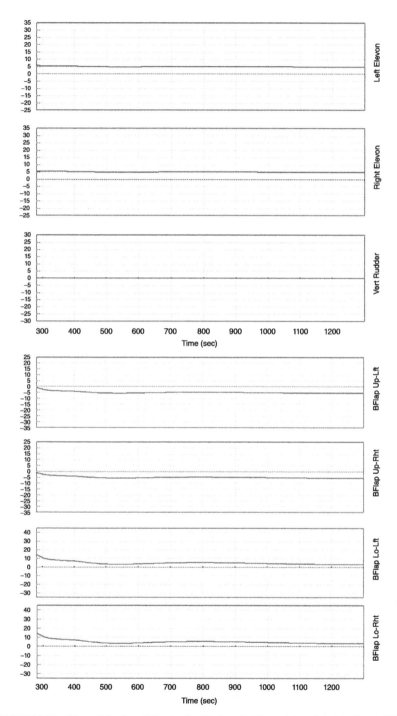

FIGURE 10.12   Control Surface Trim Angles During the Normal Acceleration Control Phase

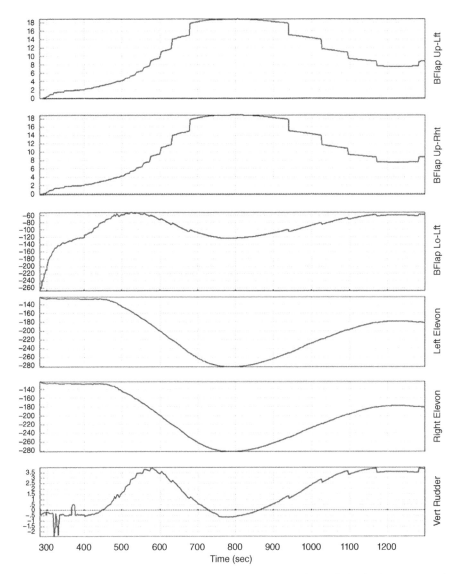

FIGURE 10.13   Moments at the Hinges of the Control Surfaces (ft-lb) in the Nz Control Phase

10.1.3.3   Performance Parameters Along the N_z-Control Phase of Trajectory
We are now able to calculate the static performance and stability parameters along the N_z-control phase of the trajectory, as described in Chapter 7, see Figure 10.14. A different mixing-logic matrix is used to combine the seven surfaces together because the aerodynamic coefficients have changed. We must also define the worst expected

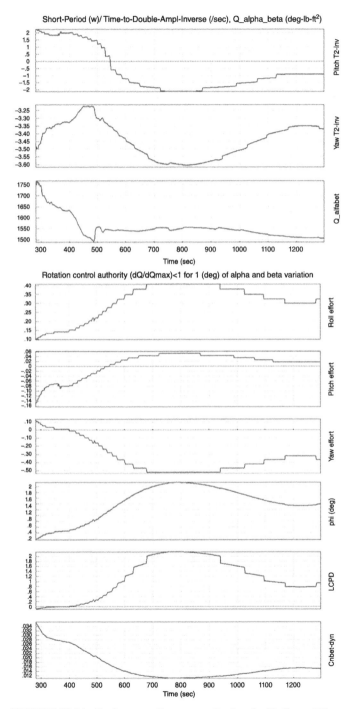

FIGURE 10.14   Performance Parameters During the N_z-Control Phase

wind disturbance in terms of α_{max} and β_{max} dispersion angles, which are expected to be less than $\pm 1°$, as before.

In the longitudinal direction the vehicle is initially statically unstable with a peak time to double amplitude 0.5 sec, but as the angle of attack is reduced it becomes statically stable, after $t = 550$ sec, with a short-period resonance of 2 (rad/sec) or 3% static margin. In the lateral direction it is statically stable with the Dutch-roll resonances peaking to 3.6 (rad/sec). The peak (Q-α, Q-β) loading due to the (α_{max} and β_{max}) dispersions is 1700 (psf-deg) which is lower than in the previous phase. The control effort to overcome the peak α_{max} and β_{max} dispersions from trim does not exceed 0.5 in all three axes, see equation 7.24. This allows sufficient control authority for other functions. The $C_{n\beta}$-dynamic is positive which means that the vehicle is directionally stable. The LCDP ratio is now positive after $t = 400$ sec and its magnitude increases to 2. This improves the roll controllability and it does not require "roll reversal". The parameter ϕ is the bank angle due to a $\beta_{max} = 1°$. It is only applicable near landing.

10.1.3.4 Contour Plots Analysis   Figure 10.15 shows contour plots for some of the critical performance parameters. Contour plots allow us to assess vehicle performance over the entire Mach versus Alpha range. The first two plots show the pitch and lateral

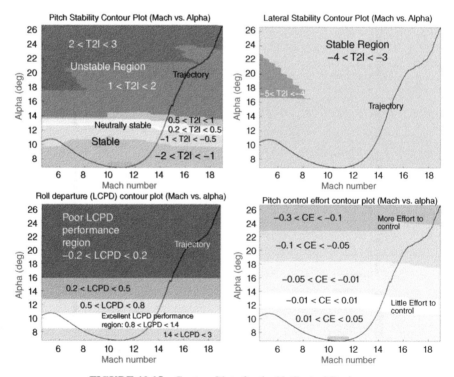

FIGURE 10.15   Contour Plots for the N_z-Control Region

stability parameter (T2-inverse) versus Mach and Alpha. The N_z-control trajectory is shown by the dark line beginning in the upper right-hand corner and ending in the lower left-hand side. In the pitch axis the vehicle is unstable at angles of attack greater than 13° because T2-inverse > 0. The rate of instability, however, is manageable. Neutral stability occurs at approximately $\alpha = 12.5°$ as seen by the thin almost horizontal white band. In the lateral direction the vehicle is statically stable across the entire region and the stability parameter is almost constant. The LCDP ratio which is a measure of dynamic roll controllability is good for angles of attack below 14°. The contour plots were calculated using a constant mixing-logic matrix and it seems that a different mixing logic should be used at high angles of attack. The pitch and yaw control authority against 1° of α_{max} and β_{max} aero disturbances is very good.

10.1.3.5   *Controllability Analysis by Using Vector Diagrams*   The following vector diagram analysis corresponds to a flight condition at $t = 800$ sec, which is in the middle of the N_z-control trajectory that corresponds to Mach 10. The aero disturbances are defined by the maximum α_{max} and β_{max} dispersions from trim which are set to $\pm1°$. A 7×3 control surface combination matrix "K_{mixM10}" was also designed for this Mach 10 flight condition. The left side of Figure 10.16 shows the pitch moment C_m plotted against the C_Z and the C_X forces (nondimensional). The big vectors show the maximum pitch moment and forces produced when the pitch control demand is maximized (before saturation). The solid vector represents the moment and forces when the pitch demand is max positive ($\delta_{+QFCS\ Max}$), and the dashed vector is due to the peak negative control demand ($\delta_{-QFCS\ Max}$). The magnitudes of the two opposite vectors are not equal. The vehicle is trimmed in pitch because $C_m = 0$ when the control $\delta Q_{FCS} = 0$. It is, however, accelerating in both $-x$ and $-z$ directions because C_X and C_Z are negative when $\delta Q_{FCS} = 0$. Notice how either a positive or negative pitch control demand has a negative effect on C_X (drag increase). Notice also that a +pitch control demand causes a positive C_Z force, reducing lift because the elevons rotate upward to increase the pitching moment. The smaller almost vertical vectors, in the upper left of Figure 10.16, represent the forces and moments generated by the disturbances $\pm\alpha_{max}$ and $\pm\beta_{max}$ which are both $\pm1°$ in this case. Increasing α makes the z-force more negative (up) shown by the solid vector pointing downwards. The rectangles represent the uncertainty in the moment and force coefficients.

On the right side of Figure 10.16 we see the vector partials in the longitudinal directions. It can be interpreted as a three-dimensional figure. The longer vectors show the pitch moment partial, the normal and axial force partials per pitch acceleration demand $\{C_{m\delta Q_{FCS}}, C_{Z\delta Q_{FCS}}, C_{X\delta Q_{FCS}}\}$. It shows that a small increase in pitch control produces a positive effect in all three directions: pitch moment, plus z and x forces. The smaller vectors are the $\{C_{X\alpha}, C_{Z\alpha}, C_{m\alpha}\}$ partials. They are two because they are calculated at two extreme values of $\pm\beta_{max}$. Negative $C_{m\alpha}$ is indicative that the vehicle is stable in this flight condition. The control vector partials are scaled to be made comparable with the disturbance partials, as already discussed. They are clearly more dominant than the disturbance partials in all directions.

In Figure 10.17, the top vector diagram is a moment partials diagram showing the variation in roll and yaw moments per acceleration demands in roll and yaw in

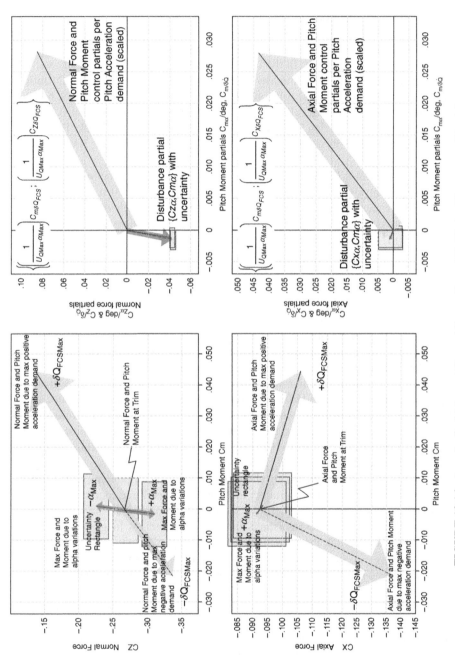

FIGURE 10.16 Longitudinal Moment C_m, and Forces C_X and C_Z Vector Diagrams and Partials

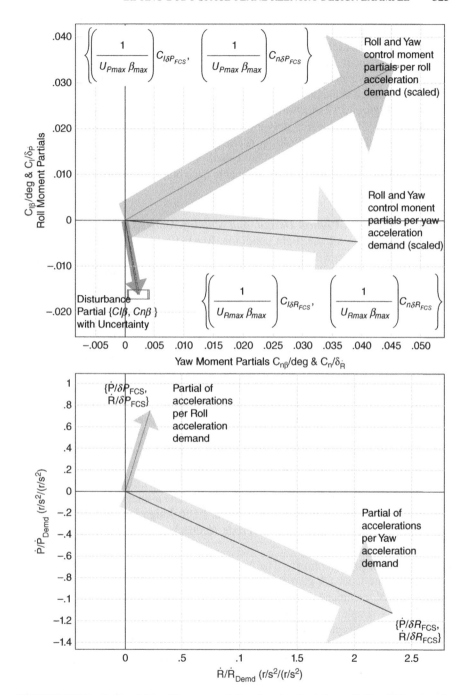

FIGURE 10.17   Roll and Yaw Moments and Acceleration Partials per Roll and Yaw Acceleration Demands

(rad/sec^2). The vector closer to the yaw axis is $\{C_{n\delta R_{\text{FCS}}}, C_{l\delta R_{\text{FCS}}}\}$ which dominates in yaw, and the other big vector is $\{C_{n\delta P_{\text{FCS}}}, C_{l\delta P_{\text{FCS}}}\}$ which affects both roll and yaw directions. The small vectors pointing downward are the $\{C_{n\beta}, C_{l\beta}\}$ partials. Notice that $C_{l\beta}$ is negative due to the dihedral and it is bigger in magnitude than $C_{n\beta}$. The rectangles centered at the tips of the vectors represent the uncertainties in the partials. The bottom Figure 10.17 shows the partials of accelerations per acceleration demands in roll and yaw. The upwards pointing vector is $\{\dot{P}/\delta P_{\text{FCS}}, \dot{R}/\delta P_{\text{FCS}}\}$, and the big vector towards the right is $\{\dot{P}/\delta R_{\text{FCS}}, \dot{R}/\delta R_{\text{FCS}}\}$. The axis units are in (rad/sec^2) per (rad/sec^2). Although the vectors are not exactly pointing in the required directions, they are, however, close to being orthogonal to each other and this is a good property for control design.

10.1.3.6   Modeling, Control Design, and Stability Analysis

We will now create a dynamic model at a fixed flight condition, Mach 10, at $t = 849$ (sec), in the N_z-controlled region. We will design control laws, analyze stability in the frequency domain, and simulate its performance in tracking the normal acceleration and ϕ commands. The process, systems, and analysis are very similar with the previous Mach 27 case, and we will skip some of the details.

LQR Control Design: The pitch N_z-control design is very similar to the α-control in the Mach 27 case. The $\{\alpha$ and α-integral$\}$ state feedback, however, is replaced with $\{N_z$ and N_z-integral$\}$ feedback, respectively. This is easy to design because in this particular flight region there is an almost proportional relationship between α and N_z. The pitch design model, initially consisting of states $\{\theta, q,$ and $\alpha\}$, is augmented to include α-integral in the state vector. The phugoid states (δh and δV) are not included in the design model because we are not interested to control those variables directly. The controller is a 1×4 state-feedback gain matrix Kq created using the LQR algorithm. The lateral design is almost identical to the Mach 27 case. It uses a design system with rates in stability axes (about the velocity vector), consisting of states $\{p_s, r_s,$ and $\beta\}$. It is augmented to include also p_s-integral and β-integral in the state vector. The stability axis model is used in the lateral LQR design for the reasons already explained. The dimension of the lateral state-feedback gain matrix Kpr is 2×5.

Control System Modification: The pitch controller consists of a 1×4, ($\theta, q, \alpha,$ α-integral) state-feedback gain Kq. The vehicle is commanded to track the normal acceleration N_z-command and the N_z-error is approximated to an α-error by the gain "$Nz2a$." This simple modification of the pitch control system from the α-control mode allows us to regulate the normal acceleration at approximately -34 (ft/sec^2) during this phase. An N_z-filter was also included to improve stability. The 2×5 lateral controller gain converts the (p_s, r_s, β, p_s-integral, β-integral) state feedback to roll/yaw acceleration demands. It is used to perform roll maneuvers by rolling the vehicle about the velocity vector (V_0). The pitch and lateral control laws are shown in Figure 10.18.

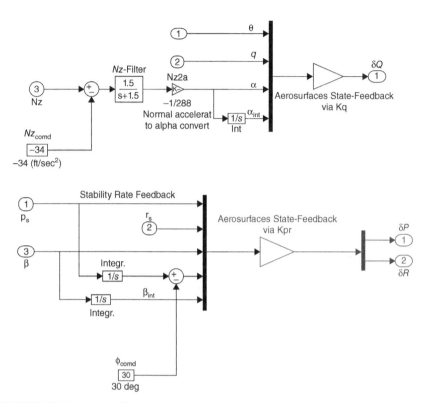

FIGURE 10.18   Pitch and Lateral State-Feedback Control Laws Derived by the LQR Method

Linear Simulation Model: The simulation model used for linear analysis is similar to Figure 10.9. The only difference is that the pitch FCS uses N_z-feedback instead of α-feedback, and the command is an N_z-command. It evaluates the system's response to roll and N_z commands and to wind disturbances. The output rates in this model are body rates and since the control system is expecting to receive rates about the velocity vector V_0, a body to stability axis transformation block is included in the simulation to convert the (p and r) body rates to stability rates (p_{stab} and r_{stab}), just like in the alpha control phase.

Stability Analysis: The linear simulation model can be configured for open-loop stability analysis and to determine the gain and phase margins by opening one loop at a time and calculating the frequency response across the opened loop with the other two loops closed. The Nichols plot in Figure 10.19 shows the phase and gain margins in the pitch axis, for the Mach #10 case. The Bode plot also shows a short-period resonance at 2 (rad/sec), which is also evident in the T2-inverse parameter, in Figure 10.14, at $t = 840$ sec.

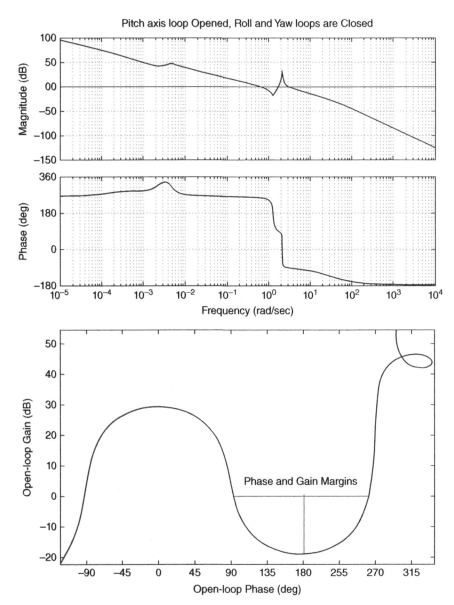

FIGURE 10.19 Nichols and Bode Plots Showing the Pitch Axis Phase and Gain Margins and a 2 (rad/sec) Short-Period Resonance

10.1.4   Flight-Path Angle Control Mode

From the N_z-control mode the FCS transitions to the flight-path angle γ control mode. Direct γ-control is commanded by the closed-loop guidance system which calculates the required flight-path angle for controlling the vehicle descent rate, as a function of range, altitude, and speed. Figure 10.20 shows some of the trajectory parameters in the γ-control region which is between Mach (5 and 0.9). The flight-path angle drops significantly toward the end of this phase, at $t = 1700$, after the vehicle performs a 30° roll maneuver to align its heading with the runway and in order to gain sufficient speed before it performs the landing flare. The dynamic pressure increases significantly in this final period as it approaches for landing, below Mach 1 at 20,000 (ft) altitude.

10.1.4.1   Control Surface Trimming   Figure 10.21 shows the aerosurface trim angles for trimming the moments along the γ-control phase of the trajectory. Note, the aerosurface trim angles are not entirely determined by the trim algorithm without any inputs from the designer, but the positions are often biased or constrained by adjusting the limits and initial surface positions prior to trimming in order to accommodate other design constraints and performance factors.

10.1.4.2   Performance Parameters in the Gamma Control Section of the Trajectory
Figure 10.22 shows the performance parameters in the gamma control section of the trajectory, calculated as described in Chapter 7. A different mixing-logic matrix K_{mixM2} was designed by the mixing-logic algorithm based on data from a fixed flight condition in this phase. The mixing-logic matrix strongly affects the control authority parameters and it was slightly modified by increasing the rudder contribution to roll (against the aileron) in order to increase the LCDP ratio because, otherwise, it would have been too low. We must also define the maximum expected disturbance due to wind shear in terms of α_{max} and β_{max} angles, which are the same as before ±1°.

The performance results show that during this phase the longitudinal axis is statically stable due to the reduction in the angle of attack. The short-period resonance varies between 1.2 and 3.2 (rad/sec) and the static margin peaks to 6%. In the lateral direction it is also statically stable with the Dutch-roll resonances peaking to 4.6 (rad/sec) during the roll maneuver. The Q-alpha, Q-beta loading is reasonably low but it is spiking to 2600 (psf-deg) during the roll maneuver. Remember that this parameter is due to α_{max} and β_{max} dispersions caused by the 1° wind shear. The control effort against wind dispersions are sufficiently small in pitch and yaw. In roll, however, the control effort (authority) was compromised to 0.65 as a trade-off in order to improve the LCDP which would have been too low otherwise. We normally like to see it at 0.5 or below. With the K_{mix} adjustment the LCDP is reasonably positive now and it does not have any sign changes that would imply roll reversals. The $C_{n\beta}$-dynamic is positive which means that the vehicle is directionally stable. The bank angle parameter ϕ is the bank angle caused due to a sideslip disturbance $\beta_{max} = 1°$, but is meaningful only near landing.

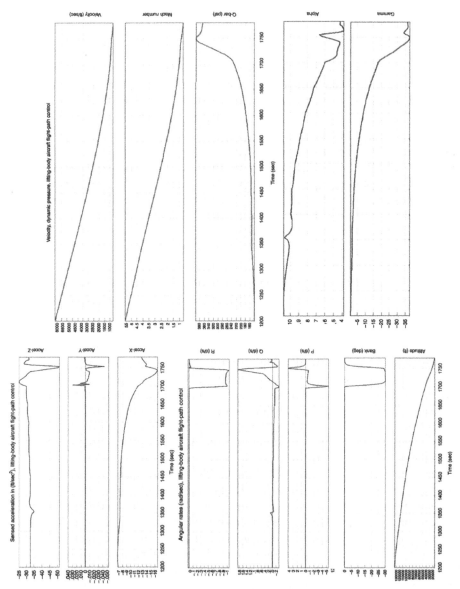

FIGURE 10.20 Trajectory Variables for the Gamma Control Phase

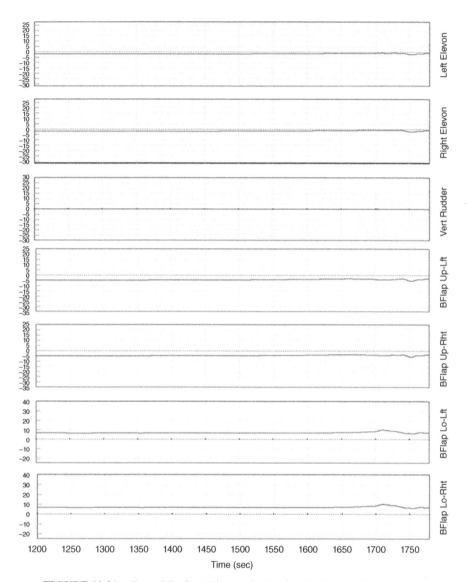

FIGURE 10.21   Control Surface Trim Angles During the Gamma Control Phase

10.1.4.3 Controllability Analysis by Using Vector Diagrams   Vector diagrams help us determine the effectiveness and orthogonality of the effectors system against shear disturbances in the steady state. The vector diagrams in Figures 10.23 and 10.24 correspond to a flight condition at $t = 1600$ sec, at Mach 1.6, which is near the middle of the γ-controlled section of the trajectory. The aero disturbance is defined by the maximum alpha and beta dispersions from trim which are defined to be $\pm 1°$. A 7×3

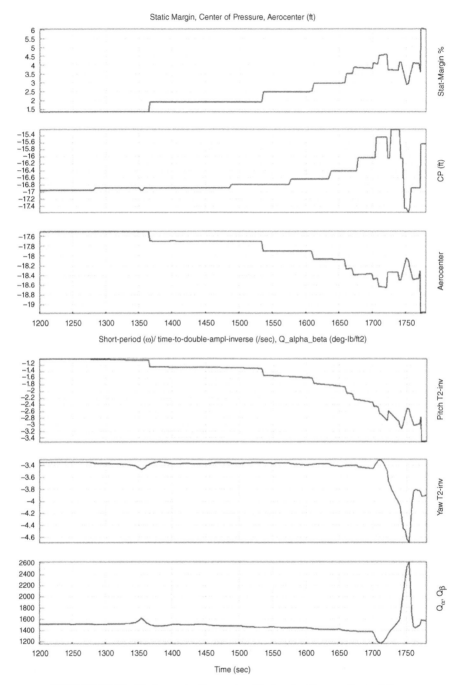

FIGURE 10.22   Performance Parameters During the Gamma Control Phase

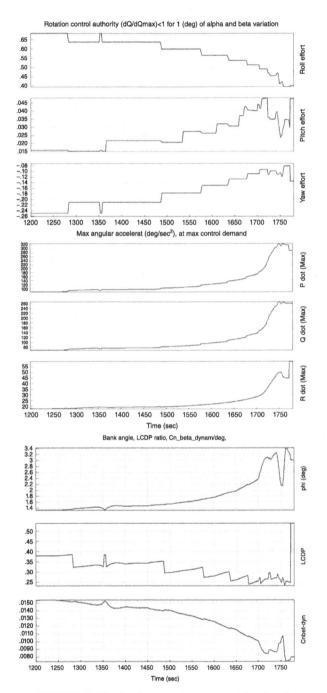

FIGURE 10.22 Performance Parameters (*Continued*)

control surface combination matrix "K_{mixM2}" was also designed for this Mach 1.6 flight condition.

The vector diagrams on the left side of Figure 10.23 show the nondimensional roll and yaw moments (C_l and C_n), and the side force (C_Y) produced when the roll and yaw FCS demands are maximized to the surface saturation limits. The solid horizontal vector pointing right shows the yaw moment produced when the yaw FCS demand is at its maximum positive position ($\delta R_{+\text{FCS Max}}$) and the dashed horizontal vector in the opposite direction shows the negative moment produced when the yaw demand is at its peak negative position ($\delta R_{-\text{FCS Max}}$). It shows that the yaw control does not produce any rolling moment. Similarly, the two control vectors in the opposite, up and down directions, are the peak moments produced when the roll FCS demand is maximized in the positive (solid), and in the negative (dashed) directions ($\delta P_{\pm\text{FCS Max}}$). The lower left figure also shows the max side force produced by varying the yaw demand to the two extreme levels. Positive yaw demand requires a negative rudder deflection which produces a negative side force. The two smaller red vectors are the roll and yaw moments generated by the angle of sideslip $\pm\beta_{\max}$ variations, and it is mainly in roll. A positive $\beta_{\max}$ generates a negative side force and a negative rolling moment because the vehicle has significant amount of dihedral effect. The rectangles at the tips of the arrows show the roll, yaw, and side force uncertainties in both the disturbance and in the control vectors. They are calculated from the uncertainties in the basic aerodynamic coefficients and in the aerosurface increment coefficients.

The vector diagrams in the upper right-hand side of Figure 10.23 is a moment partials vector diagram showing the variation in roll and yaw moments per acceleration demands in roll and yaw, which is in (rad/sec^2). The horizontal vector is the moments per yaw control demand $\{C_{n\delta R_{\text{FCS}}}, C_{l\delta R_{\text{FCS}}}\}$ which is entirely in the yaw direction. The almost orthogonal vertical vector is the moments per roll control demand $\{C_{n\delta P_{\text{FCS}}}, C_{l\delta P_{\text{FCS}}}\}$ which affects both directions but mostly roll. The shorter vectors pointing downward are the $\{C_{n\beta}, C_{l\beta}\}$ partials. Notice that $C_{l\beta}$ is negative due to the dihedral and it is bigger in magnitude than $C_{n\beta}$. The rectangle centered at the tip of the $\{C_{n\beta}, C_{l\beta}\}$ vector represents the uncertainties in the partials. The control vector partials have uncertainties too. The rectangle at the tip of the yaw control partial is due to the uncertainties in $\{C_{n\delta R_{\text{FCS}}}, C_{l\delta R_{\text{FCS}}}\}$, and the rectangle at the tip of the roll control partial is due to the uncertainties in $\{C_{n\delta P_{\text{FCS}}}, C_{l\delta P_{\text{FCS}}}\}$. The lower right diagram of Figure 10.23 shows the partials in the accelerations per acceleration demands in roll and yaw. The vertical vector is $\{\dot{P}/\delta P_{\text{FCS}}, \dot{R}/\delta P_{\text{FCS}}\}$, and the almost horizontal vector is $\{\dot{P}/\delta R_{\text{FCS}}, \dot{R}/\delta R_{\text{FCS}}\}$. The units of the axes are in (rad/sec^2) per (rad/sec^2). The two vectors show that the acceleration partials in the two directions are almost perfectly decoupled. They are pointing toward their intended directions, close to being unit vectors and orthogonal to each other which imply good controllability.

In the vector diagrams of Figure 10.24 we analyze controllability in the longitudinal direction. However, in the longitudinal axes and in this mode of operation there is only one control, which is pitch acceleration demand (δQ_{FCS}). The pitch control, in addition to the pitching moment it also generates force variations in the x and z directions, so even though we only have one longitudinal control we must still examine

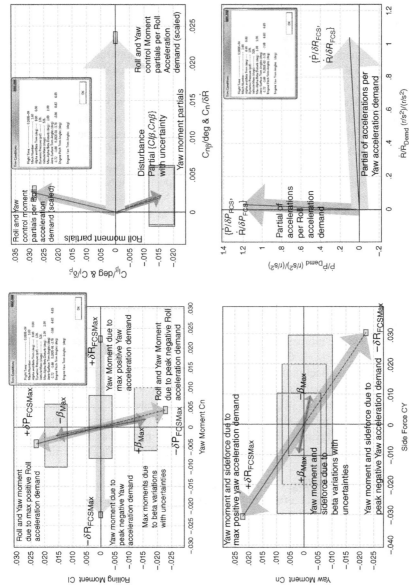

FIGURE 10.23 Roll and Yaw Vector Diagrams Showing Max Moments and Partials per Roll and Yaw Demands

its effects in all three directions. The left side of Figure 10.24 may be interpreted as a three-dimensional vector diagram with axes (C_X, C_Z, and C_m), that shows the pitching moment C_m plotted against the C_Z and the C_X forces in nondimensional form.

The vectors show the maximum pitching moment and forces produced when the pitch acceleration demand is maximized to saturation levels in both positive and negative directions. The solid vector is generated from the max positive pitch acceleration demand ($\delta Q_{+\text{FCS Max}}$), and the dashed vector is the result of the peak negative pitch acceleration demand ($\delta Q_{-\text{FCS Max}}$). Unlike the lateral directions there is no symmetry in the longitudinal axes because the peak positive control demand ($\delta Q_{+\text{FCS Max}}$) produces a larger moment and z-force variation than the peak negative control demand ($\delta Q_{-\text{FCS Max}}$). The x-force variation from trim is also different between the positive and negative pitch demands. This is because the effector system and the trim conditions are more effective in the positive pitch direction than it is in the negative. The pitch moment is balanced in pitch because $C_{m0} = 0$ when the control $\delta Q_{\text{FCS}} = 0$. However, the vehicle is accelerating in both $-x$ and $-z$ directions because C_{X0} and C_{Z0} are both negative when $\delta Q_{\text{FCS}} = 0$. Notice how either a positive or negative pitch control demand has a negative effect on C_X (drag increase). Notice also that a +pitch control demand reduces the magnitude of C_Z, reducing lift as the elevons rotate upward to increase the pitching moment. The smaller up and down vectors in the upper left diagram represent the forces and moments generated by the disturbances $\pm \alpha_{\max}$ and $\pm \beta_{\max}$ which are both $\pm 1°$ in this case, increasing α makes the z-force more negative (up). The rectangles represent the uncertainty in the moment and force coefficients.

On the right side, Figure 10.24 shows the partials in the longitudinal directions, mainly pitch moment per pitch control and pitch moment per alpha. The two figures can also be interpreted as a three-dimensional vector diagram. The long vectors represent the pitching moment, normal, and axial force partials per pitch acceleration demand $\{C_{m\delta Q_{\text{FCS}}}, C_{Z\delta Q_{\text{FCS}}}, C_{X\delta Q_{\text{FCS}}}\}$. It shows that a small increase in pitch demand produces a positive effect in pitching moment and in z-force, and a negative x-force (drag). The variations in the x and z forces are mainly due to the elevon up deflections. The smaller vectors are the $\{C_{m\alpha}, C_{Z\alpha}, C_{X\alpha}\}$ partials. Negative $C_{m\alpha}$ is indicative that the vehicle is stable in this flight condition. The control vector partials have been scaled to be made comparable with the $\{C_{m\alpha}, C_{Z\alpha}, C_{X\alpha}\}$ partials, as already described in Section 8.2.3. They are clearly more dominant than the disturbance partials.

10.1.4.4   *Vehicle Modeling, Control Design, and Stability Analysis*   We will now create a dynamic model at a fixed flight condition, at time $t = 1556$ sec, that corresponds to Mach 2 in the gamma-controlled region. We will use the dynamic model to design control laws, and a control surface mixing-logic K_{mixM2}, analyze system stability in the frequency domain, and simulate its performance when tracking γ and ϕ commands.

LQR Control Design: The control system in this phase is different from the previous two phases because the flight-path angle is directly commanded from guidance. The LQR design is based on a different state-space model that includes a γ-state instead of θ in the state vector. A γ-state and its integral are constructed by combining

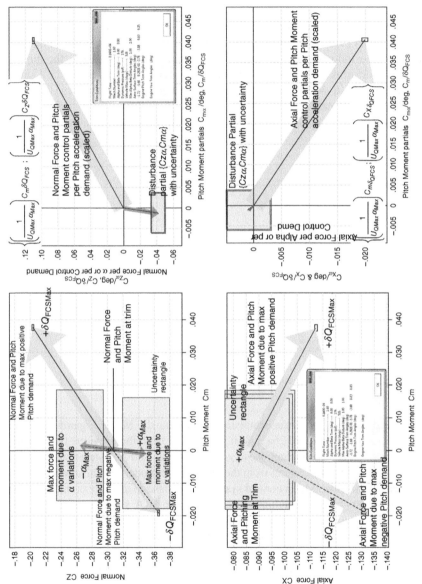

FIGURE 10.24 Pitch Axis Vector Diagrams Showing Max Moments and Partials per Pitch Demand

337

(θ and α) states. The state vector in the pitch design model is $\{\gamma, \gamma\text{-integral}, q,$ and $\alpha\}$. The γ-integral feedback helps the flight-path angle error converge to zero. α-integral feedback is not required here because we are not tracking α in this case. We also take advantage of the almost linear relationship between α and N_z in the implementation of the control system and replace the α-feedback with N_z-feedback because N_z is directly measurable from the accelerometer but not α. The phugoid states (δh and δV) are not included in the design model because they are not directly controlled. The following file in Figure 10.25 loads the design plants, the mixing matrix, and performs the control design. The pitch design calculates the 1×4 state-feedback matrix "Kq_M02_0.mat," and in the lateral design, the 2×5 state-feedback matrix "Kpr_M2_0.mat."

```
% File Init.m for Initialization and Perform Control Design
d2r=pi/180; r2d=180/pi;
[Aps, Bps, Cps, Dps] = pitch_des;          % Load Pitch aero-surf Design Model
[Als, Bls, Cls, Dls] = later_des;          % Load Lateral aero-surf Design Model
[Ave, Bve, Cve, Dve] = vehicle_sim;        % Simulation Model 6-dof
load KmixM2.mat -ascii; Kmix=KmixM2;       % Load Surfaces Mix Logic (7 x 3)

alfa0=8.791; V0=2026.5; Thet0=2.363; ge=32.174;   % Additional Vehicle Parameters
calfa=cos(alfa0*d2r); salfa=sin(alfa0*d2r);       % for Body to Stability Transform

% Convert Lateral State Vector from Body to Stability Axes, Outputs=States
[A14,B14,C14,D14]= linmod('Ldes5x');       % 5-state model (p,r,bet,pint,betint)
A15= C14*A14*inv(C14); B15= C14*B14;       % Stability axis System
C15= C14*inv(C14);     D15= D14;

% Lateral LQR Design Using Only the RCS Jets
R=[1,2]*0.5; R=diag(R);                     % LQR Weights R=[1,1]*2
Q=[10 2 0.5 10 0.01]*1; Q=diag(Q);         % LQR Weights Q=[1 0.4 0.5 0.2 0.005]*3
[Kpr,s,e]=lqr(A15,B15,Q,R)                 % Perform LQR design on Jets
save Kpr_M02_0.mat Kpr -ascii

% Pitch LQR Design Using the 7 Aero-Surfaces, States: (gamm_int,gamma,q,alfa)
[Ap4,Bp4,Cp4,Dp4]= linmod('Pdes4x');       % 4-state des model (gami,gama,q,alfa)
Ap5= Cp4*Ap4*inv(Cp4); Bp5= Cp4*Bp4;       % Convert to Output=State=(gami,gama,q,alfa)
Cp5= Cp4*inv(Cp4);     Dp5= Dp4;
R=4; Q=[2 5 2 2]; Q=diag(Q);               % LQR Weights (gami,gama,q,alfa)
[Kq,s,e]=lqr(Ap5,Bp5,Q,R)                  % Perform LQR design on Surf
save Kq_M02_0.mat Kq -ascii
```

FIGURE 10.25   M-File for LQR Control Design

Simulation Model: The linear simulation model for the Mach 2, gamma control phase, is shown in Figure 10.26a, and it is different from the previous two models because the pitch control system uses $\{\gamma\text{-integral}, \gamma, q, \text{and } N_z\}$ state feedback and its input is (γ-command) coming from the closed-loop guidance. This model is used for evaluating the closed-loop control system's response to ϕ and γ commands and also to wind gusts before applying it on a nonlinear 6-DOF simulation. The vehicle rates are body rates and a body to stability axis transformation block is included to convert the (p and r) body rates to stability rates (p_{stab} and r_{stab}) which are needed by the lateral LQR state-vector feedback, similar to the previous phases. Figure 10.26b shows the pitch control law consisting of the 1×4 state-feedback gain Kq, (α was replaced with N_z using the $Nz2a$ gain relationship). It receives a γ-command from guidance and generates the pitch demand that goes to the mixing matrix. A low-pass filter is included in the N_z loop to improve stability. The lateral controller is a 2×5 state-feedback gain Kpr, from states (p_s, r_s, β, p_s-integral, β-integral). It is commanded to perform roll maneuvers by rotating the vehicle about the velocity

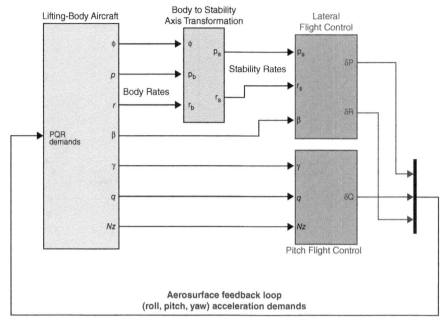

FIGURE 10.26a   Simulation Model for the Gamma Control Phase, Where the Pitch Controller Uses Gamma and N_z Feedback.

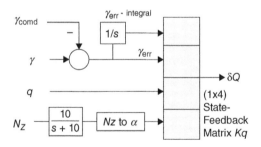

FIGURE 10.26b   Pitch Axis State-Feedback Control System for Gamma Control

vector V_0, as already described. Its outputs are roll and yaw demands to the mixing logic.

Simulation Results: We will use the linear simulation model to perform gamma and roll maneuvers simultaneously. The two commands are $\gamma_{cmd} = 2°$, and $\phi_{cmd} = 10°$. The results are shown in Figure 10.27. Both variables respond as expected to the step commands. The vehicle rolls 10° about the velocity vector creating a very small sideslip transient in β.

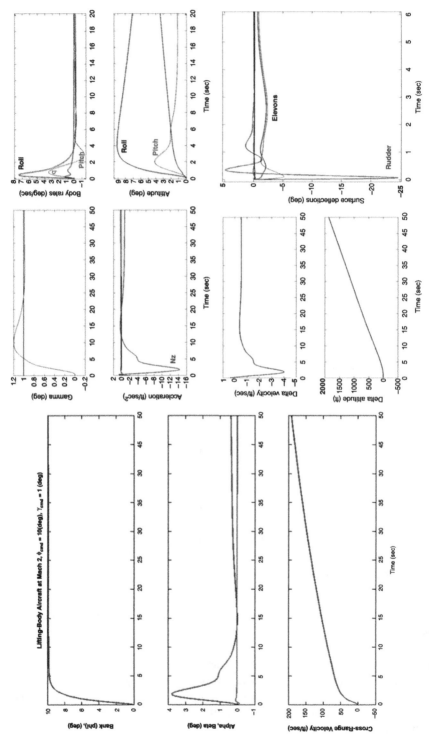

FIGURE 10.27 Linear System Response to Gamma and Phi Commands

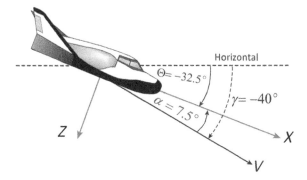

FIGURE 10.28   Approach and Landing Phase

10.1.5   Approach and Landing Phase

The approach and landing phase is different and more complex in comparison with the previous three phases because it involves additional controls and we shall analyze it in more detail. It begins at an altitude of approximately 20,000 (ft) where the vehicle dives at a steep ($\gamma = -40°$) angle, see Figure 10.28, in order to gain sufficient speed and to be able to perform its final pitch-up flare, where γ must be reduced to almost zero at landing without stalling. The closed-loop guidance system controls altitude and velocity. The FCS receives changes in altitude and velocity commands from guidance and it translates them to surface deflections. In lateral, there are no major roll maneuvers to perform during this period because the vehicle is already aligned with the runway. Small directional errors detected by the landing system due to cross-winds become commands to the heading control system and they are converted to small roll adjustments that correct the runway misalignments. The surface mixing logic used in the analysis was designed at a fixed flight condition. It includes a fourth column, in addition to roll, pitch, and yaw, which provides drag control via the speed brake. In the nonlinear simulation, however, the mixing-logic matrix is scheduled just like the control gains.

We will repeat the control surface trimming and analyze the performance parameters for this final phase of the trajectory, design and analyze the approach and landing flight control system which is significantly different from the previous phases because in the longitudinal axis we now have two separate control loops that control altitude and velocity. We will also generate uncertainty models and analyze the FCS robustness to structured parameter variations by using μ-analysis. Figure 10.29 shows some of the trajectory parameters during the approach and landing phase between Mach (0.7 and 0.25). The speed brake is partially deployed for a 65 sec period prior to landing, between $t = 1790$ and $t = 1855$ sec. The speed-brake control demand is implemented by differential deflections of the body flaps, which are controlled by the fourth column of the mixing-logic matrix. By partially deploying the speed brake it enables the velocity control system to modulate the drag coefficient, and to control speed as needed against head-wind variations. The speed brake, however, is redeployed about a minute before touchdown to enable better pitch and altitude control

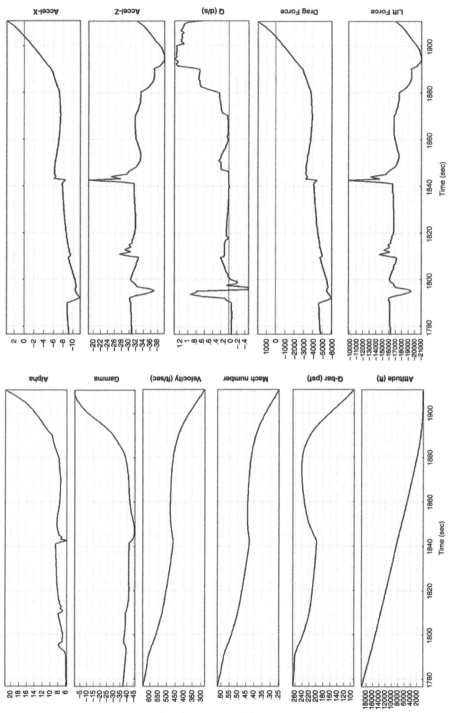

FIGURE 10.29 Trajectory Parameters During the Approach and Landing Phase

which is more critical for the final flare. Notice how the speed increases before the pitch-up flare when the speed brake is redeployed.

10.1.5.1 Control Surface Trimming We will trim the aerosurfaces along the approach and landing section of the trajectory to balance not only the three vehicle moments along the trajectory but also the axial acceleration. The x-acceleration is included in the aerosurface trimming process because a velocity control loop will be included in the FCS which will provide control in four directions, roll, pitch, yaw, and speed. In addition, a mixing-logic matrix must be designed that receives the deceleration control demand and translates it to speed-brake opening. The trimming algorithm described in Chapter 6 calculates the deflections of the aerosurfaces, see Figure 10.30, that balances the three moments and generates the expected axial acceleration along the target trajectory according to the control capability of each surface. Notice that the elevon deflections are negative (up) during the final flare in order to generate the required pitching moment to bring $\gamma = 0$ at touch down. Notice also the opening of the four body-flap deflections during the period of a partial speed-brake deployment (1790–1845 sec).

10.1.5.2 Performance Parameters Along the Approach and Landing Trajectory
The performance parameters and mainly the control authority are strongly affected by the selection of the control surface mixing matrix, that defines how the aerosurfaces combine together to allocate control in roll, pitch, yaw, and axial acceleration. The fourth column in the 7×4 surface mixing-logic matrix $K_{mixM0p4}$ determines the speed-brake deflections per axial control demand, which is mainly a differential upper and lower body-flap deflection that modulates drag. This axial acceleration control mode is of course applicable only during the period where the speed brake is partially deployed. However, in order to simplify the analysis we will keep a constant mixing-logic matrix in the entire approach and landing trajectory. The mixing-logic algorithm from Chapter 5 was used to create the surface combination matrix. The matrix was modified, however, to improve the LCDP ratio by introducing more rudder contribution in roll. We must also define the maximum wind dispersions in terms of α_{max} and β_{max} angles, which in this phase have been increased to $\pm 2°$, because the speed is reduced and it is, therefore, more vulnerable to cross-winds.

The performance parameter results in Figure 10.31 show that during this phase the vehicle is statically stable in both pitch and lateral. The short-period resonance varies between 1.8 and 3 (rad/sec) and the static margin varies between 4.5% and 9%. In the lateral direction the Dutch-roll resonance varies between 2.8 and 3.8 (rad/sec). The (Q-α, Q-β) is showing a maximum loading of 2200 (psf-deg), due to the $\pm 2°$ of α_{max} and β_{max} dispersions, which is acceptable. The control effort to counter the wind-shear disturbances is sufficiently small in pitch and yaw. In roll, however, the control effort parameter exceeds the acceptable limit. Roll authority was compromised in order to increase the LCDP magnitude and to avoid roll-reversals. It means that a strong wind gust may activate the backup RCS, since it is always available as an outer loop. The $C_{n\beta}$-dynamic is positive which means that the vehicle is directionally stable. The bank angle parameter ϕ is defined to be the bank angle to counter a cross-wind that generates a sideslip $\beta_{max} = 2°$ (in this case). The bank angle

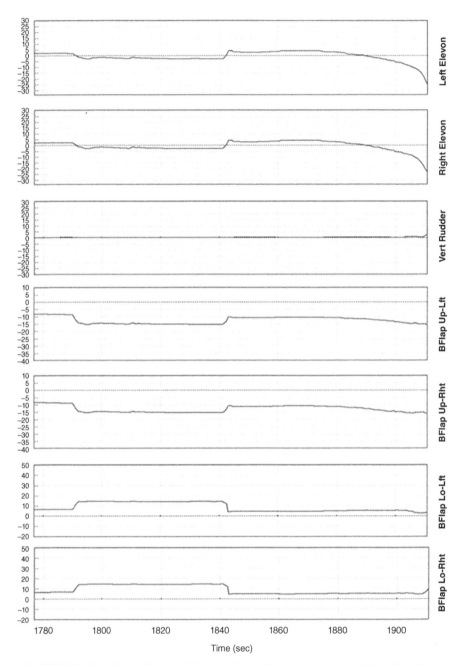

FIGURE 10.30 Control Surface Trim Angles During the Approach and Landing Phase

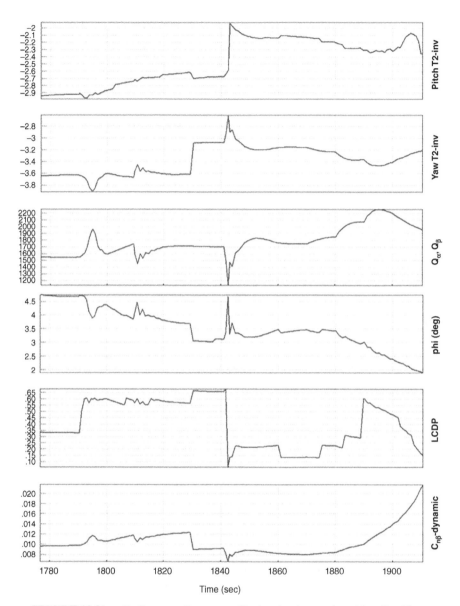

FIGURE 10.31a   Performance Parameters During the Approach and Landing Phase

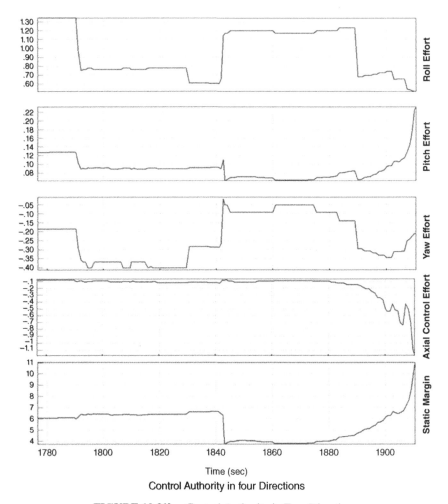

Control Authority in four Directions

FIGURE 10.31b   Control Authority in Four Directions

is less than 3° near landing, which is acceptable. The control surface hinge moments, shown in Figure 10.32, are used for sizing the actuator torques. They are calculated from equation 7.49, as a function of surface deflection, alpha, Mach number, dynamic pressure, and the hinge moment coefficients.

10.1.5.3 Contour Plots Analysis   The Mach versus alpha contour plots in Figure 10.33 show the pitch stability parameter (T2-inverse) and the LCDP ratio in the approach and land phase of the trajectory. The trajectory begins in the lower right-hand corner, where (Mach 0.65, $\alpha = 6°$) and it ends in the upper left-hand corner, where (Mach 0.25, $\alpha = 22°$). The stability parameter shows that the vehicle is statically stable in pitch with a short-period resonance between 2 and 3 (rad/sec).

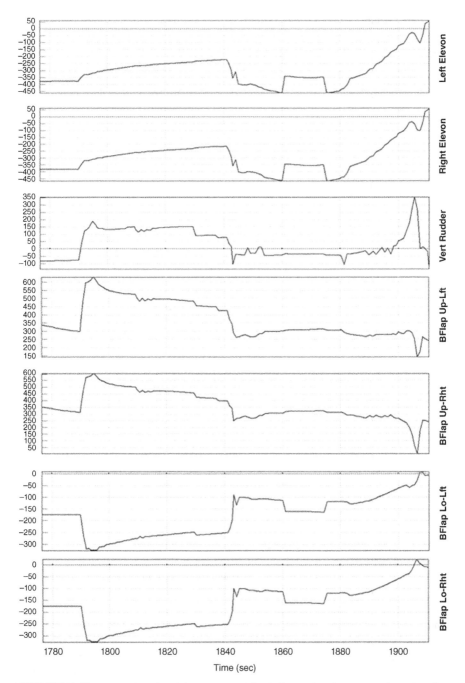

FIGURE 10.32   Control Surface Hinge Moments in (ft-lb), During the Approach and Landing Phase

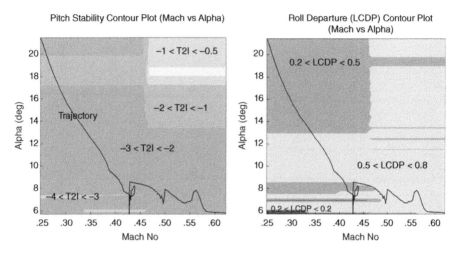

FIGURE 10.33   Contour Plots in the Approach and Landing Phase

Acceleration demands in four directions

δP	δQ	δR	δA_X	Aerosurface Commands
0.0500	-0.2000	-0.0800	0.0200	Left Elevon
-0.0500	-0.2000	0.0800	0.0200	Right Elevon
-0.1800	0	-1.0000	0	Rudder
0.0200	0	0.0100	0.0300	Upper Left Body Flap
-0.0200	0	-0.0100	0.0300	Upper Right Body Flap
0.0100	-0.0350	0	-0.0600	Lower Left Body Flap
-0.0100	-0.0350	0	-0.0600	Lower Right Body Flap

FIGURE 10.34   Control Surfaces Mixing-Logic Matrix

The LCDP ratio which measures dynamic controllability in roll is also good. The aerosurface mixing-logic matrix is shown in Figure 10.34.

10.1.5.4 Controllability Analysis by Using Vector Diagrams   The vector diagram analysis corresponds to a fixed flight condition at $t = 1839$ sec, at Mach 0.4, which is in the region where the speed brake is partially deployed and is actively controlling the speed by varying drag. The control system in this flight condition includes a speed control loop and is controlling the vehicle motion in four axes. The 7×4 control surface combination matrix is shown in Figure 10.34 and it was designed based on Mach 0.4 vehicle data. The body flaps are not only used for speed control but they are also assisting in roll, pitch, and yaw by a smaller amount. They are primarily trimming devices and slower, but they can also be used to provide some flight control assistance. The contribution amount is adjusted in the mixing-logic calculation, described in Chapter 5, by reducing the maximum deflection limit in the four body flaps to $10°$ instead of $30°$. The original mixing-logic matrix, however, was modified to improve the LCDP performance at the expense of reducing roll control authority, as already discussed.

Vector diagrams will help us evaluate the control authority of the effector system in roll, pitch, yaw, and axial acceleration, against disturbances caused by dispersions in α_{max} and β_{max}, which are expected to be less than $\pm 2°$, evaluate also the FCS authority in terms of gain capability (M_δ) against aero partials (M_α), and check the orthogonality of the effector system along the four control directions.

Pitch Controllability: Figure 10.35 shows vector diagrams in the longitudinal directions where this time the FCS is commanding two longitudinal axes independently: pitch $(\delta Q_{\pm FCSMax})$ and axial $(\delta X_{\pm FCSMax})$ accelerations. The two vector diagrams on the left side show the pitch moment, the axial and normal forces produced relative to trim when the control demands are maximized in opposite extreme directions. The figures show the pitch moment C_m plotted against the C_Z and the C_X forces in nondimensional form. The mixing logic in this flight condition uses mostly the elevons for pitch control and the speed brake to modulate drag and to vary the negative acceleration. The speed brake is implemented by differential deflections of the upper and lower body flaps with some elevon contribution.

The nearly horizontal vectors mainly along C_m show the max pitch moment and forces produced at maximum $\pm$pitch demands, $(\delta_{+QFCS\ Max}$ and $\delta_{-QFCS\ Max})$, which happens when at least one of the control surfaces saturates. The pitch control, in addition to producing a pitching moment also creates significant force variations in z, and also in x to a lesser extent. Unlike the lateral directions, the longitudinal control vectors are not symmetrical because a positive control demand $(\delta_{+\ QFCS\ Max})$ produces a larger moment and z-force variation than when a negative control demand is applied. Notice that a +pitch control demand reduces the magnitude of C_Z, reducing lift because the elevons rotate upward to increase the pitching moment. The figure shows that the vehicle is trimmed in pitch because $C_{m0} = 0$ when the control $\delta Q_{FCS} = 0$. It is, however, accelerating in both $-x$ and $-z$ directions because C_{X0} and C_{Z0} are negative when $\delta Q_{FCS} = 0$.

The almost vertical vectors on the upper left show the effects of the speed-brake control on C_X and C_m. The effect is mainly along the intended x direction. The deceleration coefficient $(C_A = -C_X)$ at trim is 0.07 in this partially deployed speed-brake position. It can be varied between 0.022 and 0.122 by the maximum and minimum speed-brake control variations. A small amount of pitching moment is also produced by the speed-brake control. The smaller vectors show the pitch moment, axial, and z forces generated by the two extreme dispersions in the angles of attack and sideslip $\pm\alpha_{max}$ and $\pm\beta_{max}$ which are $\pm 2°$ from trim positions (mostly due to $\pm\alpha_{max}$). Positive α_{max} causes a negative pitching moment because the vehicle is statically stable in this case. It also produces a negative z-force and a less negative x-force, increasing α makes the z-force more negative (up). The rectangles show the possible variations of this vector due to the uncertainties in the aero-coefficients.

The upper right plot of Figure 10.35 is a partials vector diagram that in this phase consists of two control partials. The nearly horizontal vector is the partials in pitching

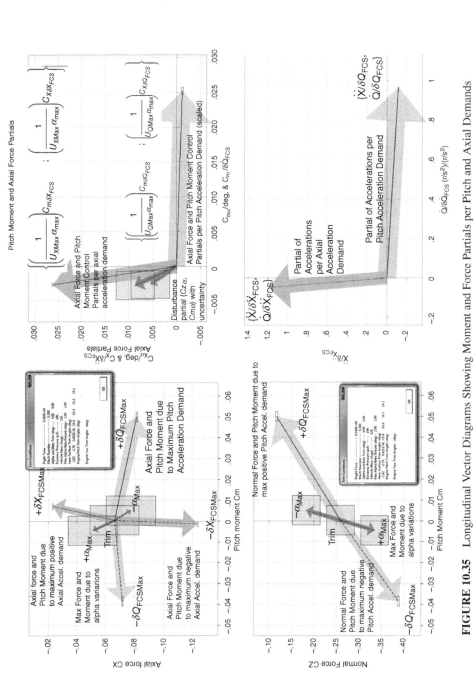

FIGURE 10.35 Longitudinal Vector Diagrams Showing Moment and Force Partials per Pitch and Axial Demands

moment and in axial force per pitch acceleration demand in (rad/sec^2), $\{C_{X\delta Q_{FCS}},$ $C_{m\delta Q_{FCS}}\}$, and the nearly vertical vector is the partials in the same directions per axial acceleration demand in (ft/sec^2), $\{C_{X\delta X_{FCS}}, C_{m\delta X_{FCS}}\}$. They are almost orthogonal to each other, pointing toward the intended directions, and they are not coupling much into each other's direction. The smaller vectors are the $\{C_{X\alpha}, C_{m\alpha}\}$ partials. They are two because they are calculated at the two extreme $\pm\beta_{max}$ positions. Notice that $C_{m\alpha}$ is negative because the vehicle is stable in this condition. The rectangle centered at the tips of the $\{C_{X\alpha}, C_{m\alpha}\}$ vectors represents the uncertainties in the two partials. Similarly the rectangle at the tip of the pitch control partial represents the uncertainties in $\{C_{X\delta Q_{FCS}}, C_{m\delta Q_{FCS}}\}$, and the rectangle at the tip of the axial control partial is the uncertainties in $\{C_{X\delta X_{FCS}}, C_{m\delta X_{FCS}}\}$.

The figure on the lower right shows the partials of pitch and axial accelerations per acceleration demands in the two longitudinal directions. The vector pointing upwards is accelerations per axial demand $\{\ddot{X}/\delta\ddot{X}_{FCS}, \dot{Q}/\delta\ddot{X}_{FCS}\}$, and the vector pointing toward the right is accelerations per pitch demand $\{\ddot{X}/\delta\dot{Q}_{FCS}, \dot{Q}/\delta\dot{Q}_{FCS}\}$. The vectors show that the two axes are almost perfectly decoupled. They are pointing toward their intended directions, close to being unit vectors and they are orthogonal to each other, which imply good controllability.

Lateral Controllability: Figure 10.36 shows the vector diagrams in the lateral directions. The diagrams on the left show the nondimensional roll, yaw moments, and the side force produced (C_l, C_n, C_Y), when the roll and yaw FCS acceleration demands are maximized. The solid almost horizontal vector shows the two moments created when the yaw FCS demand is at its maximum positive position ($\delta R_{+FCS\ Max}$) and the dashed vector in the opposite direction is the negative moment produced when the yaw demand is at its peak negative position ($\delta R_{-FCS\ Max}$). Similarly, the up and down vectors in the vertical directions show the peak roll and yaw moments produced by maximizing the roll control in both directions ($\delta P_{\pm FCS\ Max}$). The moment vectors are pointing toward their intended directions with a small amount of coupling. They are, however, perfectly orthogonal to each other which is a good property to have. The lower left diagram in Figure 10.36 also shows the max side force produced when maximizing the yaw demand. Positive yaw demand requires a negative rudder deflection which produces a negative side force. The two smaller almost vertical vectors show the roll and yaw moments produced by the aero disturbances caused by variations in the angle of sideslip $\pm\beta_{max}$ from trim and they are mainly in roll. The rectangles at the tips of the arrows show the roll, yaw, and side force uncertainties in both, the disturbance and in the control vectors.

The moment partials vector diagrams in the upper right side of Figure 10.36 show two vectors, the variation in roll and yaw moments per acceleration demands in roll and in yaw. The almost horizontal vector is the moment partials per yaw demand $\{C_{n\delta R_{FCS}}, C_{l\delta R_{FCS}}\}$ and it is pointing toward yaw. The vector pointing upward is the moment partials per roll demand $\{C_{n\delta P_{FCS}}, C_{l\delta P_{FCS}}\}$ and it affects both roll and yaw directions. The two vectors couple into each other's directions but they are nearly orthogonal to each other. The smaller vectors pointing downward are the $\{C_{n\beta}, C_{l\beta}\}$

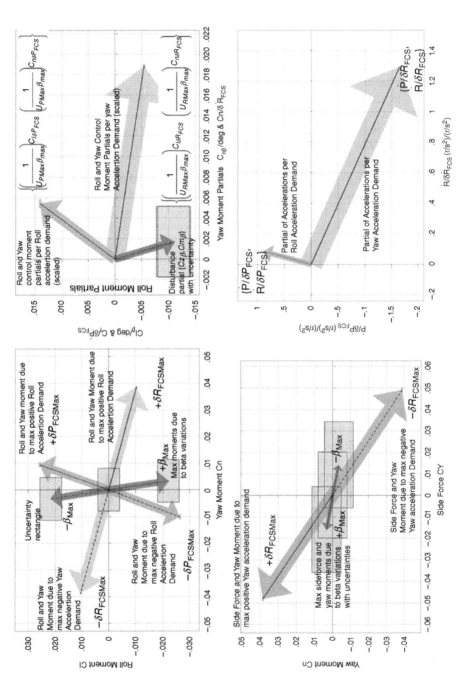

FIGURE 10.36 Roll and Yaw Vector Diagrams Showing Maximum Moments and Partials per Roll and Yaw Demands

partials. Notice that $C_{l\beta}$ is negative due to the dihedral and is bigger in magnitude than $C_{n\beta}$. The rectangle centered at the tip of the $\{C_{n\beta}, C_{l\beta}\}$ vector represents the uncertainties in the two partials. Similarly, the rectangle at the tip of the yaw control partial is due to the uncertainties in $\{C_{n\delta R_{FCS}}, C_{l\delta R_{FCS}}\}$, and the rectangle at the roll control partial is due to the uncertainties in $\{C_{n\delta P_{FCS}}, C_{l\delta P_{FCS}}\}$.

The lower right diagram in figure 10.36 shows the two vector partials of roll and yaw accelerations per acceleration demands in roll and yaw. The vertical vector is accelerations per roll demand $\{\dot{P}/\delta P_{FCS}, \dot{R}/\delta P_{FCS}\}$, and the longer vector is accelerations per yaw demand $\{\dot{P}/\delta R_{FCS}, \dot{R}/\delta R_{FCS}\}$. The axis units are in (rad/sec^2) per (rad/sec^2). They are almost orthogonal to each other, which is good property, but the roll vector is smaller indicating a reduced controllability in roll.

10.1.5.5 *Dynamic Modeling, Control Design, and Linear Analysis*   We will now create dynamic models for control design and analysis during the approach and landing phase. We select a fixed flight condition at 1839 sec, Mach 0.4, in the region where the speed brake is active and partially opened for velocity control. We will design control laws using the 7×4 control surface mixing logic already described in Figure 10.34, analyze system stability in the frequency domain, check the system's robustness to uncertainties by using μ-analysis, and create a simulation to analyze the control system's capability to track altitude and speed commands which are essential variables to control approach and landing. Two vehicle dynamic models are created: a simulation model with rates defined in the body axes, and a control design model with rates in the stability axes. The vehicle models now include the variations in altitude and velocity states (δh, and δV). A dynamic model with 60 parameter uncertainties is also created and used to analyze robustness using μ-analysis.

LQR Control Design: The longitudinal control design in this flight condition is significantly different from the previous phases because it is now using two control loops: a pitch loop to control altitude, and a deceleration control loop to control velocity. Although there is a significant amount of coupling between the two loops, it is, however, possible to achieve a certain amount of independent control in the two directions without saturating the control limits. The pitch design plant includes five states $\{\theta, q, \alpha, \delta h, \text{ and } \delta V\}$. The phugoid states ($\delta h$ and δV) are included this time because they must be directly controlled. There is no need for α-integral feedback in this case because we are not commanding α. The surface mixing matrix is also included in the design plant so the plant inputs are reduced to pitch and axial acceleration demands. The controller generated by the LQR method is a 2×5 state-feedback gain matrix. In the control law implementation, however, the α-feedback is replaced with N_z-feedback for the same reasons as before. The lateral design is similar to the previous phases. The following file loads the surface mixing matrix, and the simulation and design systems in MATLAB®, and performs the pitch and lateral LQR designs.

```
% LQR Design & Param Initialization file init.m
d2r=pi/180; r2d=180/pi;
[Aps, Bps, Cps, Dps] = pitch_des;              % Load Pitch aero-surf Design Model
[Als, Bls, Cls, Dls] = later_des;              % Load Lateral aero-surf Design Model
[As, Bs, Cs, Ds] = vehicle_sim;                % Simulation Model 6-dof
load KmixMOp4.mat -ascii; Kmix=KmixMOp4;       % Load Surfaces Mix Logic (7 x 4)

alfa0=8.531; V0=466.4; ThetO=-33.133; ge=32.174;  % Additional Vehicle Parameters
calfa=cos(alfa0*d2r); salfa=sin(alfa0*d2r);       % for Body to Stability Transform

% Convert Lateral State Vector from Body to Stability Axes, Outputs=States
[A14,B14,C14,D14]= linmod('Ldes5x');           % 5-state model {p,r,bet,pint,betint}
A15= C14*A14*inv(C14);   B15= C14*B14;          % Stability axis System
C15= C14*inv(C14);       D15= D14;

% Lateral LQR Design Using Only the RCS Jets
R=[1,10]*0.4; R=diag(R);                        % LQR Weights R=[1,1]*2
Q=[10 2 0.5 20 0.01]*1; Q=diag(Q);             % LQR Weights Q=[1 0.4 0.5 0.2 0.005]*3
[Kpr,s,e]=lqr(A15,B15,Q,R)                      % Perform LQR design on Jets
save Kpr_MOp4_0.mat Kpr -ascii                  % Lateral State-Feedback Gain

% Pitch LQR Design Using the 7 Aero-Surfaces, States: {theta,q,alfa,dH,dV}
[Ap4,Bp4,Cp4,Dp4]= linmod('Pdes5x');           % Include Kmix in design model
R=[5,1]; R=diag(R);                            % Pitch LQR Contrl Weights
Q=[0.00001 0.01 0.001 7 5]; Q=diag(Q);         % LQR State Weights {theta,q,alfa,dH,dV}
[Kq,s,e]=lqr(Ap4,Bp4,Q,R)                       % Perform LQR design on Surf
save Kq_MOp4_0.mat Kq -ascii                    % Longitudinal State-Feedback Gain

% Load Linear Sim Parameters
load THV.mat -ascii;
t=THV(:,1)'-THV(1,1); h=THV(:,2)'; v=THV(:,3)';
x0=[0 0 -2 -2 0 0 18 0 0 0]'*d2r;              % State Initialization
```

Linear Analysis Models: Figure 10.37 shows the longitudinal and lateral closed-loop control systems used for analyzing the system's response to commands and to wind gusts. The longitudinal control system commands are steps in altitude and velocity. The state-feedback loop is closed via the mixing-logic matrix which converts the control demands to deflections. This configuration is different from the previous control modes because it uses altitude and velocity feedback states which are both commanded from guidance. It also uses feedback from pitch attitude and rate, and α estimated from N_z with a low-pass filter in series to improve stability. The two longitudinal loops control the vehicle altitude and velocity by pitching and by modulating the speed brake.

In the lateral axes the control system command is heading error which is converted to roll command. The heading angle is calculated from the cross-range velocity V_{cr} and it is compared with the heading command. The error is compensated by a (PI) controller and becomes a roll command about the velocity vector. The vehicle rolls to correct its heading direction against wind disturbances. A body to stability axis transformation is included to convert the (p and r) body rates to stability rates (p_{stab} and r_{stab}) which are used by the lateral LQR state-vector feedback. The roll and yaw control demands (dP and dR) from the state-feedback matrix are converted by the mixing-logic matrix to aerosurface deflections δ_s.

This linear simulation model is used for a preliminary evaluation of the FCS performance. Figure 10.38 shows the control system's response to commands from

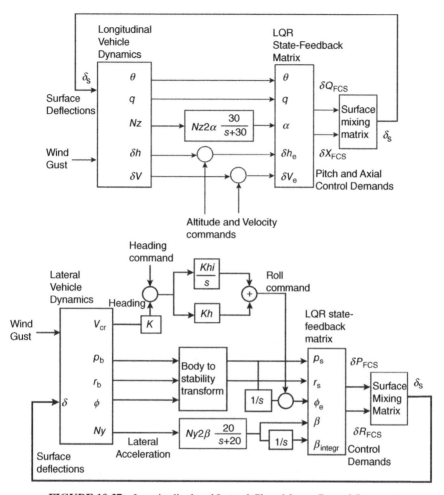

FIGURE 10.37   Longitudinal and Lateral Closed-Loop Control Systems

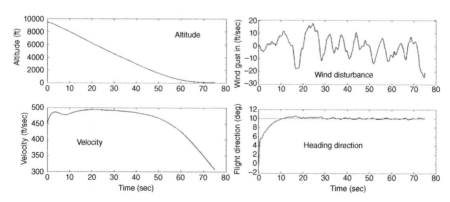

FIGURE 10.38   System's Response to Guidance Commands and to Wind Disturbances

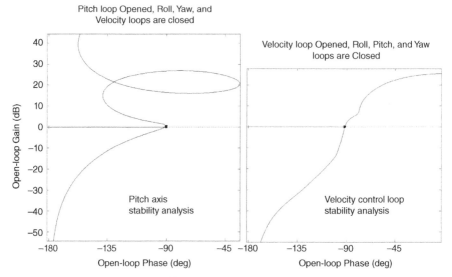

FIGURE 10.39 Nichols Plots Showing the Stability Margins of the Two Longitudinal Loops

guidance. In the longitudinal directions the system is commanded a certain altitude and velocity from a lookup table, and the lateral system is commanded a 10° change in the heading direction. A cross-wind disturbance noise is also applied. This preliminary simulation model is used for evaluating the control design. A final evaluation is obtained from the 6-DOF nonlinear simulation.

10.1.5.6 Stability Analysis The linear systems are also used for analyzing stability of the four control loops. Figure 10.39 shows the Nichols plots for the pitch axis and the velocity control loops. The loops are opened at the inputs of the mixing-logic matrix. One loop is opened at a time and the other loops are closed.

10.1.5.7 Robustness Analysis to Parameter Uncertainties Structured singular value (SSV) or μ-analysis is a very important tool for analyzing the robustness of the FCS with respect to structured uncertainties. We will, therefore, include μ-analysis in this design example. In simple words, if we know ahead of time what is the possible variation of each individual vehicle model parameter, can we use this information to analyze robustness of the closed-loop system, rather than analyzing stability using phase/gain margins? In an uncertain vehicle dynamic model some of the parameters in the model p_i are known approximately but not exactly. Their value may vary between $\pm\delta_i$, so the actual parameter value may be anywhere between ($p_i - \delta_i$ to $p_i + \delta_i$). In μ-analysis we must extract all n individual parameter uncertainties out of the vehicle model and place them in a diagonal block Δ consisting of n "fuzzy" gains (δ_1, δ_2, δ_3, ... δ_n) that connect with the nominal plant by means of n inputs and n outputs. The

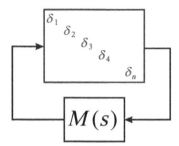

FIGURE 10.40   The Uncertainties Are Pulled Out of the Plant

gains are not exactly known but they can assume any value between $\pm\delta_i$. In order to extract the uncertainties we must first augment our vehicle model to include connections with the uncertainty block. In Section 9.8 we presented a practical method for "pulling" the uncertainties out of the plant model and creating the additional n inputs and n outputs that connect with the uncertainty Δ block, as shown in Figure 10.40. By using this method we augmented the dynamic model in our example vehicle and created the additional inputs and outputs that would connect with 60 structured uncertainties, where each uncertainty connects to an input/output pair. These are in addition to the ordinary inputs and outputs of the nominal dynamic model. The uncertainties model is dynamically identical with the original vehicle model with the exception that it includes the additional inputs and output pairs that connect with the uncertainties.

Robustness is measured by calculating the SSV across the $M(s)$ block as seen from the Δ block, without it being there. We do not need the Δ block because the uncertainty information is included in the B and C matrices. However, the block $M(s)$ represents the stabilized vehicle system including the control system with its control loops closed. The μ-analysis requires that the closed-loop vehicle model $M(s)$ should be stable and properly scaled so that the Δ block is normalized and has diagonal elements that can only vary between ±1. This is not difficult to do. The δ_i gains are absorbed in the B and C matrices and they can be replaced with ±1 variations in the Δ block. Robustness can now be measured by calculating the SSV frequency response of the $M(s)$ block across its hypothetical connections with the unitized Δ block. The great result is that the closed-loop system cannot be destabilized by any combination of the uncertainties, as long as the SSV, $\mu[M(j\omega)] < 1$, at all frequencies.

In this example we will perform separate μ-analysis for the longitudinal and lateral directions and, therefore, the dynamic model with the 60 uncertainties is separated in two subsystems. The uncertainties are mostly aerodynamic coefficients and some mass properties, and they are also separated into longitudinal and lateral uncertainties. Most uncertainties are defined by a percentage variation from their nominal values. The longitudinal model has 28 uncertainties and the lateral model has 32 uncertainties. Most of the uncertainties are rank-1, meaning that they create a single input/output pair. The X_{CG}, however, affects both longitudinal and lateral models and

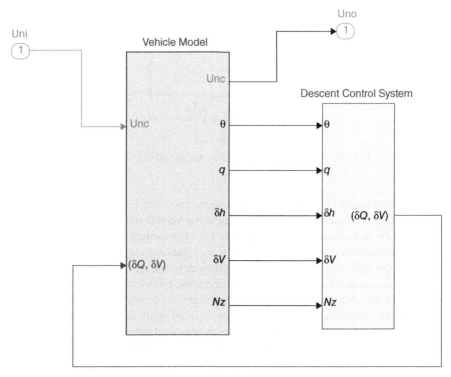

FIGURE 10.41 μ-Analysis Frequency Response is Performed Across the Connection With the Uncertainty Block Δ With the Control Loop Closed

it creates three input/output pairs. The Y_{CG} is only affecting the lateral directions. The input/output pairs are separated and are included either in the longitudinal or in the lateral models according to the states that they relate with (influence and observe). The longitudinal model in Figure 10.41 is used for calculating the SSV in the longitudinal directions. A similar model is used in the lateral axes. The stabilizing control loops are closed for the μ-analysis.

Figure 10.42 shows the longitudinal vehicle model block in detail including the 28 parameter uncertainties. It also includes the actuators and the mixing-logic matrix for the control surfaces. The pitch and axial acceleration FCS demands are converted to seven aerosurface deflections. The inputs (Uni) and outputs (Uno) vectors are theoretically connected to the normalized uncertainty block Δ, and the SSV is calculated across those inputs and outputs. The figure also shows the definition of each uncertainty and also the percentage of each parameter variation. The MATLAB® file in Figure 10.43 calculates the SSV frequency response of the pitch $M(s)$ system. A similar script calculates the SSV frequency response for the lateral $M(s)$ system using the lateral μ-analysis model.

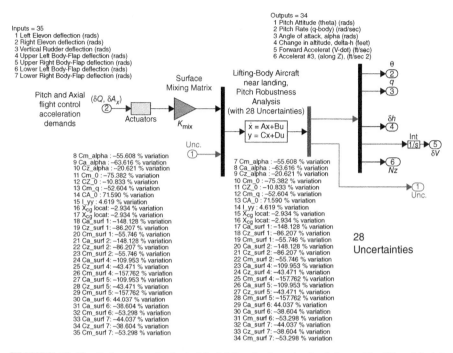

FIGURE 10.42 Robustness Analysis Model for the Longitudinal Axes, The Plant Model Includes the Vehicle Dynamics, the Surface Combination Matrix K_{mix}, Actuator Dynamics, and the 28 Connections to the Uncertainty Block

```
% Uses Mu-Analysis to Calculate System Robustness to Structured Uncertainties
d2r=pi/180; r2d=180/pi;
Npv=28; Nlv=32;                        % Number of Param Variations
[Apu,Bpu,Cpu,Dpu]= pitch_unc;          % Pitch Vehicle Model with 28 Uncertaint
[Alu,Blu,Clu,Dlu]= later_unc;          % Later Vehicle Model with 32 Uncertaint
w=logspace(-2,2,500);                  % and Frequ domain analysis

[Acp,Bcp,Ccp,Dcp]=linmod('Pitch_Robust_Anal');
sys=ss(Acp,Bcp,Ccp,Dcp);
sysf= FRD(sys,w);
blk=[-ones(Npv,1), zeros(Npv,1)];
[bnd,muinfo]= mussv(sysf,blk);
ff= get(muinfo.bnds, 'frequency');
muu=get(muinfo.bnds, 'responsedata');
muu=squeeze(muu);
muu=muu(1,:);
figure (1)
loglog(ff,muu, 'LineWidth',1.5)
xlabel('Frequency (rad/sec)')
ylabel('ssv')
Title('Pitch Mu Analysis')
```

FIGURE 10.43 MATLAB® Function That Calculates Mu of a State-Space System

Figure 10.44 shows the μ-analysis results in the longitudinal and lateral control systems. The μ-plots show that the control system is robust to the structured uncertainties in both longitudinal and lateral directions, because $\mu(\omega) < 1$ at all frequencies. The lateral axis is marginal because it touches the limit at some frequencies. It means that there is no combination of uncertainties within the specified range that can destabilize the control systems.

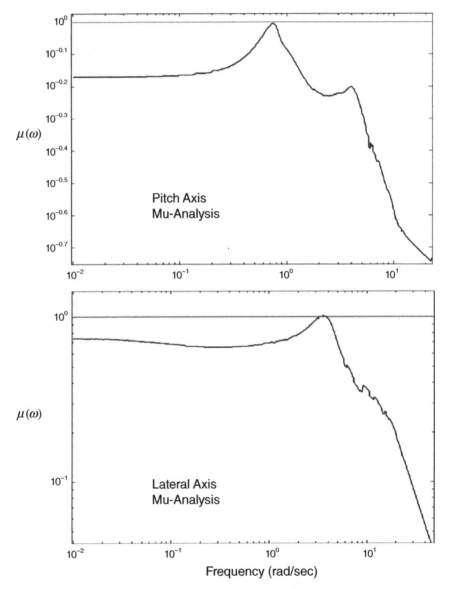

FIGURE 10.44 Mu Analysis Proves That the Control System Is Robust Against the Parameter Uncertainties in Both Longitudinal and Lateral Directions $\mu(\omega) < 1$

10.1.6   Six-DOF Nonlinear Simulation

Linear analysis is great for control design and for analyzing the vehicle stability and performance, but the ultimate validation of the control system design can only be demonstrated with a detailed (6-DOF nonlinear simulation. The simulation model in Figure 10.45 was created using MATLAB/Simulink® for the entire reentry flight, beginning after the deorbit maneuver when the vehicle is oriented at $\alpha = 30°$, and it completes 1900 sec later when the vehicle successfully lands on the runway. The guidance and control system maneuver the aircraft through the various reentry phases by employing the four types of control modes described, receiving different types of commands, and achieving different performance goals during each phase. The simulation consists of the nonlinear vehicle dynamics block and the environment block shown in detail in Figure 10.46. The block in Figure 10.47 calculates the angles of attack, sideslip, dynamic pressure, and Mach number from the velocity vector (x, y, z). The blocks in Figures 10.48 and 10.49 calculate the aerodynamic forces and moments on the vehicle as a function of the aerodynamic coefficients, Mach number, the angles of attack and sideslip, and the aerosurface deflections. Many of the blocks included in this simulation are standard off-the-shelf Simulink® functions which are available in Matlab/Simulink/Aerospace Toolkit, so they do not have to be developed from scratch.

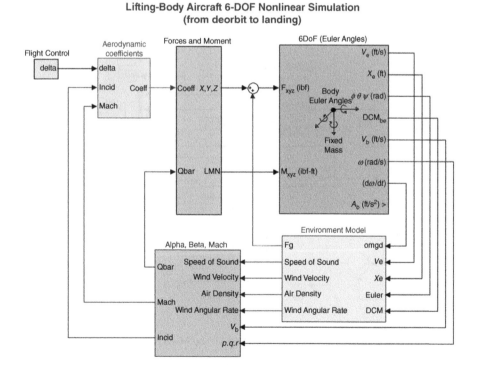

FIGURE 10.45   MATLAB® Simulation Model for the Reentry Vehicle

Environment

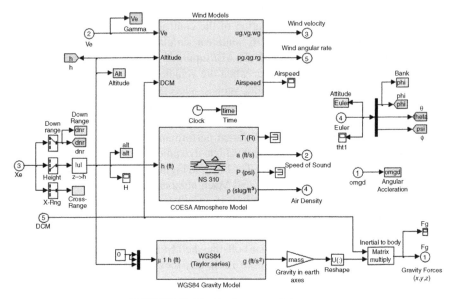

FIGURE 10.46 Environment Subsystem

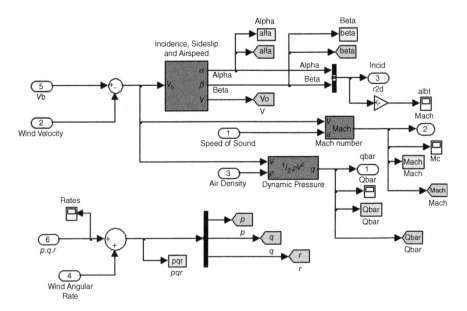

FIGURE 10.47 Angles of Attack, Sideslip and Mach Number Calculations

Moments and Forces due to the Aerodynamics

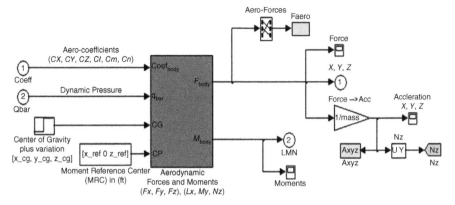

FIGURE 10.48   The Aerodynamic Forces and Moments are Functions of the Dynamic Pressure, the Aero-Coefficients, and the CG Location Relative to the Moments Reference Center

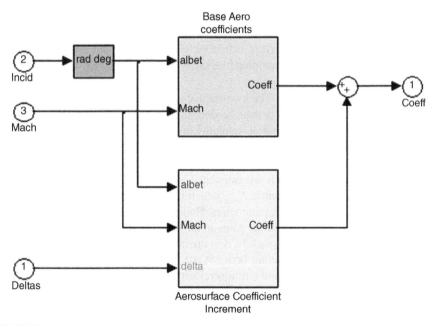

FIGURE 10.49   The Aerodynamic Coefficients Consist of the Base Body coefficients plus Increments Due to Surface Deflections; They Are Functions of the Angles of Attack and Sideslip (deg), Mach number, and Surface Deflections (deg).

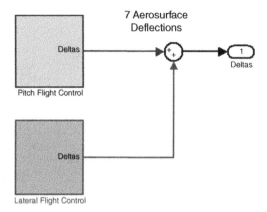

FIGURE 10.50   Flight Control Systems

10.1.6.1 Flight Control System   Figure 10.50 shows the FCS in top-level form. The longitudinal and the lateral control laws are two separate blocks generating the deflection commands to the control surfaces. They are state-feedback designs operating in different modes as already described in the control design sections. The gains were calculated and analyzed at different Mach numbers and interpolated in between Mach. The seven aerosurfaces are shared by both controllers and the deflection command signals from the longitudinal and lateral blocks are combined before being applied to the surface actuators. For simplicity the sensor feedback signals are not shown in the simulation blocks.

Pitch Flight Control System: The longitudinal control law is shown in more detail in Figure 10.51. It consists of a state-feedback gain matrix Kq converting $\{\gamma,\ \gamma\text{-integral},\ q,\ \alpha,\ \alpha\text{-integral},\ N_z,\ \text{and}\ N_z\text{-integral}\}$ error signals to pitch demands. The control law is implemented in a MATLAB® function shown in Figure 10.52. It converts the pitch state feedback to surface deflections and also interpolates the gains between the design cases which are calculated at different Mach numbers. It also interpolates the mixing-logic matrix K_{mix} which is also calculated at different Mach numbers as a function of vehicle parameters. Notice that not all variables are feeding back simultaneously but some of the gains in the feedback matrix Kq are set to zero depending on which mode the pitch FCS is operating. This type of implementation allows for a smoother transitioning between the four control modes, which are α-control, N_z-control, γ-control, and altitude/velocity control.

Guidance is implemented by means of open-loop commands which attempt to control the angle of attack, normal acceleration, flight-path angle, altitude, and speed, in different time periods along the descent flight depending on the operating mode. Initially, the pitch control mode regulates the angle of attack by commanding it to a constant value $\alpha = 30°$. This angle is gradually reduced as it transitions to the normal acceleration (N_z-control) mode where it is commanded to track a scheduled

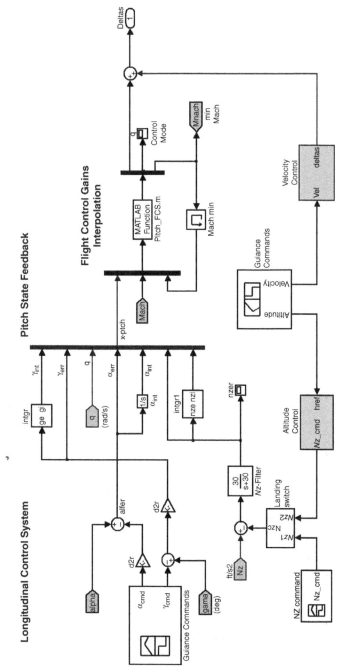

FIGURE 10.51 Longitudinal Control Law Consists of State-Feedback and Open-Loop Guidance Commands

```
function Dqpr= Pitch_FCS(u1, mach, mni)
% Implements the Pitch State-Feedback Control Laws
% u1 are the pitch state-feedback variables [gamma,gamma_int,q,alfa,alfa-int,Nz,Nz_in
global Kq Kmix r2d

Ia=int8(2); Ia2=int8(3); da=0;
mnch=min(mach,mni);

% Design Machs: [25, 20, 15, 10, 5, 4, 3, 2, 1.1, 0.9, 0.7, 0.4]
if       (mnch>=25),                      Im=int8(1);  Im2=Im;        q=1;  dm=0;
elseif (mnch>=20) & (mnch<25),  Im=int8(2);  Im2=int8(1); q=2;  dm=(mnch-20)/5;
elseif (mnch>=18) & (mnch<20),  Im=int8(3);  Im2=int8(2); q=3;  dm=(mnch-18)/2;
elseif (mnch>=17) & (mnch<18),  Im=int8(4);  Im2=int8(3); q=4;  dm=(mnch-17)/1;
elseif (mnch>=16) & (mnch<17),  Im=int8(5);  Im2=int8(4); q=5;  dm=(mnch-16)/1;
elseif (mnch>=15) & (mnch<16),  Im=int8(6);  Im2=int8(5); q=6;  dm=(mnch-15)/1;
elseif (mnch>=14) & (mnch<15),  Im=int8(7);  Im2=int8(6); q=7;  dm=(mnch-14)/1;
elseif (mnch>=13) & (mnch<14),  Im=int8(8);  Im2=int8(7); q=8;  dm=(mnch-13)/1;
elseif (mnch>=12) & (mnch<13),  Im=int8(9);  Im2=int8(8); q=9;  dm=(mnch-12)/1;
elseif (mnch>=11) & (mnch<12),  Im=int8(10); Im2=int8(9); q=10; dm=(mnch-11)/1;
elseif (mnch>=10) & (mnch<11),  Im=int8(11); Im2=int8(10); q=11; dm=(mnch-10)/1;
elseif (mnch>= 9) & (mnch<10),  Im=int8(12); Im2=int8(11); q=12; dm=(mnch- 9)/1;
elseif (mnch>= 8) & (mnch< 9),  Im=int8(13); Im2=int8(12); q=13; dm=(mnch- 8)/1;
elseif (mnch>= 7) & (mnch< 8),  Im=int8(14); Im2=int8(13); q=14; dm=(mnch- 7)/1;
elseif (mnch>= 6) & (mnch< 7),  Im=int8(15); Im2=int8(14); q=15; dm=(mnch- 6)/1;
elseif (mnch>= 5) & (mnch< 6),  Im=int8(16); Im2=int8(15); q=16; dm=(mnch- 5)/1;
elseif (mnch>= 4) & (mnch< 5),  Im=int8(17); Im2=int8(16); q=17; dm=(mnch- 4)/1;
elseif (mnch>= 3) & (mnch< 4),  Im=int8(18); Im2=int8(17); q=18; dm=(mnch- 3)/1;
elseif (mnch>=2.5)& (mnch< 3),  Im=int8(19); Im2=int8(18); q=19; dm=(mnch-2.5)/0.5;
elseif (mnch>=2.0)& (mnch<2.5), Im=int8(20); Im2=int8(19); q=20; dm=(mnch-2.0)/0.5;
elseif (mnch>=1.5)& (mnch<2.0), Im=int8(21); Im2=int8(20); q=21; dm=(mnch-1.5)/0.5;
elseif (mnch>=1.0)& (mnch<1.5), Im=int8(22); Im2=int8(21); q=22; dm=(mnch-1.0)/0.5;
elseif (mnch>=0.71)&(mnch<1.0), Im=int8(23); Im2=int8(22); q=23; dm=(mnch-0.71)/0.29;
elseif (mnch>=0.59)&(mnch<0.71),Im=int8(24); Im2=int8(23); q=24; dm=(mnch-0.59)/0.12;
elseif (mnch>=0.49)&(mnch<0.59),Im=int8(25); Im2=int8(24); q=25; dm=(mnch-0.49)/0.10;
elseif (mnch< 0.49),                     Im=int8(26); Im2=Im;       q=26; dm=0;
end

% Calculate Surface Deflections due to pitch state-feedback u1
dq1=  -Kmix(:,2,Im) *(Kq(:,:,Ia, Im) *u1);      %.. deflects at nominal (Ia, Im)
dq2=  -Kmix(:,2,Im) *(Kq(:,:,Ia2,Im) *u1);      %.. deflects at next  (Ia2,Im)
dq3=  -Kmix(:,2,Im2)*(Kq(:,:,Ia, Im2)*u1);      %.. deflects at prev (Ia,Im2)
dq= dq1 + (dq2 - dq1)*da + (dq3 - dq1)*dm;      %.. Interpolate deflects

delta= dq*r2d;                                   %.. deflects in (deg)
Dqpr=[delta; q;mnch];
```

FIGURE 10.52   Pitch Flight Control Law Converts the Longitudinal State-Vector to Surface Deflection Commands, It Also Interpolates Between the Design Cases

value. Later it transitions to gamma control mode to control the flight-path angle as commanded from guidance. Finally the FCS transitions to the approach and landing mode that controls altitude indirectly by applying N_z commands as a function of altitude error. It also controls velocity by modulating the opening of the speed brake. This indirect altitude control law (via N_z-command) in the approach and landing mode was preferred, over the direct altitude and velocity state-feedback law described earlier, because it is easier to transition from the previous gamma control mode. Figure 10.53 shows the altitude and velocity control systems in detail. It produces

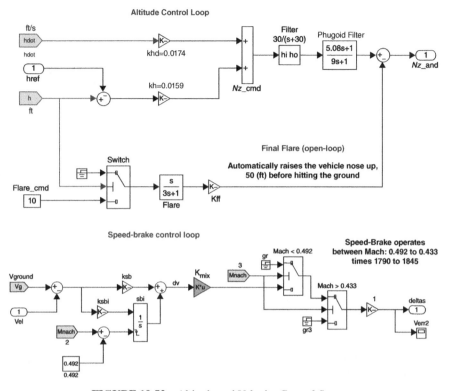

FIGURE 10.53   Altitude and Velocity Control Systems

an N_z-command as a function of altitude error and altitude rate. A lead-lag filter was added to attenuate the phugoid mode resonance. A final flare open-loop command is introduced to maximize the vehicle angle of attack during the final 50 (ft) of altitude before landing. Velocity control operates for a brief period of 70 sec during the shallow glide by modulating the opening and closing of the speed brake. It occurs at approximately 110–45 sec before landing.

Lateral Flight Control System: The lateral FCS is shown in detail in Figure 10.54. This is also a state-feedback law converting the states which are {roll and yaw stability axis rates (about the velocity vector), β, ϕ, β-integral} to deflections for the seven aerosurfaces. It also has two operating modes. During the first mode the bank angle ϕ is directly commanded open-loop, and in the second mode which is applicable prior to landing the heading direction is indirectly controlled by roll commands. The heading errors become roll commands. Notice that the rates are measured in the body axes and they are converted to stability axes as a function of the angle of attack. This reduces the sideslip angle and lateral loads when the vehicle performs

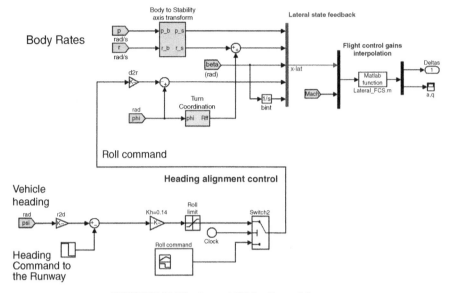

FIGURE 10.54 Lateral Flight Control System

roll maneuvers. A turn-coordination block is also included prior to the state feedback. It commands a yaw rate as a function of the bank angle (ϕ) according to the equation $R_{ff} = \frac{g}{V_0} \sin \varphi$. It uses a gravity component to counteract the centripetal side force due to turning. Notice also that the lateral state-feedback gains were designed using lateral plant models that have their output rates defined in the stability axis and the turn-coordination terms included in the vehicle dynamic model. It means that the gains know that the rates are in the stability axes and the turn-coordination logic is included.

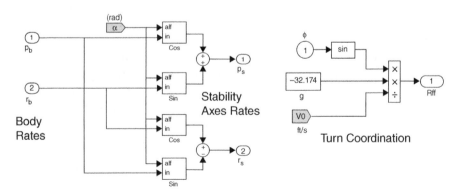

FIGURE 10.55 Body to Stability Axes Transformation and Turn-Coordination Logic

```
function Dqpr= Lateral_FCS(u2, mach)
% Implements Lateral State-Feedback Control Laws
% u2 is the lateral state-feedback variables [phi,p,r,beta,phi-integr]
global Kpr Klmix r2d
Ia=int8(2); Ia2=int8(3); da=0;

% Design Machs: [25, 20, 15, 10, 5, 4, 2, 1.1, 0.7, 0.56, 0.48]
if      (mach>=25),            Im=int8(1);  Im2=Im;       q=1;  dm=0;
elseif (mach>=20) & (mach<25), Im=int8(2);  Im2=int8(1);  q=2;  dm=(mach-20)/5;
elseif (mach>=15) & (mach<20), Im=int8(3);  Im2=int8(2);  q=3;  dm=(mach-15)/5;
elseif (mach>=10) & (mach<15), Im=int8(4);  Im2=int8(3);  q=4;  dm=(mach-10)/5;
elseif (mach>= 5) & (mach<10), Im=int8(5);  Im2=int8(4);  q=5;  dm=(mach-05)/5;
elseif (mach>= 4) & (mach< 5), Im=int8(6);  Im2=int8(5);  q=6;  dm=(mach-04)/1;
elseif (mach>= 2) & (mach< 4), Im=int8(7);  Im2=int8(6);  q=7;  dm=(mach-02)/2;
elseif (mach>= 1) & (mach< 2), Im=int8(8);  Im2=int8(7);  q=8;  dm=(mach-01)/1;
elseif (mach>=0.72)& (mach<1.0), Im=int8(9);  Im2=int8(8);  q=9;  dm=(mach-0.72)/0.28;
elseif (mach>=0.56)& (mach<0.72),Im=int8(10); Im2=int8(9);  q=10; dm=(mach-0.56)/0.16;
elseif (mach>=0.48)& (mach<0.56),Im=int8(11); Im2=int8(10); q=11; dm=(mach-0.48)/0.08;
elseif (mach< 0.48),           Im=int8(12); Im2=Im;       q=12; dm=0;
end

% Calculate Surface Deflections due to lateral state-feedback u2
dpr1= -Klmix(:,:,Im)*Kpr(:,:,Ia,Im)*u2;     %.. deflects at nominal (Ia,Im)
dpr2= -Klmix(:,:,Im)*Kpr(:,:,Ia2,Im)*u2;    %.. deflects at (Ia2,Im)
dpr3= -Klmix(:,:,Im2)*Kpr(:,:,Ia,Im2)*u2;   %.. deflects at (Ia,Im2)
dpr=dpr1 + (dpr2-dpr1)*da + (dpr3-dpr1)*dm; %.. Interpolate deflects

delta= dpr*r2d;
Dqpr=[delta; q;q];
```

FIGURE 10.56 Lateral flight Control law Converts the Lateral State Vector to Surface Deflection Commands, It Also Interpolates Between Design Cases, Which Are Fewer Than in the Longitudinal Cases

The aerosurface actuators receive deflection commands not only from the FCS, but the surface positions are prescheduled open-loop as shown in Figure 10.57. The aerosurface trim positions were calculated previously by using the trim algorithm, as already described. The flight control commands are added to the trim aerosurface deflections.

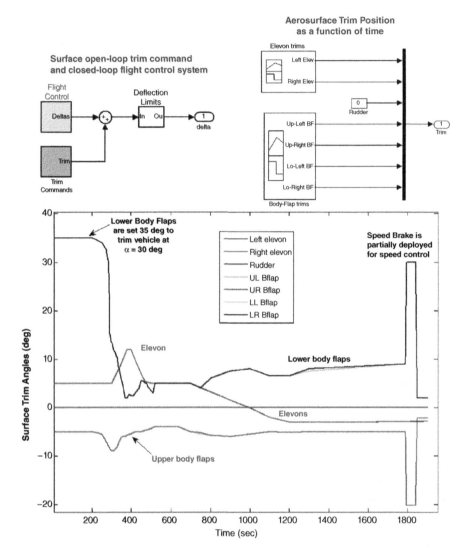

FIGURE 10.57   Aerosurface Scheduling Is Based on Previous Trim Analysis

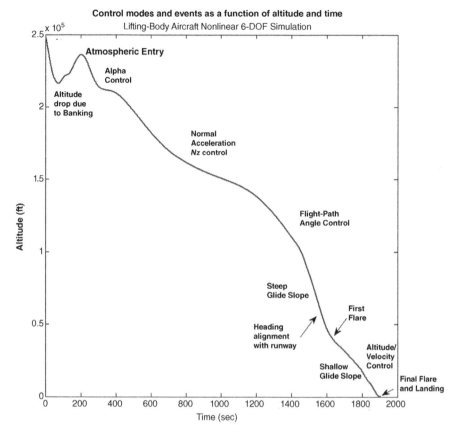

FIGURE 10.58   Altitude versus Time Including Major Events and Control Modes

10.1.6.2   Simulation Analysis

Major Events: Figure 10.58 shows the vehicle altitude versus time and highlights the control modes and the major events. The simulation begins at an altitude of 250,000 (feet) above ground where the glider vehicle enters the atmosphere with a shallow negative γ, and then it rolls a couple of times to reduce altitude and to avoid skipping back up into space. The flight control system operation begins in the alpha-control mode where the aircraft is trimmed to maintain a 30° angle of attack during this period, which optimizes its vulnerability due to friction and heating. Alpha is reduced further down and the control mode transitions to N_z-control where it maintains a comfortable and almost constant N_z acceleration for a long period.

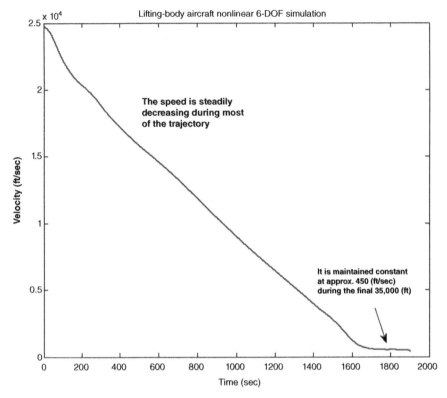

FIGURE 10.59   Velocity versus Time

The flight-path guidance is then turned on at about $t = 1400$ sec to regulate the vehicle rate of descent towards the landing site by controlling γ. In the simulation guidance is implemented with an open-loop γ-command. The γ angle is then further reduced in order to maintain sufficient speed for landing. At approximately 50,000 (feet) it rolls again in order to correct its heading and to align its direction with the runway. Figure 10.59 shows the aircraft speed versus time. The speed is steadily decreasing throughout descent and it is maintained constant at around 450 (feet/sec) during the final 35,000 (feet) of altitude by diving to reduce the glide slope. This high speed is required in order to perform the final flare before touch-down. In the final 1000 (feet) of altitude gamma begins to come up and at approximately 50 (feet) before touch down it performs the final-flare and lands with $\alpha = 13°$.

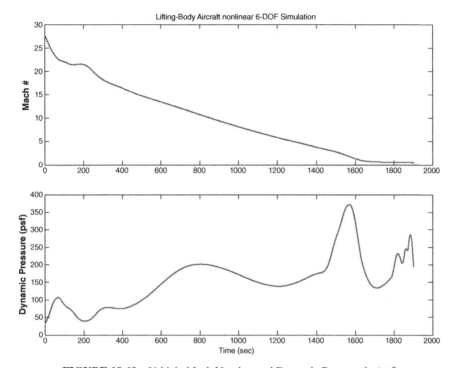

FIGURE 10.60   Vehicle Mach Number and Dynamic Pressure in (psf)

Figure 10.60 shows the variation of the Mach number and dynamic pressure as a function of time. It begins at Mach 27.5 and lands at Mach 0.4. The maximum dynamic pressure is 370 (psf) and it occurs during the heading alignment turn.

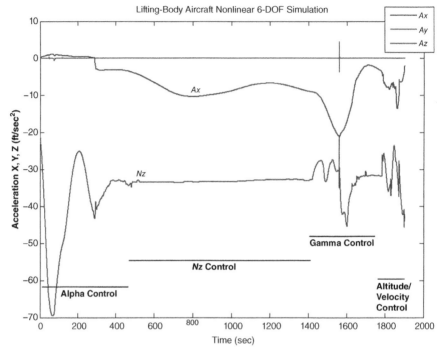

FIGURE 10.61 Vehicle Accelerations Along X, Y, and Z Highlighting the Periods of the Four Control Modes

Control Modes: Figure 10.61 shows the vehicle accelerations along x, y, and z and highlights the four control modes. The normal acceleration reaches a maximum value of little above 2 g in the beginning of atmospheric reentry during the period of alpha-control and it stabilizes at approximately 1 g during the N_z-control mode. The large aerosurface deflections that occur at $t = 1550$ sec during the heading alignment maneuver generate a short transient in the accelerations. The normal acceleration briefly increases during the heading alignment turn. The magnitude of the axial deceleration increases due to drag during the steep glide-slope dive. The normal acceleration finally peaks again before landing during the final flare.

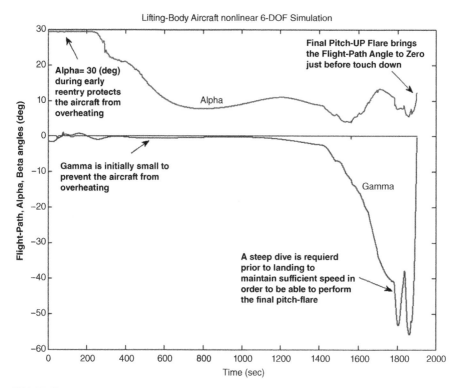

FIGURE 10.62   Angles of Attack, Sideslip, and Flight Path; γ Reaches Zero at Landing, where $\alpha = 13°$

Angles of Attack, Sideslip, and Flight-Path: Figure 10.62 shows the angles of attack, sideslip, and the flight-path angle as a function of time. The angle of attack begins at 30° in the alpha-control mode and it is gradually reduced to smaller values during the N_z-control phase and further. It is approximately 13° at landing, after the final flare that brings gamma to zero just before landing. The flight-path angle γ is initially slightly negative to optimize the atmospheric friction and heating on the vehicle. Then it comes down steeper and briefly exceeds −50° in order to maintain high velocity for the landing flare. There is a low frequency phugoid oscillation for about 1 minute during the steep glide which is attenuated further down and it does not affect landing. The sideslip angle is close to zero throughout. The small β-transients occur during the roll maneuvers.

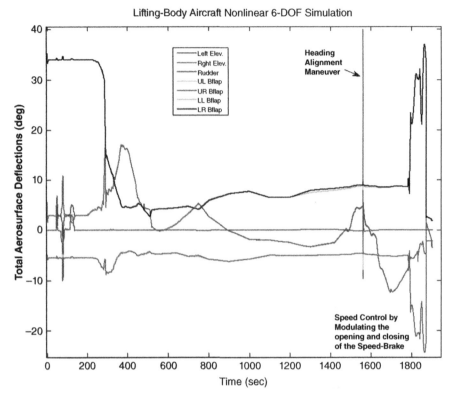

FIGURE 10.63 Surface Deflections versus Time; Includes Control plus Trim Commands

Surface Deflections: Figure 10.63 shows the aerosurface deflections as a function of time. They consist of two components: scheduled trim commands based on previous trim analysis shown in Figure 10.57, and deflection commands generated by the flight control system. It shows the rudder and the differential elevon deflections performing the two roll maneuvers. The upper body-flaps are also used in the roll heading alignment maneuver. Notice that the body-flaps are not only used for trimming but they also assist the elevons and rudder in controlling the vehicle.

Lifting-Body Aircraft Nonlinear 6-DOF Simulation

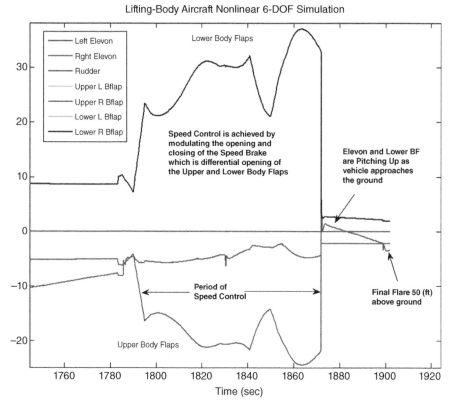

FIGURE 10.64   Surface Deflections Before Landing, Showing the Speed-Brake Activity

Speed-Brake: Figure 10.64 shows the velocity control function by means of modulating the speed-brake, which takes place a couple of minutes before landing and it lasts approximately 70 seconds. The speed-brake operates by differentially deflecting the upper and lower body-flaps. During this period the speed-brake is partially opened (trimmed) at approximately 30° for the lower flaps and −20° for the upper flaps, see Figure 10.57. The additional opening and closing of the upper and lower body-flaps is adjusted as shown by the velocity control system that attempts to control the vehicle speed by adjusting the deceleration. The ratio of upper to lower body-flap deflections is determined by the surface mixing-logic. The velocity command in this simulation is scheduled from a look-up table. The speed-brake is given enough time to regulate the landing speed and it closes about a minute before landing in order to maximize the accuracy and performance of the altitude control system.

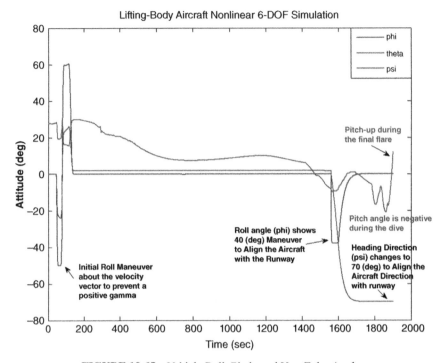

FIGURE 10.65   Vehicle Roll, Pitch, and Yaw Euler Angles

Heading Alignment Maneuver: Figure 10.65 shows the vehicle attitude in terms of Euler angles as a function of time. It demonstrates mainly the second roll maneuver that aligns the direction of the vehicle with the runway. The maneuver is performed by the heading alignment control system, shown in Figure 10.54, which applies a roll command proportional to the alignment error. The heading angle ψ is approximately 1.9° after the first roll maneuver. It is modified to −70° after the second roll maneuver to align the heading of the aircraft with the runway. The roll angle ϕ reaches a peak value of −40° during the second roll maneuver. The pitch attitude θ that comes down to −20° during the steep dive in order to maintain high speed, but it goes up after the flare and reaches a maximum 13° before touch down.

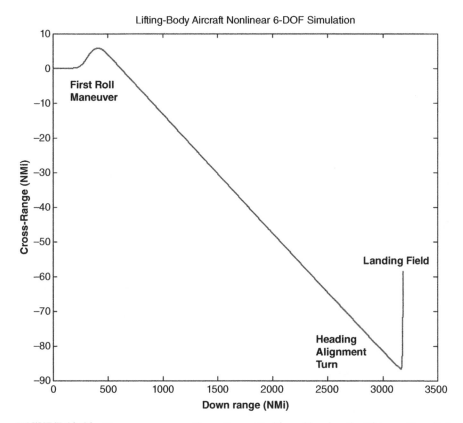

FIGURE 10.66   Downrange versus Cross-Range Positions Showing the Distance Travelled and the Heading Adjustment Maneuvers

Downrange versus Crossrange: Figure 10.66 shows the downrange versus crossrange vehicle positions beginning from the atmospheric reentry point, all the way to landing. The first roll maneuver which occurs during early reentry, points the aircraft heading direction towards the landing site. The second maneuver which occurs near the end of flight, further adjusts the heading angle and aligns the aircraft with the runway before landing.

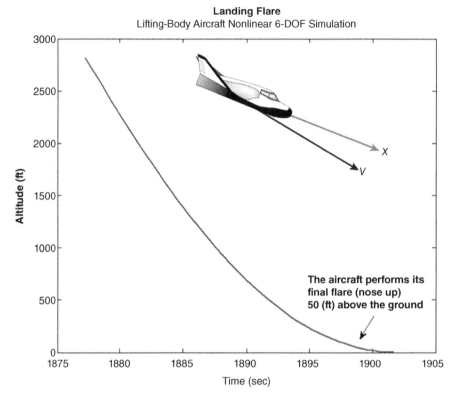

FIGURE 10.67   Aircraft Altitude During Final Flare and Landing

Final Flare and Landing: Figure 10.67 shows the altitude versus time during the final 25 seconds of flight where the vehicle performs its pitch up flare and lands. It shows that the direction of the velocity becomes horizontal ($\gamma = 0°$) after the flare which occurs at approximately 50 (feet) above the ground. The success of the flare depends on the landing speed which should be higher than 350 (feet/sec) before pitching up. Ground effects were not included in the simulation. The proper speed versus altitude is regulated by the guidance system which prevents it from exceeding the maximum allowable dynamic pressure while maintaining a sufficient speed for landing at the proper location. However, the purpose of this example is not to discuss the entry guidance system but to present the dynamic modeling and control issues. The simulation files and data for this example can be downloaded from the website Flixan.com/Examples/Lifting-Body Aircraft/.

10.2   LAUNCH VEHICLE WITH WINGS

In this example we will analyze a launch vehicle with wings similar to the one shown in Figure 10.68. It takes off vertically like a typical launch vehicle, reaches high altitudes and returns to land like a glider without engine power. During ascent it uses six engines that provide a total 350,000 (lb) of thrust. The engines are not the same but of different types and thrusts, selected intentionally for the purpose of demonstrating how to combine different types of engines. The engines' configuration is shown in Figure 10.69. The first two engines are gimbaling ±14° in pitch and ±12° yaw and they have constant thrusts of 70,000 (lb) each. Engines #3 and #4 are also gimbaling but only ±5° in pitch and yaw and they can also vary their thrusts between 9200

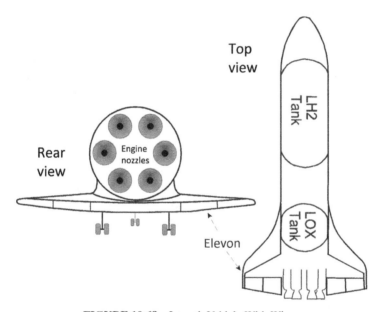

FIGURE 10.68   Launch Vehicle With Wings

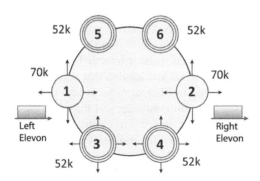

FIGURE 10.69   Six Main Engines and Two Elevon Are Used During Ascent

and 100,000 (lb) each. Engines #5 and #6 can also vary their thrusts from 9200 to 100,000 (lb) each, but they are fixed and they do not gimbal. The launch vehicle also uses two elevons which are located at the wings. During ascent the elevons are used to control roll because the gimbaling engines are too close together and the moment arm is too short for roll control. Notice, that this vehicle does not have a vertical tail and rudder. During descent the engines are not active but the vehicle uses the two elevons for pitch and roll control and two lateral thruster jets for yaw control which are located near the front.

This launch vehicle is a great example for demonstrating the effectors combination methodology that was described in Chapter 5. The mixing-logic matrix optimally combines multiple types of effectors together and optimizes control authority because it takes into consideration the control capability of each effector in the control directions. In the upcoming sections we will analyze the vehicle performance during ascent along a predefined trajectory. We will evaluate its performance and stability parameters, its capability to trim along the trajectory and against disturbances. We will also examine the effects due to lack of lateral symmetry caused by the Y_{CG} offset, engine and actuator failures, create linear models, design flight control laws, and analyze its dynamic performance. The analysis files for this example and a more detailed documentation can be downloaded from Flixan.com/Examples/Space Plane.

10.2.1 Trajectory Analysis

We must first take a look at the trajectory that this vehicle is intended to fly, which is shown in Figure 10.70. It takes off vertically and reaches a maximum dynamic pressure of 500 (lb/ft^2) and an altitude of 240,000 (ft). The thrust increases to 360,000 (lb) and then it is gradually reduced to 180,000 (lb) at the end of the ascent trajectory before it is cut off. The initial vehicle weight is 241,000 (lb) and it comes down to 64,000 (lb) when the propellants are depleted. The CG location also varies during flight. The angle of attack is mostly zero but it reaches $-2°$ as the vehicles maneuvers.

10.2.2 Trimming along the Trajectory

Our first step is to determine the effector trim angles along the trajectory. This vehicle uses all six engines and the two elevons to trim during ascent. Some of the engines are only gimbaling, some are only throttling, and some are both gimbaling and throttling. Therefore, in addition to the three moments we will also include the axial acceleration among the directions to be trimmed. The engine throttling capability should be able to achieve the axial acceleration that is defined in the trajectory. The trimming algorithm, described in Chapter 6, takes into consideration the locations of the engines, thrusts, maximum gimbaling and throttling capability, and also the effectiveness of the aerosurfaces, in order to calculate the trim deflections and the thrust variations required to balance the moments and the axial acceleration along the target trajectory.

Figure 10.71 shows the trim positions of the effectors along the trajectory. There are a total of 14 effectors on this vehicle (four engines gimbaling in two directions,

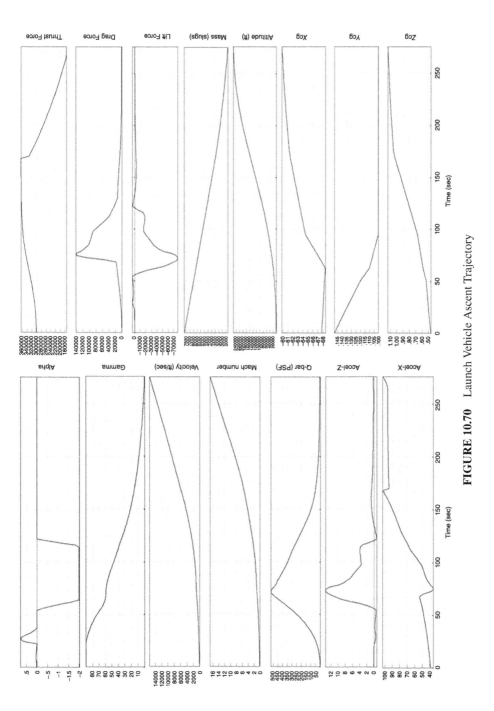

FIGURE 10.70 Launch Vehicle Ascent Trajectory

383

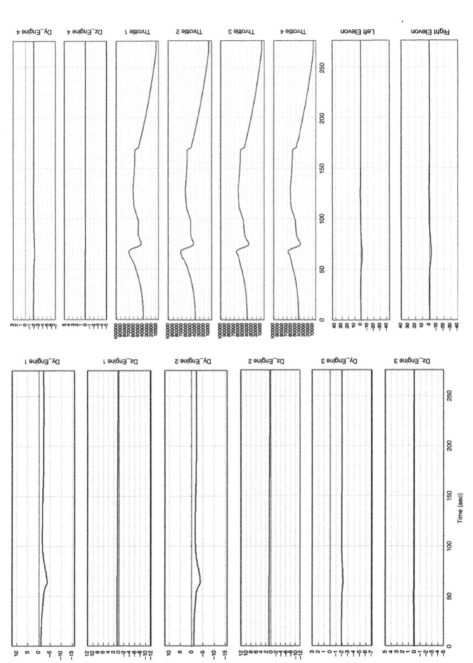

FIGURE 10.71 Trim Positions of the Effectors Along the Trajectory

384

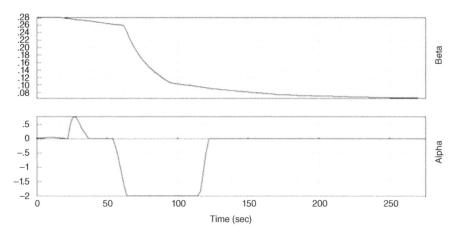

FIGURE 10.72   Angles of Attack and Sideslip. β Is Nonzero Because of the Y_{CG} Offset

four throttling engines, and two elevons). Notice in Figure 10.72, that the $+Y_{CG}$ offset causes the vehicle to yaw creating a sideslip β for balancing the yawing moment. Some of the TVC engines are slightly tilted in yaw ($\delta_Z = 0.3°$) in order to counteract the yawing moment due to β. The TVC engines are pitching in the negative direction in order to balance the pitching moment along the input trajectory. The pitch deflections of the TVC engines δ_Y are different between the left and right sides, and also the two Elevon deflections are slightly different. It is because they are trimming the rolling moment generated due to the lack of lateral symmetry. The thrusts of the four throttling engines are also varying during trim as they try to match the acceleration defined in the trajectory.

10.2.3   Trimming with an Engine Thrust Failure

Engine failures are not uncommon in launch vehicles with multiple engines and in most cases they should be able to complete their missions in the event of at least one engine failure. Failures may be due to a loss of thrust or a TVC actuator hard-over. In the analysis that follows we will demonstrate that we are able to trim and control the vehicle in the event of thrust loss in one of the six engines. We will fail one of the throttling engines (#6), and allow the remaining five engines and the control surfaces to adjust and continue trimming the vehicle along the same trajectory. An engine can easily be made to fail in the trimming algorithm by setting its thrust and its max deflection or throttle capability to zero during trimming, which is easier done than changing the setup to use fewer engines.

Figure 10.73 shows trim positions of the effectors with and without the engine thrust failure. The original configuration with the six engines is compared against the failed engine configuration that uses only five thrusting engines. The target trajectory is still the same but the deflections and throttles of the remaining functioning engines have changed because they are now attempting to compensate for the failed engine thrust. Only the throttle values instead of actual thrusts are shown for the thrust

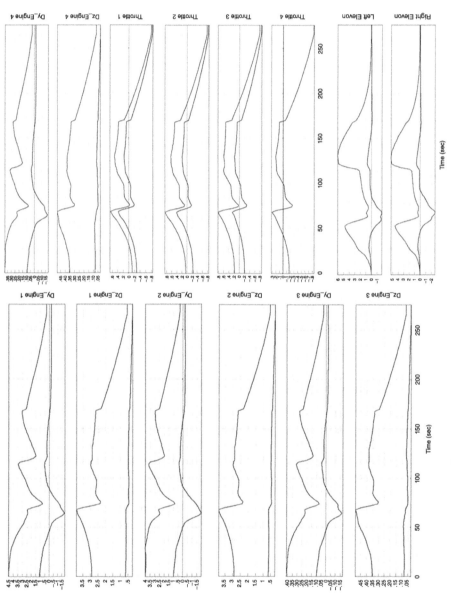

FIGURE 10.73 Effector Trim Positions Before and After Engine #6 Thrust Failure

varying engines. Here are some interesting points to notice between the nominal and the failed engine cases.

- In the failed engine thrust case, the throttle value and also the thrust of Engine #6 (which is the 4th throttling engine) are zero.
- The new trimming with the failed engine increased the throttle levels on the remaining three throttling engines (Throttle #1 to 3), as expected, in order to make up for the failed engine thrust as necessary for achieving the expected trajectory acceleration.
- The yaw deflections $(\delta_{Z1}, \delta_{Z2}, \delta_{Z3}, \delta_{Z4})$ of the four gimbaling engines are bigger in the failed thruster case, in order to provide a negative yawing moment and to compensate against the positive yawing moment produced by the absence of engine #6 thrust.
- The pitch deflections $(\delta_{Y1}, \delta_{Y2}, \delta_{Y3}, \delta_{Y4})$ in the four TVC engines are also increased in order to compensate for the lack of symmetry.
- The aerosurface deflections are also increased in the failed thruster case. The difference between left and right Elevons is greater in order to counteract the rolling moment produced due to the lack of lateral symmetry.

10.2.4 Analysis of Static Performance along the Trajectory

Figure 10.74 shows the static performance parameters of the launch vehicle calculated along the trajectory. An effector combination matrix must first be created, using the mixing-logic algorithm, in order to efficiently combine the effectors together and to achieve the required accelerations. Notice that the four trim directions are also directions to be controlled, so the mixing matrix has four inputs that correspond to the 4 control/trim directions (3 rotations plus x-acceleration), and 14 outputs that correspond to the 14 effectors. The vehicle controllability is measured by its control authority against wind-shear disturbances which are defined by the maximum expected dispersions of the angles of attack and sideslip from trim. In this case they are set to $\alpha_{max} = \pm 4°$, and $\beta_{max} = \pm 4°$.

The static stability analysis of this vehicle along the trajectory shows that it is unstable in pitch during the first 130 sec and later it becomes stable. During the unstable period the shortest time to double amplitude (T2) is 0.62 (sec) and occurs at 70 sec. In the lateral direction the vehicle is always stable with a positive $C_{n\beta}$-dynamic. The yaw T2-inverse parameter is negative (stable), and the Dutch-roll resonance peaks to 1.4 (rad/sec) at 80 sec. The (Q-alpha, Q-beta) loading at (α_{max} = 4°, β_{max} = 4°) is also acceptable. It peaks at about 1000 (psf) in the high dynamic pressure region. The control effort is good in all four directions (roll, pitch, yaw, and x-acceleration), because it requires less than 50% of full control authority to counteract disturbances due to the (α_{max} = 4°, β_{max} = 4°) dispersions from trim. This allows sufficient control authority for maneuvering and for reacting against wind-gust disturbances. It means that changes in the accelerations due to wind-shear disturbances are easily counteracted by the controls. The LCDP is positive and very close to 1, which implies that the vehicle has good roll controllability and turn coordination through the entire ascent trajectory. It means that we can rely on the elevons

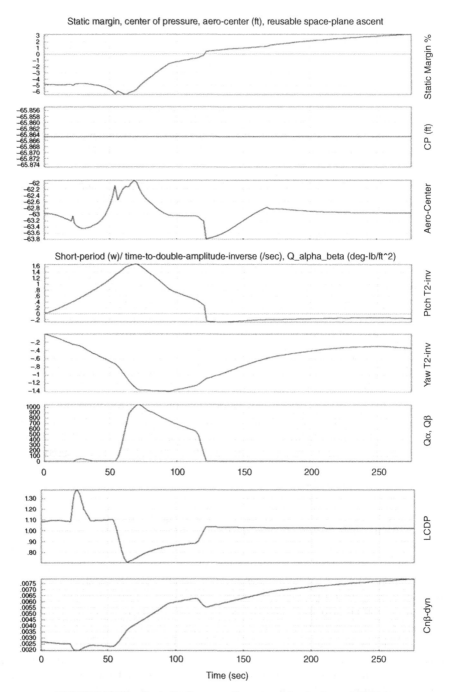

FIGURE 10.74   Static Performance Parameters for the Launch Vehicle

Rotational and Translational Control Authority in the Four Directions (parameter must be less than 0.5)

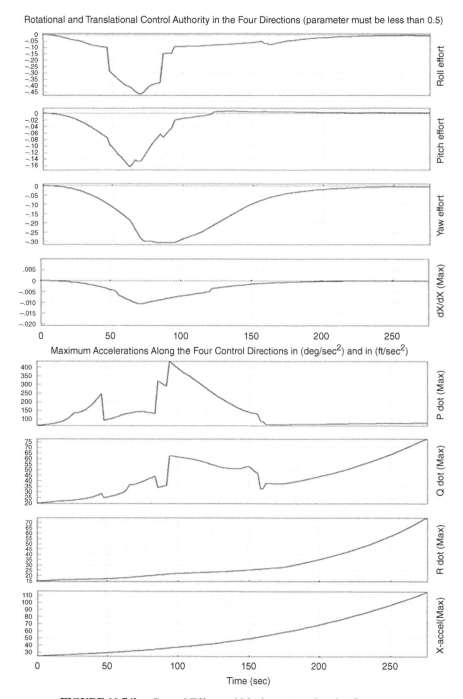

Maximum Accelerations Along the Four Control Directions in (deg/sec^2) and in (ft/sec^2)

Time (sec)

FIGURE 10.74b   Control Effort and Maximum Acceleration Parameters

and the TVC to control roll without any assistance from RCS jets. The $C_{n\beta}$-dynamic parameter is positive which implies that the vehicle also has good directional stability.

10.2.5   Controllability Analysis Using Vector Diagrams

We will now examine the control authority of the vehicle against aerodynamic disturbances by using vector diagrams. The disturbance is a wind shear defined in terms of maximum alpha and beta dispersion angles from trim. The 14×4 effector mixing matrix is shown in Figure 10.75. It converts the control demands in four directions (which are roll, pitch, yaw, and axial accelerations) to TVC and aerosurface deflections and to throttle commands. The deflections then produce the accelerations in the required directions. Notice that the mixing matrix calculation is not exactly symmetric because of the Y_{CG} offset.

Figure 10.76 shows the maximum moments, axial force, and the accelerations produced along the four control directions when the control demands are maximized. The two plots on the left side compare the roll and yaw moments and the accelerations produced when the roll control $\delta P_{\pm FCSMax}$ and the yaw control $\delta R_{\pm FCSMax}$, reach their saturation levels. The small vectors represent the roll/yaw moments and accelerations generated by the $\pm\alpha_{max}$ and $\pm\beta_{max}$ dispersions from trim. The plots show that both roll and yaw directions are well controllable because the two control vectors are a lot more powerful than the disturbance moments. They are also very well decoupled from each other and are pointing toward their intended directions.

Similarly, the two vectors on the upper right show the maximum pitch moment and axial force produced by maximizing the pitch and axial acceleration demands $\delta Q_{\pm FCSMax}$ and $\delta X_{\pm FCS\_Max}$. The units are nondimensional. The diagram on the lower right shows the max control accelerations in pitch (rad/sec^2) and in axial acceleration

Acceleration demands in four directions

dP	dQ	dR	dA_X	Effector commands
↓	↓	↓	↓	
0.1465	-0.5698	0.0300	0.0003	Engine 1 Ptch Deflect dy1 (rad)
-0.1393	-0.4969	-0.0289	0.0003	Engine 2 Ptch Deflect dy2 (rad)
0.0071	-0.0552	0.0014	0.0000	Engine 3 Ptch Deflect dy3 (rad)
-0.0064	-0.0517	-0.0013	0.0000	Engine 4 Ptch Deflect dy4 (rad)
0.0101	-0.0015	-0.7207	-0.0002	Engine 1 Yaw Deflect dz1 (rad)
0.0101	-0.0015	-0.7207	-0.0002	Engine 2 Yaw Deflect dz2 (rad)
-0.0103	0.0028	-0.1007	-0.0000	Engine 3 Yaw Deflect dz3 (rad)
-0.0103	0.0028	-0.1007	-0.0000	Engine 4 Yaw Deflect dz4 (rad)
-0.0324	0.9381	0.9775	0.0292	Engine 3 Throttle Input dTh3/Th3
0.0063	0.9312	-0.9814	0.0286	Engine 4 Throttle Input dTh4/Th4
-0.0076	-0.9227	0.9943	0.0302	Engine 5 Throttle Input dTh5/Th5
0.0317	-0.9297	-0.9907	0.0296	Engine 6 Throttle Input dTh6/Th6
0.1680	-0.7860	-0.0353	-0.0002	Left Elevon Deflection (rad)
-0.2574	-0.6777	0.0220	-0.0001	Right Elevon Deflection (rad)

FIGURE 10.75   Mixing-Logic Matrix

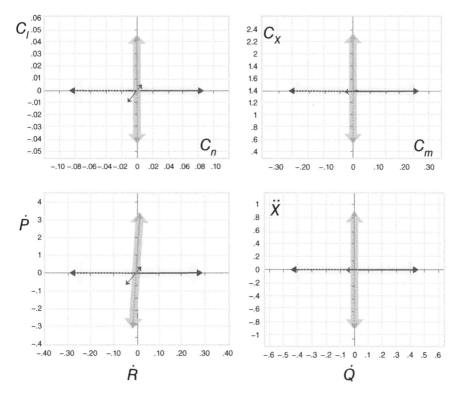

FIGURE 10.76 Maximum Moment and Force/Acceleration Vector Diagrams in Four Control Directions

in (g). The pitch axis is controlled by gimbaling the first four TVC engines in pitch. The axial acceleration is controlled by throttling the last four engines. The horizontal vectors in opposite directions represent the variation in pitch moment and pitch acceleration by maximizing the pitch demand $\pm \delta Q_{\text{FCSMax}}$ in both directions. Similarly, the vertical vectors represent the variation in the axial control force and acceleration by maximizing the axial acceleration demand $\pm \delta X_{\text{FCS\_Max}}$ in both directions. The axial force coefficient at trim is biased positively, centered at $C_{X0} = 1.4$, because the vehicle is accelerating due to the main engine thrust. The throttle control is able to vary C_X between 0.5 and 2.3. The small vectors show the effects of the wind disturbance which is mainly due to $\pm \alpha_{\max}$. The disturbance affects mainly the pitch direction C_m, not C_X. A negative α_{gust} creates a negative pitching moment because the vehicle is unstable.

Figure 10.77 shows the vector moment and force partials at the top and the acceleration partials at the bottom. The vector diagram on the upper right compares the roll and yaw aeromoments partials with respect to beta $(C_{l\beta}, C_{n\beta})$ against the moments partials per roll and yaw flight control demands. The vector pointing upward is the moment partial per roll demand $(C_{l\delta P_{\text{FCS}}}, C_{n\delta P_{\text{FCS}}})$, and the vector pointing toward the right is the moment partial per yaw demand $(C_{l\delta R_{\text{FCS}}}, C_{n\delta R_{\text{FCS}}})$. Both vectors are

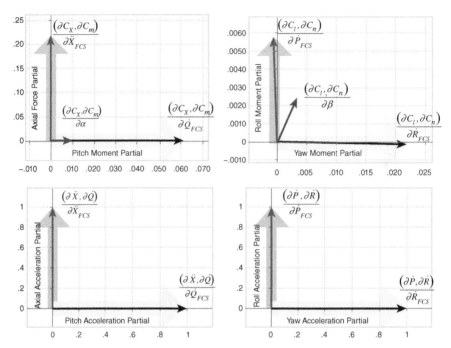

FIGURE 10.77 Moment, Force, and Acceleration Vector Diagram Partials in All Four Control Directions

pointing in the intended directions with almost zero cross-coupling. The two control vectors are scaled in order to be made comparable to the disturbance vectors for controllability analysis, as described in equations 8.3 through 8.6. The $(C_{l\beta}, C_{n\beta})$ disturbance partial vector is smaller in comparison with the control partials, so there is no controllability issue in the roll and yaw directions.

The longitudinal results are similar. The upper left-hand diagram shows the axial force and pitch moment partials. The vector pointing upward is the x-force and moment produced per axial acceleration demand $(C_{X\delta X_{FCS}}, C_{m\delta X_{FCS}})$ and it is pointing the proper direction. The vector pointing toward the right is the pitch moment and x-force per pitch acceleration demand $(C_{X\delta Q_{FCS}}, C_{m\delta Q_{FCS}})$ and it is also in the right direction. The smaller red vector is the x-force and pitch moment partials due to alpha $(C_{X\alpha}, C_{m\alpha})$ and it is small in comparison with the control partials.

All four partials are orthogonal to each other and they are pointing toward their intended directions. The decoupling between axes is accomplished by the mixing-logic matrix, shown in Figure 10.75 that is calculated as described in Chapter 5. The decoupling is even more evident in the acceleration per acceleration demand partials, shown in the lower half of Figure 10.77, where all four accelerations per acceleration demand partials are unit vectors pointing exactly in the demanded directions. This ideal situation of attaining perfect open-loop diagonalization is easily achievable in this example because of the multitude of control effectors.

10.2.6   Creating an Ascent Dynamic Model and an Effector Mixing Logic

We now come to the most interesting part which is the control design. We used the Flixan program to generate a dynamic model for our launch vehicle using the equations of motion described in Chapter 3. The flight condition of the model is at $t = 50$ sec after liftoff. This vehicle blends multiple types of effectors together and the dynamic model has 14 control inputs and each input corresponds to an effector. It has eight gimbal deflection inputs for the four TVC engines (four pitch and four yaw gimbals), four throttling inputs for the four engines that have a variable thrust, and two control surface inputs. Among the six main engines; engines #1 and #2 are only gimbaling, engines #5 and #6 are only throttling, and engines #3 and #4 are both gimbaling and throttling. The engine nozzles are not mounted parallel to the vehicle x-axis but they are tilted $-2°$ in pitch, which is also their trim position. The max deflections of the first two engines are $\pm 14°$ in pitch and $\pm 12°$ yaw. Engines #3 and #4 are only able to gimbal $\pm 5°$ in pitch and yaw. The four throttling engines have a nominal thrust of 55,300 (lb) and they can vary their thrusts $\pm 95\%$, that is, from 9200 (lb) when the throttle input is -1 to 101,400 (lb) when the throttle input is $+1$. Zero throttle control input corresponds to nominal thrust. There are two additional inputs that correspond to the two elevon deflections. The last input to the dynamic model is a wind-gust velocity disturbance in (ft/sec). The wind-gust direction is defined perpendicular to the vehicle x-axis and it is exciting both pitch and lateral axes, as it was described in Section 3.7.

There is one additional feature that makes this vehicle model even more interesting. It includes two tanks that contain sloshing propellants and the slosh dynamics for the two tanks are included in the dynamic model, as it is described in the propellant sloshing Section 3.8. There is a liquid oxygen (LOX) tank located near the back of the vehicle and it is much heavier than the liquid hydrogen (LH2) tank which is located closer to the front of the vehicle, see Figure 10.68.

10.2.7   Ascent Control System Design, Analysis and Simulation

The FCS is expected to control the same four directions that were included in the mixing-logic matrix design, that is, roll, pitch, yaw, and axial velocity. The mixing matrix translates the 4 FCS demands to 14 effector commands and it is included in the FCS. We begin by using the dynamic models to design state-feedback control laws for this flight condition. Then we will perform stability analysis in the frequency domain, develop a simulation model in MATLAB/Simulink®, and simulate the system's response to flight control commands and also to wind gusts.

10.2.7.1 Flight Control Design   The flight control gains are calculated by using the following MATLAB® file. It loads the pitch and lateral design systems and uses the LQR control design algorithm to calculate the state-feedback matrices Kq and Kpr for the pitch and lateral control systems. The slosh resonances are not included in the design models because the slosh states are not measurable for feedback, plus we do not want to complicate the control system and to include additional states for estimating slosh. The state vector for the pitch design model consists of states [θ, q, α, θ-integral], and the state vector of the lateral design model consists of states [ϕ, p, ψ, r, β].

```
% Pitch LQR Design for Space-Plane
[Ap,Bp,Cp,Dp]= vehi_pitch;                    % Load Pitch Design Model
[Api,Bpi,Cpi,Dpi]= linmod('Pitch_Design');    % Augment Pitch Simulink model
sys1=SS(Api,Bpi,Cpi,Dpi);

% Weights[theta, q, alpha, theta_int]
Q= diag( [5.5,  3.5, 0.01, 0.1]);             % Weights(thet,q,alpha,thet_int)
R=1;                                          % Control Weights R=2
[Kq,S,E]= LQR(sys1,Q,R);
save Kq.mat Kq -ascii                         % Save the LQR gains in Kq.mat

% LQR Lateral Design for Space-Plane
[Al,Bl,Cl,Dl]= vehi_later;                    % Load the Lateral Design Model
sys2=SS(Al,Bl,Cl,Dl);
% states: [phi, p, psi,r, beta]
Q=diag([ 0.5, 0.2, 0.5, 0.2,  0.001]);
R=diag([1  1]*1);                             % Control Weights
[Kpr,S,E]= LQR(sys2,Q,R);
save Kpr.mat Kpr -ascii                       % Save the LQR gains in Kpr.mat
```

10.2.7.2 Simulation Model   Figure 10.78 shows a closed-loop simulation model that was developed in Simulink® and it will be used for analyzing the system's performance. It has four control loops roll, pitch, yaw attitudes, plus a velocity control loop. The 14 × 4 effector combination matrix is included in the vehicle dynamic block. The dynamic model is excited by a wind-gust disturbance pulse that is shaped by a low-pass filter. The pitch state-feedback control loop via matrix *Kq* is shown in the upper section of Figure 10.78, and the roll/yaw state-feedback control loop via matrix *Kpr* is shown in the lower section. There is also a velocity control loop that receives a velocity command from guidance. It measures axial acceleration and velocity and controls the velocity by commanding the throttle input. The simulation model receives three commands from guidance which are represented with step inputs in this simple implementation. There is a pitch attitude command θ_{comd}, a roll command ϕ_{comd}, and a delta velocity command δV_{cmd}. It can also receive a yaw command but it is not shown. We will use this model to simulate two cases by commanding one input at a time: (a) a wind-gust excitation, (b) a pitch attitude step command. It is interesting to observe how the mixing matrix allocates control to the various effectors in a coordinated action in order to achieve the desired guidance commands.

Wind-Gust Simulation: In the first simulation, shown in Figures 10.79 and 10.80, we apply a square wind-gust velocity pulse of 30 (ft/sec) for a period of 4 sec. The pulse is smoothed by the low-pass filter that makes it look more like a real gust. Its direction is perpendicular to the vehicle *x*-axis, exciting both the pitch and yaw directions and creating positive alpha and beta transients, as described in the wind-gust modeling Section 3.7. It generates a roll/yaw transient also. Notice how both

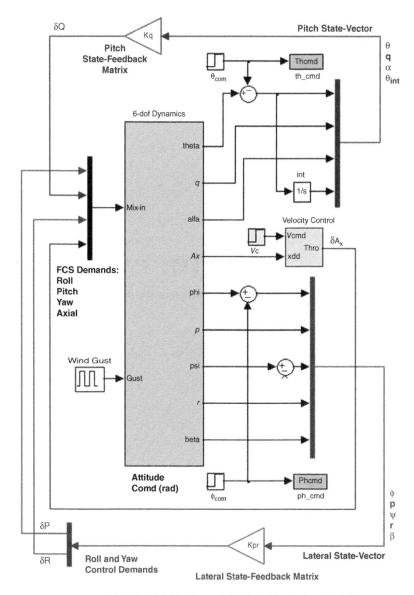

FIGURE 10.78   Launch Vehicle Simulation Model

the TVC engines and aerosurfaces react against the gust disturbance as they stabilize the vehicle.

Pitch Attitude Command Simulation: The simulation model of Figure 10.78 is also used to calculate the system's response to a pitch attitude command $\theta_{comd} = 1$ (deg) with all other inputs set to zero, see Figure 10.81. This maneuver is performed by

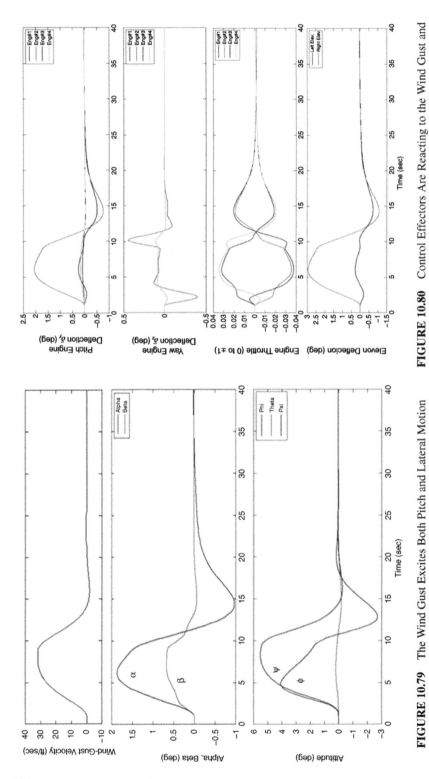

FIGURE 10.80 Control Effectors Are Reacting to the Wind Gust and Stabilizing the Vehicle

FIGURE 10.79 The Wind Gust Excites Both Pitch and Lateral Motion

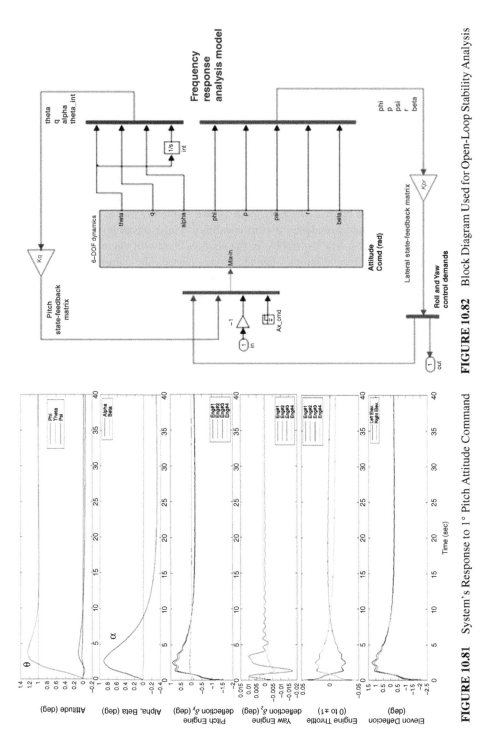

FIGURE 10.81 System's Response to 1° Pitch Attitude Command

FIGURE 10.82 Block Diagram Used for Open-Loop Stability Analysis

deflecting the TVC engines and also the two aerosurfaces together in pitch. The small difference between the left and right elevon deflections is because of the Y_{CG} offset. The small oscillations are caused by the propellant sloshing and it is damping out.

10.2.7.3 Stability Analysis   Figure 10.82 is a block diagram used for open-loop stability analysis. Since the system has four control loops, only one loop is opened at a time to calculate the frequency response across the opened connection, with the other three loops closed. In the case shown, it is configured for yaw analysis. Figure 10.83 shows the Nichols charts in the pitch and yaw axes where the slosh resonances are more observable and highlights the phase and gain margins.

The LOX mode corresponds to the heavier slosh mass and it is much stronger than the LH2. It is phase stable in both pitch and yaw directions with plenty of margins and does not require baffles. The LH2 mode is also stable in yaw. In the pitch axis, however, the LH2 has a tendency toward instability (phase unstable) and when the damping coefficient is low the bubble in the Nichols chart will encircle the critical point and cause slosh oscillations when the loop is closed. The damping coefficient was increased in order to reach an acceptable gain margin. It means that baffles may have to be included in the LH2 tank in order to dampen out the fuel motion and to provide an acceptable stability margin.

In some cases we may get by without having to use baffles, even if the slosh mode is unstable, simply because the mass is small in comparison with the vehicle weight. We may find out that the performance is acceptable without baffles, and even if the slosh mode becomes unstable, the amplitude of the oscillations will be limited by the diameter of the tank, plus the wave will eventually break and the disturbance effect on the vehicle will be a small limit cycle that may not have any significant impact on performance.

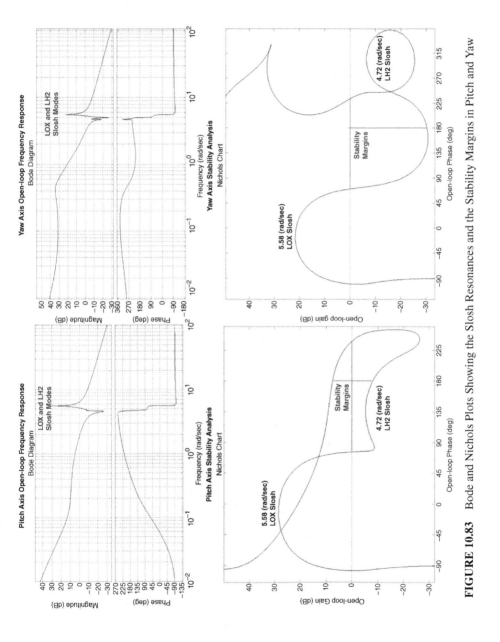

FIGURE 10.83 Bode and Nichols Plots Showing the Slosh Resonances and the Stability Margins in Pitch and Yaw

10.3 SPACE STATION DESIGN EXAMPLE

In this example we will design and analyze the control system of a large space station concept in orbit around the earth. The space station is a single-keel configuration similar to the picture in Figure 10.84, consisting of a truss structure with some attached modules for the crew, equipment, experiments, etc. which are attached on the vertical truss structure. There is also a horizontal boom with two rotating solar arrays which are always pointing toward the sun and complete one rotation per orbit relative to the spacecraft. For attitude control the spacecraft uses reaction control jets (RCS and CMGs, either independently or together. The CMGs supply the control torque, but they cannot provide a control torque indefinitely in the same direction because they have limited momentum capacity (which is torque times time, and when the momentum reaches saturation level the control torque drops to zero. Therefore, the torque direction must be reversed prior to saturation in order to reduce the CMG momentum. When this does not happen naturally during normal operations it must be forced into torque reversal by firing the RCS jets in the proper directions. A more efficient method is to utilize the gravity-gradient torque for desaturation. This requires, however, knowledge of gravity-gradient dynamics in the control law.

The Space Station is flying in a circular orbit and most of the time at a nearly constant attitude with respect to the local vertical local horizontal (LVLH) frame. The LVLH x-axis is in the velocity direction, the z-axis is toward the earth, and the y-axis toward the right solar array. The attitude control system (ACS) has several modes

FIGURE 10.84 Space Station Single-Keel Configuration

of operation for maneuvering, attitude hold, momentum management using RCS or CMG control, docking, etc. One of the control modes attempts to fly the station at the torque equilibrium attitude (TEA) which is the attitude where the average aerodynamic torques balance against the gravity-gradient torques. In this mode the average torques are zero and the momentum does not saturate but it cycles below its maximum limit. The torque also cycles below its maximum limit. One approach to keep the spacecraft at the average TEA and to avoid secular CMG momentum build up is to use feedback from the CMG momentum and to have some knowledge of the gravity-gradient torque in order to use it properly in bringing the momentum down. When this is properly designed the spacecraft automatically converges towards the average TEA, without having to be commanded there, and the CMG momentum continuously cycles, but it never exceeds its maximum magnitude capability. The CMG's function in this mode is not to maneuver the space station attitude around but to stabilize it at the TEA and to attenuate the attitude oscillations which are caused by cyclic aerodynamic disturbances.

The aerodynamic disturbances consist of steady torques and also cyclic components that induce attitude oscillations. The steady torques are caused by the steady aerodynamic torque due to the difference between the center of pressure and the center of mass. There are two frequencies associated with the cyclic disturbance torques: at orbital rate ω_0, and at twice the orbital rate $2\omega_0$. The components at orbital rate are due to the difference in atmospheric density between the sunny and the dark sides of the earth, and the components at twice the orbital rate are caused by the variation in drag due to the rotation of the solar array. The purpose of the control system is to point the station at an attitude where the secular torques cancel and only the cyclic remain. We will, therefore, design a control system that will stabilize the attitude at the TEA and manage the combined CMG momentum by preventing it from saturating. We will also analyze the stability and performance of the station's control system by means of simulations and classical frequency domain methods.

The purpose of this example is to teach the student how to create simple rigid-body spacecraft models and use them to design attitude control laws, and also how to develop 6-DOF simulations to test spacecraft performance. The work-files, data, and analysis models for this example can be downloaded from Flixan.com/Examples/ Large Space Station/.

10.3.1 Control Design

In the TEA mode of operation the space station orients itself to an almost steady LVLH attitude that balances the average aerodynamic torques with the average gravity-gradient torques. In addition to attitude and rate feedback the ACS uses feedback from the CMG momentum to prevent it from building up. The LQR control design method is used to stabilize this coupled system by means of state feedback. The method requires a linear model of the spacecraft attitude and CMG dynamics operating relative to the LVLH frame and a performance index in order to calculate the state-feedback gain matrix.

The dynamic model in equation 10.1 will be used for attitude control and CMG momentum management design. It is derived from equation 3.58 after replacing the body rates with LVLH rates, and it is augmented with additional states as shown in Figure 10.85. It includes the CMG momentum dynamics in the body frame and also the gravity-gradient torque which is a function of attitude. It is the gravity-gradient torque that will be used to stabilize the CMG momentum. The CMG momentum integral is also included in the design model to help bound the CMG momentum closer to zero. The states of the LQR design model are spacecraft rates, attitude, CMG momentum, and CMG momentum integral. The resulting LQR state feedback stabilizes not only attitude but also the CMG momentum. The control law points the station attitude toward a TEA by balancing the aerotorques with gravity-gradient torques. The CMG momentum cycles because it responds to the cyclic aerodynamic disturbances but it does not diverge to saturation. This is a continuous momentum desaturation method that relies on sufficient knowledge of the mass properties and orbit information. It is much more attractive than RCS desaturation because it does not use propellant, plus it is not vulnerable to RCS disturbances that interfere with micro-gravity experiments.

$$
\begin{bmatrix} I_{XX} & I_{XY} & I_{XZ} \\ I_{XY} & I_{YY} & I_{YZ} \\ I_{XZ} & I_{YZ} & I_{ZZ} \end{bmatrix} \begin{pmatrix} \ddot{\varphi} \\ \ddot{\theta} \\ \ddot{\psi} \end{pmatrix} = \omega_o \begin{bmatrix} 0 & 2I_{YZ} & I_{XX}-I_{YY}+I_{ZZ} \\ -2I_{YZ} & 0 & 2I_{XY} \\ -I_{XX}+I_{YY}-I_{ZZ} & -2I_{XY} & 0 \end{bmatrix} \begin{pmatrix} \dot{\varphi} \\ \dot{\theta} \\ \dot{\psi} \end{pmatrix}
$$

$$
+ \omega_o^2 \begin{bmatrix} 4(I_{ZZ}-I_{YY}) & 3I_{XY} & -I_{XZ} \\ 4I_{XY} & 3(I_{ZZ}-I_{XX}) & I_{YZ} \\ -4I_{XZ} & -3I_{YZ} & (I_{XX}-I_{YY}) \end{bmatrix} \begin{pmatrix} \varphi \\ \theta \\ \psi \end{pmatrix} + \omega_o^2 \begin{pmatrix} -4I_{YZ} \\ 3I_{XZ} \\ I_{XY} \end{pmatrix} + T_d + T_c
$$

$$
\begin{pmatrix} \dot{h}_X \\ \dot{h}_Y \\ \dot{h}_Z \end{pmatrix}_{CMG} = \omega_o \begin{pmatrix} h_Z \\ 0 \\ -h_X \end{pmatrix}_{CMG} - T_c
$$

$$(10.1)$$

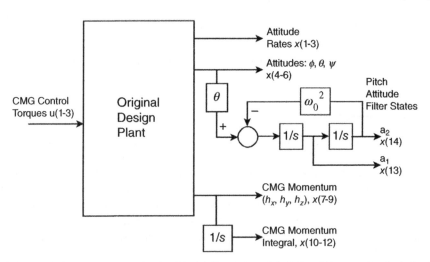

FIGURE 10.85 Augmented Design Model Consisting of Additional States to be Optimized by the LQR

The dynamic model in equation 10.1 when used without augmentation for state-feedback design does not provide sufficient CMG momentum attenuation. In addition, the cyclic aerodynamic disturbances generate unacceptably large pitch attitude oscillations at orbital frequency ω_0. It is, therefore, necessary to augment this model by including some additional states in order to make it more useful for LQR design.

The new design model includes the integral of the CMG momentum in all three axes, plus a filter that amplifies the pitch attitude θ oscillation at ω_0. An effective method for attenuating the effect of a disturbance at a known frequency on the state θ is to introduce a resonance of the same frequency in the design model. The resonance is implemented by two additional states (α_1, α_2) and it is excited by the pitch attitude oscillations θ that we would like to attenuate. The augmented design model is shown in Figure 10.86, and its inputs are the CMG control torques about x, y, and z body axes. The augmented design model state-vector consists of the following states which are also outputs.

1. Roll Rate LVLH (rad/sec)
2. Pitch Rate LVLH (rad/sec)
3. Yaw Rate LVLH (rad/sec)
4. Roll Attitude LVLH (rad)
5. Pitch Attitude LVLH (rad)
6. Yaw Attitude LVLH (rad)
7. Roll CMG Momentum (ft-lb-sec)
8. Pitch CMG Momentum (ft-lb-sec)
9. Yaw CMG Momentum (ft-lb-sec)
10. Roll CMG Momentum integral (ft-lb-sec^2)
11. Pitch CMG Momentum integral (ft-lb-sec^2)
12. Yaw CMG Momentum integral (ft-lb-sec^2)
13. Filter state a_1 function of (theta)
14. Filter state a_2 function of (theta)

Before using the LQR algorithm, however, we must first set the weighting matrices Q and R, which are used for adjusting the states versus controls penalization in the LQR optimization. They are set as shown in the MATLAB® script below. The script calculates the 3×14 state-feedback matrix K_{pqr}, shown in Table 10.1. The state-feedback must achieve an acceptable trade-off between attitude errors and CMG control magnitudes.

```
Q0=diag([1.e-3,   1000,    1.e-3, ...        % rate (wl) weights
         1.e-5,   1000,    1.e-5, ...        % LVLH attitude weights
         1.e-10, 1.e-12,  1.e-10, ...        % CMG Momentum weights
         1.e-12, 3.e-13,  1.e-12, ...        % CMG Moment-Integral weights
         1.e-5,  1.e-5]);                    % Disturb Accomod Filter weights
R0=diag([1, 1, 1]);                          % CMG Control Torque Weights
Kpqr = LQR(A,B,Q,R);
```

	x1	x2	x3	x4	x5	x6	x7	x8
x1	-1.923e-005	-4.575e-006	0.001741	2.819e-006	-1.786e-007	2.116e-008	0	0
x2	5.171e-008	1.693e-007	-5.397e-005	-2.375e-007	-8.52e-007	-5.689e-011	0	0
x3	-0.001889	9.501e-005	1.907e-005	8.389e-008	5.624e-009	8.68e-007	0	0
x4	1	0	0	0	0	0	0	0
x5	0	1	0	0	0	0	0	0
x6	0	0	1	0	0	0	0	0
x7	0	0	0	0	0	0	-0.0011	0
x8	0	0	0	0	0	0	0	0
x9	0	0	0	0	0	0	0	0
x10	0	0	0	0	0	0	1	1
x11	0	0	0	0	0	0	0	1
x12	0	0	0	0	1	0	0	0
x13	0	0	0	0	0	0	0	0
x14	0	0	0	0	0	0	0	0

	x9	x10	x11	x12	x13	x14
x1	0	0	0	0	0	0
x2	0	0	0	0	0	0
x3	0	0	0	0	0	0
x4	0	0	0	0	0	0
x5	0	0	0	0	0	0
x6	0	0	0	0	0	0
x7	0.0011	0	0	0	0	0
x8	0	0	0	0	0	0
x9	0	0	0	0	0	0
x10	0	0	0	0	0	0
x11	0	0	0	0	0	0
x12	1	0	0	0	0	0
x13	0	0	0	0	0	-1.21e-006
x14	0	0	0	0	1	0

A

	u1	u2	u3
x1	-8.415e-009	-9.919e-010	-9.559e-011
x2	-9.919e-010	-2.487e-008	-7.893e-011
x3	-9.559e-011	-7.893e-011	-9.078e-009
x4	0	0	0
x5	0	0	0
x6	0	0	0
x7	1	1	0
x8	0	0	1
x9	0	0	0
x10	0	0	0
x11	0	0	0
x12	0	0	0
x13	0	0	0
x14	0	0	0

B

	x1	x2	x3	x4	x5	x6	x7	x8	x9	x10	x11	x12	x13	x14
y1	1	0	0	0	0	0	0	0	0	0	0	0	0	0
y2	0	1	0	0	0	0	0	0	0	0	0	0	0	0
y3	0	0	1	0	0	0	0	0	0	0	0	0	0	0
y4	0	0	0	1	0	0	0	0	0	0	0	0	0	0
y5	0	0	0	0	1	0	0	0	0	0	0	0	0	0
y6	0	0	0	0	0	1	0	0	0	0	0	0	0	0
y7	0	0	0	0	0	0	1	0	0	0	0	0	0	0
y8	0	0	0	0	0	0	0	1	0	0	0	0	0	0
y9	0	0	0	0	0	0	0	0	1	0	0	0	0	0
y10	0	0	0	0	0	0	0	0	0	1	0	0	0	0
y11	0	0	0	0	0	0	0	0	0	0	1	0	0	0
y12	0	0	0	0	0	0	0	0	0	0	0	1	0	0
y13	0	0	0	0	0	0	0	0	0	0	0	0	1	0
y14	0	0	0	0	0	0	0	0	0	0	0	0	0	1

C

	u1	u2	u3
y1	0	0	0
y2	0	0	0
y3	0	0	0
y4	0	0	0
y5	0	0	0
y6	0	0	0
y7	0	0	0
y8	0	0	0
y9	0	0	0
y10	0	0	0
y11	0	0	0
y12	0	0	0
y13	0	0	0
y14	0	0	0

D

FIGURE 10.86 Augmented Design Model in State-Space Form

TABLE 10.1   State-Feedback Matrix

$K_{pqr} = 10^5$

-6.0168	0.9835	0.0604	-0.0073	-0.0001	-0.0037	0	0	0	0	0	0	0	0
0.3908	0.2948	0.0535	0.0003	-0.0109	-0.0003	0	0	0	0	0	0	0	0
-2.3654	1.0020	-4.1478	-0.0081	-0.0004	-0.0014	0	0	0	0	0	0	0	0

10.3.2   Simulation and Analysis

Figure 10.87 shows the linear simulation model used for evaluating the state-feedback gain and for adjusting the LQR optimization weighting matrices Q and R. The augmented state vector is fed back via the state-feedback gain K_{pqr} to produce the CMG control torque T_c. In Figure 10.87, the two states resonance used in the design model becomes a disturbance filter which is part of the control law. This is not a noise attenuation filter such as those used in filtering out flex modes but it is the opposite. It amplifies the control torque at the disturbance frequency and phases it to counteract the disturbance. The control design is finally tested with a 6-DOF simulation that uses the nonlinear equation 2.23. It is implemented in a Simulink® model shown in Figure 10.88. The space station 6-DOF equations are coded in a MATLAB® function "RigBod_Dynal_LVLH.m," shown in Figure 10.89. The model does not have an attitude command. Its inputs are aerodynamic disturbance torques in all three axes estimated from other models. The disturbances consist of steady-state plus cyclic components at the two frequencies (ω_0 and $2\omega_0$). The simulation results in Figure 10.90 demonstrate the performance of the CMG momentum management design under the stimulus of aerodynamic disturbances and gravity-gradient torques that cause the attitude to drift toward the TEA.

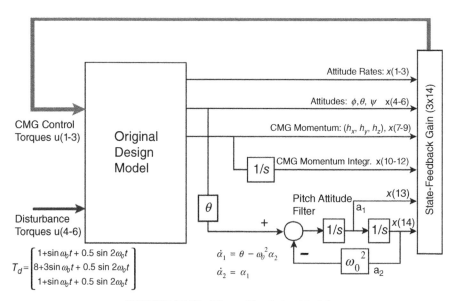

FIGURE 10.87   Linear Simulation Model

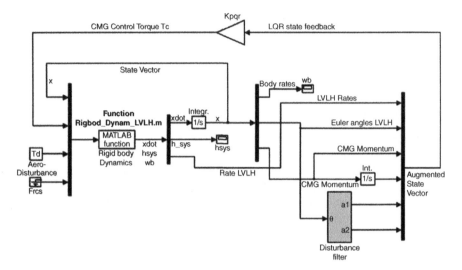

FIGURE 10.88   Nonlinear 6-DOF Simulation Model

```
function dot= Rigbod_Dynam_LVLH(x,Tc,Td,f)                    % s/c Dynamics with CMG
global J JI wo
global nt Jloc Jdir CB2L

% dot= Rigbod_Dynam2_LVLH(x,Tc,Td,f)
% Space Station Dynamics in the Local Vertical Local Horizontal plane
% State Variables (x)
% x(1-3)  = body rates (wb)
% x(4-6)  = Attitude LVLH
% x(7-9)  = CMG Momentum in body (hcmg)
% Inputs:
% Tc(3)   = Control Torque (ft-lb)
% Td(3)   = Disturbance Torque (ft-lb)
% f(nt)   = RCS jet thrusts (lbf)

dot= zeros(15,1);
wb  = x(1:3);                                                 % body rates
the = x(4:6);                                                 % Vehicle Atttitude LVLH
hcmg= x(7:9);                                                 % CMG Momentum

cphi= cos(the(1)); sphi= sin(the(1));
cthe= cos(the(2)); sthe= sin(the(2));
cpsi= cos(the(3)); spsi= sin(the(3));

c= [-sin(the(2))*cos(the(3)); ...                            % Gravity Gradient terms
    cos(the(1))*sin(the(2))*sin(the(3)) + sin(the(1))*cos(the(2)); ...
    -sin(the(1))*sin(the(2))*sin(the(3)) + cos(the(1))*cos(the(2))];

CB2L= [cpsi, -cphi*spsi,  sphi*spsi; ...                     % Body to LVLH Transform
       0,     cphi,       -sphi; ...
       0,     sphi*cpsi,  cphi*cpsi]/cpsi;

Tr= zeros(3,1);                                              % Calculate Thruster Moments
for i=1:nt
    Tr= Tr + f(i)*cross(Jloc(:,i), Jdir(:,i));              % Total RCS Moment (ft-lb)
end

dot(1:3)= JI*(-cross(wb,J*wb) + 3*wo^2*cross(c,J*c)-Tc +Td +Tr);  % Rate of Change of body rates wb-dot
dot(4:6)= CB2L*wb + [0, wo, 0]';                            % rate of Euler angls wrt LVLH
dot(7:9)= -cross(wb,hcmg) + Tc;                            % Rate of CMG Momentum in body
dot(10:12)= J*wb + hcmg;                                    % System Momentum
dot(13:15)= CB2L*wb + [0, wo, 0]';                          % Rates in LVLH
```

FIGURE 10.89   MATLAB® Function That Implements the Nonlinear Equations

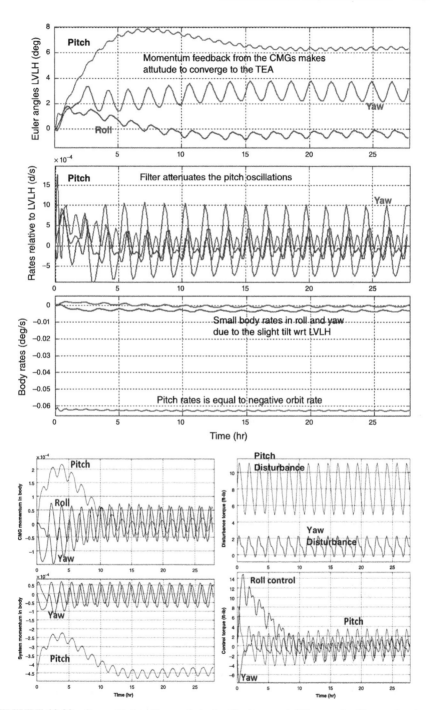

FIGURE 10.90 Space Station Responds to the Environmental Torques by Converging to a New Attitude While Managing the CMG Momentum

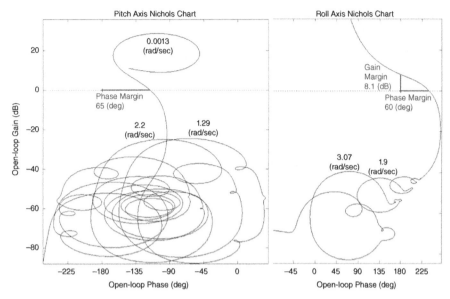

FIGURE 10.91 Frequency Response Analysis With Structural Flexibility in Pitch and Roll Showing the Phase and Gain Margins

The simulation plots in Figure 10.90 show the space station's response to the external torques. Its attitude is initialized at zero in the LVLH frame and it has an initial body rate $(0, -\omega_0, 0)'$. The spacecraft eventually reorients to a different LVLH attitude at which the external torques balance in all directions. At steady state the CMG momentum oscillates at around zero without a bias. This is because of the momentum plus momentum integral feedback from the CMGs. The pitch attitude oscillations at orbital rate are attenuated by the disturbance filter. Notice how the pitch control torque is bigger in size and oscillates at ω_0 to cancel the cyclic orbital component of the disturbance. The only remaining oscillation in pitch attitude is generated by the disturbance torque at $2\omega_0$, which is caused by aerodynamic drag variations from the rotation of the solar panels. The roll and yaw attitude oscillates mostly at orbital frequency ω_0. The pitch attitude converges to 6.3° and the yaw attitude converges to 3° relative to the LVLH. The roll attitude is small. At steady state the body rate is mainly in pitch at negative orbital rate. There is a small component of rate coupling in the negative roll direction because the yaw attitude is slightly tilted positive at 3°. We could have included a second similar filter tuned at $2\omega_0$ to eliminate the small pitch oscillation, or a yaw oscillation filter for better attenuation in yaw, but we shall leave this as an exercise to the reader.

Figure 10.91 is the Nichols plots in the Pitch and Roll axes. It shows the phase and gain margins with flexibility included. For further analysis in this Space Station example, including RCS design, and for accessing the models and data files the reader is recommended to go to the website: Flixan.com/Examples/Large Space Station.

BIBLIOGRAPHY

[1] Abramson, H.N. (1966) "The Dynamic Behavior of Liquids in Moving Containers with Applications to Space Vehicle Technology," NASA SP-106.

[2] AGARD Conference Proceedings No. 235 on Dynamic Stability Parameters, AD A063851, January 1979.

[3] AGARD-CP-260 (1978) "Stability and Control."

[4] Anderson, B.D.O. and Moore, J.B. (1989) *"Optimal Control, Linear Quadratic Methods,"* Prentice-Hall.

[5] Ashley, H. and Zartarian, G. (1956) "Supersonic GAFD Theory. The Piston Element Method. Second Order Theory," *American Institute of Aeronautics and Astronautics Journal*, 23 (12).

[6] Baumgartner, R. I. and Elvin, J. D. (1995) "Lifting Body – An Innovative RLV Concept," Space Programs and Technology Conference, AIAA 95-3531, September 1995.

[7] Betts, K. M., Rutherford, C. R., McDuffie, J., Johnson, M. D., Jackson, M., and Hall C. (2007) "Stability Analysis of the NASA ARES I Crew Launch Vehicle Control System, AIAA GN&C Conference, AIAA 2007-6776, August 2007, SC.

[8] Blakelock, J.H. (1991) *"Aircraft and Missiles,"* John Wiley & Sons Ltd.

[9] Brinker, J. S. and Wise, K. A. (2001) "Flight Testing of Reconfigurable Control Law on the X-36 Tailless Aircraft," *Journal of Guidance, Control, and Dynamics*, 24 (5):903–909.

[10] Britt, R. T., Jacobson, S. B., and Arthurs, T. D. (2000) "Aeroservoelastic Analysis of the B-2 Bomber," *Journal of Aircraft*, 37 (5):745–752.

Performance Evaluation and Design of Flight Vehicle Control Systems, First Edition. Eric T. Falangas.
© 2016 by Eric T. Falangas. Published 2016 by John Wiley & Sons, Inc.

[11] Calhoun, P. (2000) "An Entry Flight Controls Analysis for a Reusable Launch Vehicle," 38th Aerospace Sciences Meeting & Exhibit, AIAA 2000-1046, January 2000,Reno, NV.

[12] Dailey, R. L. (1995) "Lecture Notes for the Workshop on H-infinity and Mu Methods for Robust Control," American Control Conference, June 1995, Seattle, WA.

[13] Doyle, J.C., Francis, B.A., and Tannebaum, A. (1991) *"Feedback Control Theory,"* Macmillan.

[14] Doyle, J.C., Glover, K., Khargonekar, P.P., and Francis, B.A. (1989) "State-space solutions to standard H2 and H-infinity control problems," *IEEE Transactions on Automatic Control*, 34 (8):831–847.

[15] Dykman, J.R. and Rodden, W.P. (1999) "Structural Dynamics and Quasistatic Aeroelastic Equations of Motion," *Journal of Aircraft*, 37 (3):538–542.

[16] Etkin, B. (1972) *"Dynamics of Atmospheric Flight,"* John Wiley & Sons Inc.

[17] Etkin, B. and Reid, L.D. (1996) *"Dynamics of Flight: Stability and Control,"* John Wiley & Sons Ltd.

[18] Fisher, J., Lawrence, D., and Zhu, J. (2002) "Auto-Commander—A Supervisory Controller for Integrated Guidance and Control for the 2nd Generation Reusable Launch Vehicle," AIAA Paper 2002-4562, August 2002, Monterey.

[19] Fossen, T.I. (2011) *"Mathematical Models for Control of Aircraft and Satellites*, 2nd Edition, Norwegian University of Science and Technology.

[20] Freymann, R. (1984) "New Simplified Ways to Understand the Interaction Between Aircraft Structure and Active Control Systems," AIAA Guidance and Control Conference, AIAA 84-1868, New York.

[21] Gainer, T.G. and Hoffman, S. (1972) "Summary of Transformation Equations and Equations of Motion Used in Free-Flight and Wind-Tunnel Data Reduction Analysis," NASA SP-3070.

[22] Giesing, K. and Rodden, W.P. (1972) "Subsonic GAFD Theory. The Doublet Lattice Method. Subsonic Unsteady Aero For General Configurations Computer Program N5KA," Technical Report: AFFDL-TR-71-5, Part II, Vol. II, April 1972.

[23] Glover, K. and Doyle, J.C. (1988) "State-space formulae for all stabilizing controllers that satisfy an H∞ norm bound and relations to risk sensitivity," *Systems and Control Letters*, 11:167–172.

[24] Greensite, A.R. (1970) *"Analysis and Design of Space Vehicle Flight Control Systems,"* Spartan Books.

[25] Haiges, K., Chiang, R.Y., Madden, K., and Safonov, M. (1991) "Robust Control Law Development for Modern Aerospace Vehicles," WL-TR-91-3105, August 1991.

[26] Hall, C.E., Gallaher, M.W., and Hendrix, N. D. (1998) "X-33 Attitude Control System Design for Ascent, Transition, and Entry Flight Regimes," AIAA-98-4411.

[27] Hall, C., Lee, C., Jackson, M., West, M., Whorton, M., Brandon, J., Hall, J., Jang, J., Bedrossian, N., Compton, J., and Rutherford, C. (2008) "Ares I Flight Control System Overview," AIAA GN&C Conference, AIAA 2008-6287, August 2008, Honolulu, HI.

[28] Heiberg, C. Bailey, D., and Wie, B. "Precision Pointing Control of Agile Spacecraft using Single Gimbal CMG," AIAA-97-3757.

[29] Heiberg, C., Bailey, D., and Wie, B. (2000) "Precision Spacecraft Pointing Using Single Gimbal CMGs," *Journal of Guidance, Control, and Dynamics*, 23 (1).

[30] Hughes, P.C. (1986) *"Spacecraft Attitude Dynamics,"* John Wiley & Sons Inc.

[31] Hyde, R.A. (1995) "H-infinity Aerospace Control Design," Springer-Verlag London Ltd.

[32] Ingalls, S. and Mrozinski, R. (1994) "Preliminary Performance and Static Stability of a Single-Stage-to-Orbit Rocket Variant WB001," AIAA.

[33] Junkins, J.L. (1989) "Mechanics and Control of Large Flexible Structures, Progress in Astronautics and Aeronautics," *American Institute of Aeronautics*, 129.

[34] Kay, J., Mason, W.H., Durham, W., Lutze, F., and Benoliel, A. (1993) "Control Authority Issues in Aircraft Conceptual Design," AIAA Paper 93-3968, August 1993.

[35] King, A., Ryan, R.S., and Scoggins, J.R. (1967) "Use of Wind Shears in the Design of Aerospace Vehicles," *Journal of Spacecraft and Rockets*, 4 (11):1526–1532.

[36] Kwakernaak, H. and Sivan, R. (1972) *"Linear Optimal Control Systems,"* John Wiley & Sons Inc.

[37] Lukens, D., Schmitt, A., and Broucek, G. (1961) "Approximate Transfer Functions for Flexible Booster and Autopilot Analysis," Convair Report No. AE 61-0198, Convair Astronautics Division, General Dynamics, Flight Control Laboratory, April 1961, San Diego, CA.

[38] Lukens, D.R., Turney, R.L., Feffemmm, R.L., Kittle, J.W., and Reed, T.E. (1967) *"Dynamic Stability Of Space Vehicles,"* General Dynamics Corporation, San Diego, CA.

[39] Maciejowski, J.M. (1989) *"Multivariable Feedback Design,"* Addison-Wesley Publishing Co.

[40] McRuer, D., Ashkenas, D., and Graham, A.I. (1973) *"Aircraft Dynamics and Automatic Control,"* Princeton University Press, New Jersey.

[41] NASA SP-8036 (1970) "Effects of Structural Flexibility on Launch Vehicle Control Systems," February 1970.

[42] Nelson, R.C. (1998) *"Flight Stability and Automatic Control,"* McGraw-Hill Int.

[43] Noll, R.B., Zvara, J., Deyst, J.J. (1970) "Effects of Structural Flexibility on Launch Vehicle Control Systems," NASA Space Vehicle Design Criteria, NASA SP-8036, February 1970.

[44] Owens, D.H. (1978) *"Feedback and Multivariable Systems,"* IEE Control Engineering Series, Peter Peregrinus LTD, IEE.

[45] Peterson, L.,D., Crawly, E.F., Hansman, R.J. (1989) "Non-Linear Fluid Slosh Coupled to the Dynamics of a Spacecraft," *AIAA Journal*, 27 (9):1230–1240.

[46] Raterink, K.E., Pesek, D., Fernandes, S. (1991) "Analysis of STS Ascent Digital Autopilot Load Relief Feature in Conjunction with Pitch and Yaw Table Wind Biasing," TM-6.23-07-32, November 1991, Houston.

[47] Rodden, W.P. and Johnson, E.H. (1994) "User's Guide, MSC/NASTRAN® Aeroelastic Analysis – Version 68," October 1994.

[48] Rodden, W.P. and Giesing, J.P. (1970) "Application of Oscillatory Aerodynamic Theory to Estimation of Dynamic Stability Derivatives," *AIAA Journal of Aircraft Engineering Notes*, 7(3):272–275.

[49] Safonov, M.G. (1982) "Stability Margins of diagonally perturbed multivariable feedback systems," *IEE Proceedings Part D*, 129(6):251–256.

[50] Safonov, M.G., Limebeer, D.J.N, and Chiang, R.Y. (1989) "Simplifying the H∞ theory via loop-shifting, matrix-pencil and descriptor concepts," *International Journal of Control*, 50 (6):2467–2488.

[51] Schmidt, D.K. and Newman, B. (1998) *"Modeling, Model Simplification and Stability Robustness with Aeroelastic Vehicles,"* AIAA 88-4079-CP.

[52] Skogestad, S. and Postlethwaite, I. (2005) *"Multivariable Feedback Control: Analysis and Design,"* John Wiley & Sons Ltd, West Sussex, England.

[53] Stephen, B. P. and Craig, B. H. (1991) *"Linear Control Design, Limits of Performance,"* Prentice Hall Information and System Sciences, Englewood Cliffs, NJ.

[54] Stevens, B.L. and Lewis, F.L. (2003) *"Aircraft Control and Simulation,"* 2nd Edition, John Wiley & Sons, Inc, Hoboken, NJ.

[55] Tewari, A. (2007) *"Atmospheric and Space Flight Dynamics: Modeling and Simulation with Matlab and Simulink,"* Birkhauser Basel, Boston.

[56] Uy-Loi L. (1997) "Stability and Control of Flight Vehicles."

[57] Wehmueller, K.A. and Hauser, F.D. (1991) "Robust Load Relief Autopilot Design for Advanced Launch System (ALS) Vehicles," AIAA-91-2817-CP.

[58] Weissman, R. (1973) "Preliminary Criteria for Predicting Departure Characteristics and Spin Susceptibility of Fighter Type Aircraft," *Journal of Aircraft*, 10 (4):214–219.

[59] Weissman, R. (1975) "Status of Design Criteria for Predicting Departure Characteristics and Spin Susceptibility," *Journal of Aircraft*, 12 (12):989–993.

[60] Wie, B. (1985) "Thrust Vector Control Design for a Liquid Upper Stage Spacecraft," *Journal of Guidance*, Control, and Dynamics, 8(5):566–572.

[61] Wie, B. (1998) *"Spacecraft Vehicle Dynamics and Control,"* AIAA Education Series, AIAA, Inc., Reston, VA.

[62] Wie, B., Bailey, D., and Heiberg, C. (2000) "Rapid Multi-Targeting Acquisition and Pointing Control of Agile Spacecraft," AIAA GN&C Conference, AIAA-2000-4546, August 2000.

[63] Wie, B., Bailey, D., and Heiberg, C. (2001) "Singularity Robust Steering Logic for Single-Gimbal Control Moment Gyro," *Journal of Guidance, Control, and Dynamics*, 23 (5):865–872.

[64] Wie, B., Du, W., and Whorton, M. (2008) "Analysis and Design of Launch Vehicle Flight Control Systems," AIAA GN&C Conference, AIAA 2008-6291, August 2008.

[65] Wie, B., Du, W., and Whorton, M. (2008) "Dynamic Modeling and Flight Control Simulation of a Large Flexible Launch Vehicle," AIAA GN&C Conference, AIAA 2008-6620, August 2008.

[66] Wie, B., Heiberg, C., and Bailey, D. (2000) "Singularity Robust Steering Logic for Redundant Single-Gimbal Control Moment Gyros," AIAA GN&C Conference, AIAA-2000-4453, August 2000.

[67] Zipfel, P.H. (2007) "Modeling and Simulation of Aerospace Vehicle Dynamics," 2nd Edition, AIAA Educational Series.

INDEX

Performance Evaluation and Design of Flight Vehicle Control Systems, First Edition. Eric T. Falangas.
© 2016 by Eric T. Falangas. Published 2016 by John Wiley & Sons, Inc.

Printed and bound by CPI Group (UK) Ltd, Croydon, CR0 4YY

16/04/2025

14658579-0005